W0080846

Natural Environments and Human Health

Dedication

This book is dedicated to all those individuals who strive to protect our natural environments and promote the idea of integrating experiences in natural settings into the lives of our citizens. We believe that good things happen when people and natural environments meet.

Natural Environments and Human Health

Alan W. Ewert, PhD

Indiana University, USA

Denise S. Mitten, PhD

Prescott College, USA

Jillisa R. Overholt, PhD

Warren Wilson College, USA

www.cabi.org

CABI is a trading name of CAB International

CABI	CABI
Nosworthy Way	745 Atlantic Avenue
Wallingford	8th Floor
Oxfordshire OX10 8DE	Boston, MA 02111
UK	USA
Tel: +44 (0)1491 832111	Tel: +1 (617)682-9015
Fax: +44 (0)1491 833508	
E-mail: info@cabi.org	E-mail: cabi-nao@cabi.org
Website: www.cabi.org	

© CAB International 2014. All rights reserved. No part of this publication may be reproduced in any form or by any means, electronically, mechanically, by photocopying, recording or otherwise, without the prior permission of the copyright owners.

A catalogue record for this book is available from the British Library, London, UK.

The Library of Congress has cataloged the hardcover edition as follows:

Ewert, Alan W., 1949- author.
  Natural environments and human health / Alan Ewert, Denise Mitten, Jillisa Overholt.
    p. ; cm.
  Includes bibliographical references and index.
  ISBN 978-1-84593-919-9 (hbk)
  I. Mitten, Denise, author. II. Overholt, Jillisa, author. III. C.A.B. International, issuing body. IV. Title.
  [DNLM: 1. Environment. 2. Environmental Health. 3. Environmental Exposure. 4. Public Policy. WA 30.5]

  RA566.27
  362.1969'8--dc23

2013025140

ISBN-13: 978 1 78639 529 0 (PB)

Commissioning editor: Rachel Cutts
Editorial assistant: Emma McCann
Production editor: Shankari Wilford

Typeset by SPi, Pondicherry, India
Printed and bound by CPI Group (UK) Ltd, Croydon, CR0 4YY
First printed in hardback in 2014. Transferred to POD paperback in 2019.

Contents

Preface

Guard well your spare moments. They are like uncut diamonds. Discard them, and their value will never be known. Improve them, and they will become the brightest gems in a useful life.

Ralph Waldo Emerson

The story of humans and their interactions with natural environments is long and varied. A large part of this story has encompassed how, and in what ways, natural environments have influenced human health. From a historical perspective, this influence has involved both individuals responding to natural environments and also using natural environments for health-related reasons, as well as the ways in which societies have served to provide for these interactions. Two questions underlie the provision and use of natural environments for health. First is the question of whether or not natural environments can actually create positive health outcomes such as reduced stress, lower blood pressure, or a heightened sense of well-being. Second is the question of how natural environments effectuate or cultivate these impact factors related to health. Is the impact due to the natural setting simply being somewhat of a novelty for many individuals, and this 'change of pace' offering an offset to modern living? Or is there something about natural environments that helps foster a change in health status or the maintenance of that status? From another perspective, visitors to natural environments often come with others, such as family and/or friends. Does the very presence of these other people serve as a facilitating factor in enhancing health?

Natural settings often have different types and levels of environmental characteristics such as levels of pollution, crowding or high visitor use, and proximity to external influencing agents such as traffic, visual disturbances, and litter. Other characteristics of natural environments that often have some direct correlations to positive health factors include higher air quality, modified temperatures, reduced levels of noise and signs of urban disturbances, and the presence of vegetation, water, beautiful scenery, and quiet. Both the literature and our own experiences often suggest to us, that it is a combination of these and other variables related to the environment that often provides us with a powerful antidote to the stress of modern-day life and influences our health status.

Whether it be the verdant healing gardens developed in ancient Babylon, the herbs and plants gathered by our ancestors for medicinal purposes, or the vision quests practiced by First Nations, all the way to the development of greenspaces or nature preserves in our modern urban environments, societies of all eras and regions have recognized that natural environments play an important role in the fostering of positive health benefits. Whether these benefits are accrued through a quiet walk along a stream, or hiking up a challenging mountain, enjoying the camaraderie of friends out for a picnic in a municipal park, or even breathing in the air while in a cool, green forest, health comes to us, while in nature, and through nature.

This book is about the myriad of ways that nature and natural environments serve to foster health. Whether the natural environment provides a setting, an experience or some intrinsic quality, in this book we examine the long relationship people have had with natural settings, how this relationship can result in improved health, some of the theories and concepts that frame our thinking regarding the human/nature interaction, and how research has informed our thinking about how natural landscapes impact our health.

In addition, we have examined how different groups have responded to the health/natural environments interaction. For example, indigenous and First Nations peoples have used natural medicines and healing rituals involving natural landscapes for thousands of years, with this traditional ecological knowledge (TEK) carrying over into modern times. In another example, we examine the role that children and different life stages play in health-related issues and natural environments.

Along with issues of life stage or membership in a particular group, either historical or contemporary, we believe that the health-related benefits that natural environments provide to people can be augmented by the development of specifically designed programs or experiences. Towards this end, structured programs, such as those offered through organizations engaged in adventure education, and similar types of programs can be used to promote the health benefits associated with natural environments.

Finally, providing evidence-based knowledge and then developing subsequent policies that support the use of natural environments for enhancing health is an ongoing process. Like other research processes, developing research efforts specifically targeted towards understanding the health/natural environment connection often face a number of issues that need to be addressed. In addition, evidence-based policy needs to account for the myriad of situations and needs of the variety of people who engage in natural environments, whether deliberately or vicariously for health-related outcomes. As suggested by Ralph Waldo Emerson's quote at the beginning of this Preface, what often matters most is what individuals do with their 'spare time'. For many, that spare time is often spent in a natural environment and increasingly, for health-related outcomes.

Alan W. Ewert
Denise S. Mitten
Jillisa R. Overholt

Acknowledgements

The authors would like to extend a special thanks to Emma McCann from CABI for guiding this book through the many travails of publishing a textbook, as well as to Brian Forist (PhD student at Indiana University) and Chiara D'Amore, Janet Ady and Betsy Wier (PhD students at Prescott College) for their diligent assistance in helping us with a number of different technical and literature tasks associated with this book.

1

Overview

To me a lush carpet of pine needles or spongy grass is more welcome than the most luxurious Persian rug.

Helen Keller

Why This Book and Why Now

Today, literally hundreds of millions of people around the globe will engage in some form of contact with natural environments through work, living circumstances, or play and recreation. Some of these contacts will be in the form of high adventure activities, such as white-water rafting or mountain climbing. Others will involve a quiet walk down a winding path or along a wind-swept coastline. Still others will engage in a natural environment through tending to their garden, woodlot, or a local municipal park. In the US alone, there are about 300 million annual visits to national parks with nearly 50% of all Americans participating in at least one form of outdoor recreation in 2010. Many of these participants did so for health-related reasons or incidentally received health benefits while recreating (Outdoor Foundation, 2011). And for many Indigenous and First Nations peoples, the natural environment has provided and continues to provide a critical core to the very rhythm and existence of their lives, with some people practicing subsistence living by

attending to the various cycles and rhythms of the natural environment. Many others, some by choice and many without choice, live close to the land in countries where securing water, for example, consumes hours of their days. Without a doubt, there are countless ways in which people and natural environments meet, whether for needed resources such as water, minerals, animals and plants, or simply the enjoyment of interacting, within a leisure or recreational context, with the natural setting.

Whatever their underlying needs, values, and reasons, the interaction between human health and natural environments involves a myriad of experiences, settings, and beliefs and it is this myriad that constitutes the subject of *Natural Environments and Human Health*. This interaction in terms of health and well-being has never been more important than in this current time. Both individually and collectively, health has become a significant issue of concern for much of the world's population. Some of these health concerns are because of toxins in nature and others are because of humans' lack of contact with nature, particularly in many Westernized countries (Pergams and Zaradic, 2008). A sample of the broad reach of these issues involves both psychological as well as ecological factors. Table 1.1 lists major health concerns that are related to natural environments, loosely divided into two major categories: Ecology-based and Physiological/Psychological-based.

© CAB International 2014. *Natural Environments and Human Health*
(A.W. Ewert, D.S. Mitten and J.R. Overholt)

Table 1.1. Major health concerns related to natural environments.

Ecology-based	Physiological/Psychological-based
Pesticides	Levels of physical activity
Air/water quality	Perceived general health
Toxic contaminants	Levels of obesity
Ozone depletion/acid rain	Sense of well-being
Excessive noise	Quality of life
Loss of biodiversity	Mood
Environmental degradation	Rates of recuperation
Global warming	Environmental injustice
Exotic disease distribution	Ability to provide focus and attention

Thus, one way of thinking about the intersection of natural environments and human health is the way in which the connection impacts health and well-being. Humans' relationship with the environment is complex and multidimensional. Humans impact the environment and the environment affects humans. In this book a systems approach or lens is used, meaning that everything in the universe (and perhaps beyond) is connected to and affects everything else, and that everything known to humans is in effect one living system of which humans are a part.

In considering our understanding of the relationship between natural environments and human health, two overarching questions serve to guide our thinking about the issue. First, is nature *beneficial* to human health? That is, can contact with natural environments increase health and well-being? Second, if natural environments are beneficial to human health, *how* are nature and natural environments beneficial? Other questions that flow from whether and how natural environments impact health include:

- From a research perspective, what issues and concerns should we take into account when developing our understanding about natural environments and health?
- What is the dosage or minimum threshold of effect from different exposures to natural environments?
- Are these effects due to vicarious or confounding issues such as novelty?
- Are the effects of natural environments on human health influenced by variables such as time, perceptions, and background?

- Can structured programs be developed that multiply the effect of natural environments upon human health?

Past research and literature provide a substantial amount of information to answer these questions related to health and natural environments. For example, we now have information about a number of health-related issues especially in relation to time in nature lowering blood pressure, increasing social connections, and increasing longevity. In addition, as illustrated in Table 1.2, there are varying levels of confidence concerning the level and quality of this information. While Table 1.2 presents only a partial listing, what becomes apparent is the growing breadth of information concerning human health and natural environments as well as the depth of that information. In Chapter 7, we provide a more detailed examination of the benefits and outcomes associated with natural environments and human health.

Accordingly, this book seeks to address these influences both by providing an overview of what is currently known about a given phenomenon (such as physical activity in natural environments) as well as discussing some of the past and current theories that seek to explain how these connections actually work. Thus, the book provides a bridge between what we do (individually and collectively) in natural settings and how that action can impact our health, both individually and collectively as the human species. Our hope is that the information in this book will spur students and professionals to want to know more about the connections between human health and the environment on a personal and professional level. We want

Table 1.2. Health-related information concerning human health and natural environments.

Information	Quality of information/confidence in information
Natural environments can be restorative	High: substantial database
Genetic predisposition may play a role in the effect of natural environments on health	Low: little research
Initial responses to natural environments are usually affective rather than cognitive	Medium: heavy reliance on anecdotal reports
Positive emotional states/blocked negative toned feelings	High: substantial research and anecdotal base
Decreased levels of stress and increased resilience and hardiness	Medium to high and developing
Building social networks/shared experiences	Medium
Physical fitness and levels of physical activity	High
Values (personal growth, self-awareness, reflection)	Developing

readers to have useful information and to be part of the rich dialogue occurring in many disciplines as we find ways to increase health and well-being for all people. By extension the book modestly addresses how human understanding of the importance of the natural environment to our health and well-being can influence our relationship with the natural world.

Throughout the book we offer examples of research. We describe and discuss research and problems that may arise from research design or interpretation. We want readers to think critically about research and be able to analyze and interpret results. The bottom line based on the research for this book and the experience of the authors is that nature has been and continues to be essential and incredibly positive for human life and that appropriate connections with nature will positively influence human development, health, and well-being.

Who is This Book For?

Understanding the connection between human health and natural environments has implications for a broad range of people, both individually and collectively. Moreover, a number of often disparate disciplines have an interest in these connections. For example, health professionals have an obvious interest in how natural landscapes and time in them can provide positive health benefits or serve as a mediating variable for health problems. Likewise, landscape designers and policy planners have both an

interest in and specific roles to play in this relationship (Cheng and Monroe, 2012). This issue, however, extends beyond the purview of the expert or researcher; other groups such as recreational professionals, conservation groups, social scientists, and educators also have a stake in understanding how natural environments influence health. For example, greenspace environments and their relationship to health have received a substantial amount of research focus through a variety of academic and professional disciplines (de Vries *et al.*, 2003). So, who is this book intended for? Simply put: a wide range of students, academics, researchers, planners, and practitioners from a variety of disciplines and backgrounds. These disciplines include, but are not limited to: Parks and Recreation, Adventure Therapy, Anthropology, Applied Health, Architecture, Biology, Business, Cognitive Science, Conservation Psychology, Developmental Psychology, Ecology, Ecopsychology, Education, Environmental Psychology, Evolutionary Psychology, Health & Wellness, Landscape Architecture, Law, Natural Resource Management, Medicine, Nursing, Psychological Sciences, Political Science, Public Health, Religion, Social Psychology, Social Work, Sociology, Sustainability, and Urban Planning.

Readers specifically can expect to take four concrete pieces from this book. Through knowing the health benefits of being in nature, people will:

1. more readily be predisposed to integrating nature into their personal and professional lives;

2. be more prepared to implement action whether that is in programming or through legislation;

3. understand some of the critiques and techniques of researching the health benefits of nature so that they will be able to keep current and assess future research; and

4. renew their appreciation of nature and want to go outside and take everyone with them.

Underlying Assumptions

This book has a foundational assumption that interaction with the natural world is most often a positive, health-enhancing experience. While some literature discusses potential negative outcomes from interaction with natural environments (Andrews and Gatersleben, 2010; Bruni *et al.*, 2012), the vast majority of research and scholarship supports the overall assumption in this book that natural environments can provide health-promoting experiences and behaviors for individuals and groups (Johansson *et al.*, 2011; Kline *et al.*, 2011). For example, Russell (2003) and Mitten (1994) propose that therapy done in natural settings can be effective in creating more positive and health-enhancing behaviors. In addition, Ewert and Galloway (2012) and Mitten (2009) suggest that programs such as those that are adventure-based can provide opportunities for achieving health and wellness within society. In a similar fashion, O'Brien *et al.* (2011) draw connections between outdoor education and skills developed in woodlands and greenspaces and human health and well-being. In a socio-ecological approach to health, health results from an interwoven relationship between people and their environment. Natural environments play a key role in a socio-ecological approach to health because these environments encourage and enable people to relate to each other and the natural world (Maller *et al.*, 2006). Along the same thinking, there are a number of theories and practices related to the restorative nature of natural environments, namely attention restoration, friluftsliv, and psycho-evolutionary connections, as discussed later in the book.

We acknowledge that nature can be harsh; many environments because of their geographic location, including altitude, are inclement for humans. Storms, earthquakes, and volcanoes can have devastating effects; still nature is not evil or against humans. We are part of the global and perhaps cosmic ecosystem with both benign and catastrophic forces.

The second primary assumption made in this book is that the effects of natural environments upon human health can actually be felt, observed, and/or measured. Larivière *et al.* (2012) suggest that while experiences in natural environments often result in positive anecdotal accounts, a substantial amount of variation exists from a variety of empirically derived studies and perspectives. Understanding these potential and real outcomes is made even more urgent by authors such as Lyytimaki (2012) and Louv (2005) who believe that members of industrialized societies are becoming more alienated and separated from nature experiences and more accustomed to indoor and technology-oriented environments; they argue that this separation from nature has adverse effects on individuals' health and well-being.

The third primary assumption is that in order for the natural environment to positively impact human health we need to care for the environment. With an understanding of our need for nature for health and well-being, people are more likely to be advocates to ensure environmental or ecological quality. People will have a deeper understanding of the human–nature connection and when we feel that on a visceral level we will develop a strong ecological conscience leading to positive action (see Chapter 8).

Defining the Terms

A number of terms are used extensively in this book and are defined in the following way.

- *Health* is a state of complete physical, mental, and social well-being and not merely the absence of disease or infirmity (World Health Organization, 1948).

Health and well-being is seen as spanning across six dimensions of a person's existence. The dimensions involve both the micro (immediate, personal) and macro (global, planetary) environments (Blonna, 2011). The six dimensions of health and well-being include the following:

○ *Emotional health* is being in touch with feelings, having the ability to express them, and being able to control them when necessary. Optimal functioning involves understanding that emotions assist us get in touch with what is important in our lives. Our emotions help us feel alive and provide us with a richness of experience that is uniquely human.

○ *Environmental health* includes aspects of human disease and injury that are influenced by variables in the environment. This includes the study of both direct and pathological effects of various chemical, physical, and biological agents, as well as the effects on health of the physical and social environments such as parks, greenspaces and undeveloped landscapes (adapted from the US Department of Health and Human Services, 2000).

○ *Intellectual health* is the ability to process information effectively. Intellectual wellness involves the ability to use information in a rational way to problem solve and grow. It also includes factors such as creativity, spontaneity, and openness to new ways of considering situations.

○ *Physical health* is how well the body performs its intended functions. Absence of disease, though an important influence on physical wellness, is not the sole criterion for health. The physical domain is influenced by factors such a genetic inheritance, nutritional status, fitness level, body composition, and immune status.

○ *Social health* is being connected to others through various types of relationships. Individuals who function optimally in this domain are able to form friendships, have intimate relationships, and give and receive affection. They are able to give of themselves and share in the joys and sorrows of being part of a community.

○ *Spiritual health* is often described as feeling connected to something beyond oneself. Spiritual wellness is expressed through an inner peace and understanding of one's place in the greater universe. People can express spirituality through participation in organized religious activities, often involving the belief in a supreme being or supernatural force as well as a formalized code of conduct by which to live, or in many other ways such as spending time in nature. The crucial underlying feeling is a perception of life as having meaning beyond the self, often enhanced by being part of a community and helping others.

• *Connectedness to nature:* An individual's sense of connection or relationship to the natural world. This concept is often linked to predisposing factors such as personality, past experience with natural environments, and a specific setting and/or situation.

• *Constructivism* as a learning theory is the active mental constructions of children resulting in the ways in which children organize and act on their knowledge and values.

• *Cure* is to eradicate a disease condition or symptom(s) that the patient has. Curing happens at the level of the body.

• *Environment* is the natural, physical, and societal surroundings that affect individuals' functioning on both the micro and macro levels. On a macro level this might include natural-area buffering from potential storms, floods, or other environmental challenges. The well-being of our micro environment includes the level of functioning in our school, home, worksite, and neighborhoods. Our social support system includes family and friends and is also part of our micro environment; it affects our personal safety by influencing whether or not we are at risk of and fear such issues as theft, crime, and violence. Air and water quality, noise pollution, overcrowding, and other factors that influence our stress level are also affected by our micro environment (Blonna, 2011).

- *Ethic of care* presumes that there is moral significance in the fundamental relationships and dependencies in life and affirms the importance of caring motivation, emotion, and the body in moral deliberations. The ethic builds on the concept of empathy and is inspired by memories of being cared for and of the idealizations of self (Mitten, 1994).

- *Healing* is process that leads to a sense of well-being which includes optimism and calmness. Healing happens to the whole person and is about confidence and trust in life. If a disease has no cure, the person may still experience healing. For example, the focus of hospice and palliative care is healing (Mitten, 2004).

- *Indigenous* refers to people who are native to a particular environment and have maintained living in a natural area or environment insomuch as is realistic in the 21st century. These people are usually ethnic minority groups affected negatively by colonization and often marginalized. In this book examples from indigenous people do not mean that all indigenous people are the same nor do we intend to mythologize indigenous people. Throughout the book we have tried to include specific examples when referring to indigenous people.

- *Green exercise*: Physical exercise performed in relatively natural settings.

- *Natural environments*: Surroundings or the geophysical space that encompasses all living and non-living features and systems that are dominantly influenced by environmental processes with minimal human disturbances and not managed by humans for human purposes. In this book we use natural environments, nature, natural settings, the natural world, natural landscapes, greenspaces, and outdoor places interchangeably.

- *Place-based education* is an educational philosophy and method focused on enhancing people's relationship with their personal community and surrounding land. This experiential pedagogy links curricula, the local classroom, or informal educational settings to the students' local community by connecting multidisciplinary topics to cultural, political, economic, and ecological concepts.

- *Praxis* is the practical application of a theory or a branch of knowledge.

- *Stress*: A resultant situation where an individual perceives the situation as exceeding her or his abilities or 'a holistic transaction between an individual and a potential stressor resulting in a stress response' (Blonna, 2011, p. 12). Chronic stress is linked to factors such as obesity, high systolic blood pressure, and elevated heart rate.

- *Stress response* is a change in the body's internal environment (i.e. leaving homeostasis) in response to a stressor. When people are stressed, hormone levels change (e.g. cortisol, adrenaline, and noradrenaline increase) resulting in, among other things, blood pressure rising and heart rate increasing.

- *Systems theory* is a coherent scientific framework that understands that everything in the universe (and perhaps beyond) affects everything else and that everything known to humans is in effect one living system of which humans are a part.

- *Quality of life* is a person's ability to enjoy normal life activities and an overall sense of well-being. Quality of life often has a strong connection to individual health perceptions and life functioning.

- *Wellness* is the realization of the fullest potential of an individual—physically, psychologically, socially, spiritually, and economically—and the fulfillment of expectations associated with one's roles in the family, community, spiritual life, workplace, and other settings (Smith *et al.*, 2006, p. 5).

While there are many more terms that could be included, the above list represents those that are most essential to a shared understanding of the relationship between human health and natural environments.

Structure of this Book

Historically, there has been much discussion and accompanying literature focused on the role that nature plays in human welfare. Only recently, however, with both the debate over climate change and the 2005 publishing of Louv's book *Last Child in the Woods*, has the discussion over the importance of natural environments for human well-being been rekindled. This issue is both timely and of critical importance to our society as we continue to define the role that natural environments will play for the health and well-being of future generations. Increasingly, there is recognition that these environments play a critical role in providing benefits such as stress reduction, healthy childhood development, mental health, enjoyment, aesthetics, and catharsis, rather than only in commodity production. This book examines the long history of natural environments being used in these non-commodity production ways and traces the development of the connection of humans to environments within the context of how they impact our personal and collective health. This information is presented in 11 chapters.

Chapter 1 provides an overview of the book by focusing on the intended audience, the underlying assumptions, and definition of commonly used terms. In Chapter 2, 'Human Perceptions of Nature', the discussion revolves around the ways that people's attachment to the natural world is driven by perceptions as well as evolution and biology. Salient worldviews of natural environments and health, and how these worldviews impact individual and collective behaviors are identified. Finally, we reverse the direction of this discussion by examining the impact of a lack of natural settings upon health and, in particular, focus on modernity from the perspectives of society lacking a relationship of personally significant connection with the natural world, as well as the growing awareness of an ecological crisis and its impact on human health. Transitioning then to nature and health throughout history, Chapter 3, 'The Historical Connection between Natural Environments and Health', provides an overview of the history of the connection between human health and natural environments and how this connection has evolved and changed over time. Following this, we identify particular characteristics and attributes associated with human health that have been linked to natural settings.

In Chapter 4, we discuss a number of salient concepts and theories closely identified with human health and natural environments. This chapter is divided into five sections: (i) genetic theories; (ii) other evolutionary-grounded theories; (iii) psychological theories; (iv) restoration and restorative environments; and (v) intentionally designed experiences (IDEs). Genetic theories include naturalistic intelligence, the biophilia hypothesis, and indigenous consciousness. In the psychological theories section, identity and the natural environment, psychologically deep and extraordinary experiences, flow, the peak experience, and transcendent experiences are examined. The restoration and restorative environments section looks at psycho-evolutionary theory (PET) and attention restoration theory (ART).

The discussion on IDEs attempts to combine some of the extant theories such as PET with how specifically designed programs can increase the impact that natural environments can have on health-related issues such as stress or anxiety. That is, how can we develop health-enhancing experiences and programs using the psychological theories discussed in this chapter?

Chapter 5 focuses on the topic of 'Human Development and Nature'. Specific attention is paid to the developmental process and myriad of ways that nature and natural environments can be used to enhance human development, particularly with a view towards positive health behaviors. Human development from birth through end-of-life stages is covered, combining developmental theory with the research about human health and natural environments. In Chapter 6, 'Adaptations and Applications', we examine health-related concepts such as folk biology, traditional ecological knowledge (TEK), friluftsliv, ecofeminism, and the socio-ecological approach to human health. Applications included in this chapter are nature medicines, conservation and the development of a land ethic, spirituality, and citizen science. A section of Chapter 6 describes the growing interest in ecotherapy and the broader field of ecopsychology.

Chapter 7 centers on the outcomes and benefits typically linked to human health and natural environments. Within this chapter, a broad range of activities, fields of study and settings are covered. Of major concern in this chapter is the research conducted in the area of human health and the natural environment and the specific outcome measures and documented benefits of spending time in nature. A sampling of documented benefits includes physical, psycho-emotional, spiritual, and social well-being. Included are suggestions for future research.

Chapter 8 discusses sense of place and the role of education in developing issues such as resilience, connection to nature, and building an environmental conscience. This chapter attempts to merge the reality of the positive impacts of natural environments on human health with the involvement of individuals, groups, and communities. Knowing that science provides a growing body of knowledge pointing to the value of natural environments for human health is important but insufficient if the individual does not either understand these effects or know how to apply them.

Chapter 9 looks at adventure and outdoor education programming as an example of an innovative approach for integrating natural environments and health. Chapter 10 expands on the ways to develop the connection between health and natural environments by focusing on how public policy might be developed in order to facilitate ways in which individuals can better access and benefit from natural landscapes. For example, these future actions might involve policy changes such as greenspace development, providing parks and other natural environments, and moving to make natural environments personally relevant. This chapter also examines the challenges and issues facing future research such as fidelity concerns, who will actually use the research findings, and how the issues of confounding variables will be dealt with.

The book ends with Chapter 11, 'Resources', and provides a collection of web sites, extant literature and other sources that speak to the connection between human health and natural environments.

References

Andrews, M. and Gatersleben, B. (2010) Variations in perceptions of danger, fear and preference in a simulated natural environment. *Journal of Environmental Psychology* 30(4), 481.

Blonna, R. (2011) *Coping with Stress in a Changing World*, 5th edn. McGraw-Hill, Boston, Massachusetts.

Bruni, C.M., Chance, R., Schultz, P.W. and Nolan, J. (2012) Natural connections: bees sting, snakes bite, but they still are nature. *Environment and Behavior* 44(2), 197–215.

Cheng, J. and Monroe, M.C. (2012) Connection to nature: children's affective attitude toward nature. *Environment and Behavior* 44(1), 31–49.

de Vries, S., Verheij, R.A., Groenewegen, P.P. and Spreeuwenberg, P. (2003) Natural environments – healthy environments? An exploratory analysis of the relationship between greenspace and health. *Environment and Planning A* 35(10), 1717–1732.

Ewert, A. and Galloway, G. (2012) Take a park, not a pill: promoting health and wellness through adventure programming. In: Martin, B. and Wagstaff, M. (eds) *Controversial Issues in Adventure Programming*. Human Kinetics, Champaign, Illinois, pp. 130–137.

Johansson, M., Hartig, T. and Staats, H. (2011) Psychological benefits of walking: moderation by company and outdoor environment. *Applied Psychology: Health and Well-Being* 3(3), 261–280.

Kline, J.T., Rosenberger, R.S. and White, E.M. (2011) A national assessment of physical activity in US National Forests. *Journal of Forestry* 109(6), 343–351.

Larivière, M., Couture, R., Ritchie, S.D., Côté, D., Oddson, B., et al. (2012) Behavioural assessment of wilderness therapy participants: exploring the consistency of observational data. *Journal of Experiential Education* 35(1), 290–302.

Louv, R. (2005) *Last Child in the Woods: Saving Our Children from Nature-Deficit Disorder*. Algonquin Books of Chapel Hill, Chapel Hill, North Carolina.

Lyytimaki, J. (2012) Indoor ecosystem services: bringing ecology and people together. *Human Ecology Review* 19(1), 70–76.

Maller, C., Townsend, M., Pryor, A., Brown, P. and St. Leger, L. (2006) Healthy nature healthy people: 'contact with nature' as an upstream health promotion intervention for populations. *Health Promotion International* 21(1), 45–54.

Mitten, D. (1994) Ethical considerations in adventure therapy: a feminist critique. In: Cole, E., Erdman, E. and Rothblum, E.D. (eds) *Wilderness Therapy for Women: The Power of Adventure*. The Haworth Press, Binghamton, New York, pp. 55–84.

Mitten, D. (2004) Adventure therapy as a complementary and alternative therapy. In: Bandoroff, S. and Newes, S. (eds) *Coming of Age: The Evolving Field of Adventure Therapy*. Association of Experiential Education, Boulder, Colorado, pp. 240–257.

Mitten, D. (2009) Under our noses: the healing power of nature. *Taproot Journal* 19(1), 20–26.

O'Brien, L., Burls, A., Bentsen, P., Hilmo, I., Holter, K., *et al.* (2011) Outdoor education, lifelong learning and skills development in woodlands and green spaces: the potential links to health and well-being. In: Nilsson, K., Sangster, M., Gallis, C., Hartig, T., de Vries, S., *et al.* (eds) *Forests, Trees and Human Health*. Springer, New York, pp. 343–372.

Outdoor Foundation (2011) *Outdoor Recreation Participation Report 2011*. Outdoor Foundation, Boulder, Colorado.

Pergams, O.R.W. and Zaradic, P.A. (2008) Evidence for a fundamental and pervasive shift away from nature-based recreation. *Proceedings of the National Academy of Sciences USA* 105(7), 2295–2300.

Russell, K.C. (2003) An assessment of outcomes in outdoor behavioral healthcare treatment. *Child and Youth Care Forum* 32(6), 355–381.

Smith, B.J., Tang, K.C. and Nutbeam, D. (2006) WHO health promotion glossary: new terms. *Health Promotion International* 21(4), 340–345.

US Department of Health and Human Services (2000) *Healthy People 2010*. Government Printing Office, Washington, DC.

World Health Organization (1948) Preamble to the Constitution of the World Health Organization as adopted by the International Health Conference. *Official Records of the World Health Organization* 2, 100.

2

Human Perceptions of Nature

At the heart of commons-based cultural systems is the recognition that diversity, biological, linguistic and cultural, must be protected as a means of survival.
R. Martusewicz (2005, p. 340)

To better understand humans' connection to nature and the relationship between human health and natural environments, it is important to consider how both individuals and societies perceive the natural environment. Human actions are driven by perceptions regardless of the reality of our evolutionary, biological, and psychological connections with nature. Perceptions of nature are driven by worldviews, which are impacted by culture (social interactions) and our interaction with the environment at large. These perceptions ultimately account for the ways that humans interact with, value, protect, benefit from, and use the natural environment.

The overarching goal of this chapter is for people to learn that the natural environment is personally significant to all of us. We first make a case that humans are innately connected with nature evolutionarily, biologically, emotionally, spiritually, and socially. This connection is reciprocal; we are influenced by nature and we influence nature. Next we address the evolution of worldviews and human perceptions of nature leading to the modern, predominantly Western tendency to place humans outside ecosystems; and the

subsequent sense of and reality of disconnection from nature. Finally, we examine potential concerns rising from estrangement from nature and the ramifications of such separation, concluding that a worldview that includes a positive affiliation with nature will lead to increased health and well-being for humans.

What does Connected Mean? Are We Connected to Nature?

Connected means that we are attached or united, we are joined; we are related as in family ties. In a state of connection there is a link or a bond; there is cause and effect. We are connected to nature; meaning that we are in relationship with other living beings and these relationships impact our social, psychological, spiritual, and biological selves, which in turn impact our health and well-being. Early humans were well-connected and interwoven with nature, co-evolving for millennia.

Our social and physical environments influence our perceptions. Therefore, daily interactions and close dependence on nature influenced perceptions of early humans as they lived lives intertwined with natural patterns such as solar and lunar cycles, salmon spawning cycles, whale and bird migrations, insect activity, and berry ripening. Most likely

© CAB International 2014. *Natural Environments and Human Health*
(A.W. Ewert, D.S. Mitten and J.R. Overholt)

this reliance and close relationship offered many positive health and developmental benefits while also presenting hardship and challenge. Evidence of this relationship is depicted in 40,000-year-old cave paintings in areas such as Chauvet in southern France and Ulm, Germany, where horses are drawn with great perspective and accuracy. Archeologists interpret the drawings to represent a close connection to these animals in spiritual and physical ways. Prior to discovery of the cave paintings this sort of relationship with animals was believed to have developed much later in human evolution. Some societies continue to live in close contact with nature, even to the extent that they mimic the behavior of animals and plants—a practice that science now knows as biomimicry. McGregor (2010) talks about the Evenk or Reindeer people from the Siberian taiga in northern Russia who mirror reindeer behavior, historically following the migrations and relying on them for most of their needs including their shamanic spiritual and healing tradition. They believe that their soul connection to the reindeer allows them to 'see the future, understand the unknowable, heal individuals, and advise the entire community' (p. 13). They think of this as Bayanay or an all-knowing, all-feeling spirit, or a shared consciousness (Vitebsky, 2005; Klokov, 2007) that gives them a sense of belonging to a world larger than themselves. Originally coastal dwellers, the Evenk combine pastoralism (reindeer and horses) with fishing, hunting, and gathering; they were able to migrate into the more mountainous taiga region only because of their mutualistic relationship with the reindeer as pack animals. Today the destruction of the pasture land severely limits the reindeer herding, though local officials and Evenk are trying to revitalize their pastoral lifestyle. This disruption in connection to their natural environment has negative physical, mental, and spiritual health consequences for the Evenk people.

Human behaviors influence nature, further demonstrating our connection. A mutualistic relationship between the Evenk people and reindeer is evidenced in the behavior of herds used by the people for milk and packing burdens (though not for meat). These herds now naturally stay close by due to protection people offer from predators and by smudging biting insects.

Of course, not all human influence on animal behavior is positive or mutualistic. In the past century as human technology developed and wildlife habitat decreased, hunting pressure increased, resulting in changed animal behavior. Bears under hunting pressure now avoid hunters by shifting activity from day to night (Miller, 2012). This behavior change lessens an individual bear's chance of being killed; however, the bear population as a whole is more vulnerable to starvation. Hunting season begins in late August and ends in late October, coinciding with the bears' need to eat copious amount of berries and accumulate fat storage for winter. Daytime eating is more efficient for bears and the cost of not having enough fat stored for hibernation is low birth rates and death (Ordiz et al., 2012). Hunters' use of deer feeding stations has also shifted deer activity from day to night to avoid hunting pressure.

Hunting is a connection with nature that humans have had since ancestral times. Many cultures exhibit emotional and spiritual ties with the animals they hunt. The Zunis people do not destroy the bones of animals they kill as an act of honoring the animal (Earhart, 2001). Swan (1992) asserts that hunting is a close relationship with the hunted animal. He cites examples of American Indian, Inuit, and West African conceptions of the willingness of animals to be killed and the misconception that hunting exhibits dominance over animals. In this systems way of thinking, an animal consents to be taken and a power greater than both animals and humans allows each to survive.

The previous examples illustrate evolutionary and social connections with the natural world; these interwoven relationships can be explained by systems theory, defined in Chapter 1. The following sections offer additional examples that demonstrate specific areas of connection.

Evolutionary connections

The evolution of the human species was made possible by the evolution of plants, especially trees, which process sunlight into energy through photosynthesis and release oxygen.

Before the abundance of oxygen released by photosynthetic plants the Earth's atmosphere was anaerobic and humans would not have evolved in the anaerobic environment. The oxygen-rich atmosphere formed the ozone layer, thus blocking ultraviolet solar radiation, and enabling more complex, oxygen-dependent forms of life, including humans, to evolve. The effect of this co-evolution, an absolute necessity for survival, continues today in terms of humans needing to breathe oxygen that plants make.

The results of co-evolution can be seen in many relationships in nature and this connection ties human health to the health of the natural environment. Humans eat plants which convert inorganic compounds to organic compounds (Pollan, 2002). Plants provide people with essential nutrients and human activity disperses plant seeds. Human guts also host intestinal microbes. For the past 5 years the National Institutes of Health-sponsored Microbiome Project has studied the bacteria, fungi, one-celled archaea, and viruses that live within the human digestive system. The conclusions were that humans can be considered a superorganism because of the extent and complexity of intestinal microbial life found. Microbes outnumber human cells at least ten to one and perform essential digestive and immune system functions. This relationship between humans and digestive tract microbiota mirrors examples of mutualistic evolution found at the global ecological system levels. These findings have implications for the healthcare and nutrition fields, and highlight the importance of human consumption of certain plants to promote growth of intestinal organisms that help human digestion and promote health.

Similarly, scientists have discovered the essential role of microbiota in plant health, a commensal relationship developed over 400 million years. Microbiota digest nutrients and protect plants from pathogens in a symbiotic relationship with fungi that creates mycorrhizae at plant roots. Discovery of the disruption and misunderstanding of the ecological role microbiota play in human gut and soil systems has caused some people to speculate a relationship between the loss of the organisms and the increase in immune system diseases in humans. Arden Andersen (2004), soil scientist and physician, claims that human health, gut microbiota, and soil health are directly correlated and have evolved concurrently. This evolutionary connection ties human health to the health of the natural environment.

Another connection humans share with nature is similarities in aspects of our response evolution. As one example, Sarah Earp, undergraduate music and neuroscience major, and Donna Maney, a neuroscientist at Emory University, report similar neural responses and pathways activated for birds and humans when listening to music, possibly demonstrating similar evolution of emotional responses to music and bird songs (Earp and Maney, 2012). They found that both music and bird song elicit responses in interconnected regions of human and bird brains thought to regulate emotion. The response of the mesolimbic reward system indicates the same neuroaffective mechanisms (meaning the way our emotions are tapped) in the bird and human listeners; namely, that some music or song results in dopamine release for birds and humans, and certain other music or song results in the activation of the amygdala or fear. This discovery may demonstrate that music shares many similar social functions in humans and in birds such as facilitating social contact, reducing conflict, helping to maintain personal attachments, and communicating emotional states (Koelsch, 2010). This study gives credence to the argument that humans, as well as other species, interact emotionally with their environment, including the natural environment.

Plants, like humans, have been shown to respond to music and emotions. Peter Tompkins and Christopher Bird discuss this phenomenon in *The Secret Life of Plants* (Tompkins and Bird, 1974) and Cleve Backster, past Interrogation Specialist for the Central Intelligence Agency, has presented his work with polygraph instruments and plants demonstrating emotions or reactions at the Institute of Noetic Sciences and the Institute for Transpersonal Psychology in Palo Alto, California. Backster (2003) reported that plants communicate with humans on an energetic or intuitive plane, demonstrating connection.

He measured stress increases in plants in relationship to the stress of their human caretakers thousands of miles away. Backster found the connection to humans' emotions still strong even if the plants were in lead containers, e.g. as human stress increased plant stress also increased.

Many animals seem to be aware of environmental changes before humans. Aware humans piggyback on the animals' reactions by using their abilities in sensing and noticing the physical and biological signs for dangerous or other situations. For example, birds becoming noisy or quiet can signal predators. Rupert Sheldrake (2005) wrote about the many animals that survived the 2004 tsunami in Southeast Asia. He noted that in the 1970s authorities in earthquake-prone areas in China relied on cues from animals in order to evacuate towns. More research is needed in this area; although in a retrospective research project Marapana *et al.* (2012) found compelling evidence that animals that were not caged or tied were mostly able to escape from the 2004 tsunami. In the Yala National Park in Sri Lanka, no animal deaths were reported, even though it is on the coast.

These previous examples show the evolutionary connections between humans and other elements of the natural world. We need the microbes in our gut to survive and plants need the microbes in the soil to survive precisely because we have evolved together. In some cases co-evolution created mutualistic relationships, connecting humans to natural elements, and prompting the web of life theory that in fact all life is one living system. Other examples, such as human response to music being similar to bird response, show some of the similarities in our evolutionary development with other animals. A final aspect of co-evolution is the mapping of biodiversity and linguistic diversity. Skutnabb-Kangas *et al.* (2003) mapped approximately 7000 languages and found that high areas of biodiversity predicted high areas of linguistic diversity.

Biological connections

This section provides information about how humans' biological functions, including physical health, depend on interactions with nature. Many people, if asked whether humans are connected to nature, would say 'of course': we breathe the air, drink water, eat plants and animals, and use minerals and trees for shelters. From a purely utilitarian standpoint, the idea that humans are connected to nature is widely recognized. At the same time, a remarkable number of people in Western countries might at first respond to the question with a shrug or a disinterested attitude. While we depend on nature for our survival it is also true that in Western countries people can go for long periods of time without being in natural environments, which can cause us to lose our cognitive sense of our connection with nature and result in a disconnect from nature that can be harmful for both people and nature. If humans believe that they do not need nature they ignore their impact on nature. This physical separation may lead to the false sense that we are no longer connected to nature or even that we can survive without nature by relying on technology and human ingenuity.

However, humans' physical connection with nature is ever present though it may be experienced differently, depending on economic status and geographic location. For example, during droughts people with adequate financial resources can afford to eat more expensive imported food. People not able to import food on demand because of economic or other reasons may starve. Ban Ki-moon (2012) tells us: 'Droughts, such as we have recently seen in the United States, Kazakhstan, Russia, Brazil and India, also raise prices in the marketplace—with potential economic, political and security ramifications'. In 2012, 15 million children worldwide starved primarily due to drought conditions and their parent(s) being unable to buy food. In 2011 in Kenya, Ethiopia, Somilia and Djibouti millions of people starved primarily because of drought (BBC News Africa, 2011). In ecosystems called drylands, such as in the Horn of Africa where Somalia is located, the effects of climate change are particularly evident. The intensifying cycles of extreme drought and flooding in this area caused the need for emergency relief for 10 million people in 2011 (Ki-moon, 2012). These numbers illustrate the very real biological connection we have to the natural world and how we are affected by it.

Climate change is altering the geographic distribution of plants and animals. Mosquitoes, biologically connected to humans because they feed on human blood, have increased their range in recent years leading to an increase in human exposure to malaria, yellow fever, and dengue (Reiter, 2001). Cities such as Nairobi and Aursha were purposefully located at an altitude where the climate was unfavorable for mosquitoes, thus decreasing the risk of infection. Due to a warming climate in that area, over 4 million people who once were not at great risk for malaria now are at risk. Additionally, because there are more human-built heated indoor spaces, mosquitoes now have indoor resting sites and can live longer at higher altitudes. The result has been an increase in the mosquito's ability to transmit malaria in the East African Highlands (Reiter, 2001). In this case, humans are changing the habitat and behaviors of the insects to our health detriment.

Florence Nightingale, Ellen Swallow, and many others raised concerns about the impact of the environment on human health and disease, especially as it relates to clean water and diseases such as cholera. They helped humans modify their behavior and environment, and improved health. Our biological health is absolutely connected to nature. Humans need clean air to breathe, clean water to drink, and nutritious food to eat. Even Hippocrates talked about the significant effect that 'airs, waters, and places' have on human health over time (Philo, 2009). Our connection to nature is such that changes humans make to the natural environment can have positive or negative physical effects on health.

Psychological/emotional and spiritual connections

It may be harder to see and feel emotional connections with nature than it is to understand our biological connection, but the examples of evolutionary connections described above also point to psychological, emotional, and spiritual connections to nature. If a plant many miles away exhibits the stress of its caretaker, it implies an emotional or psychological connection. Our psychological connections with nature are mostly demonstrated by: (i) people's self-report of feeling better when in natural environments; (ii) people performing better on stress indicators or cognitive tasks after spending time in nature; (iii) nature having an immunizing effect on people's psychological health or people having psychological problems attributed to a lack of nature; and (iv) showing that people's environmental behavior is related to their emotional attachment to nature or animal and human interactions.

There are many examples of humans simply feeling better after contact with nature. Marghanita Laski (1961) studied spiritual experiences and found that ecstasy usually takes place shortly after making contact with something valuable or beautiful or both. Nature, including bodies of natural water and beautiful natural settings, is the most common trigger that inspires ecstasy experiences. Her research was a combination of literary analysis and survey using a questionnaire. Francis and Cooper-Marcus (1991) asked a sample of university students in the San Francisco area what settings they sought when feeling stressed or depressed. Seventy-five per cent of the students cited outdoor places—wooded urban parks, places offering scenic views of natural landscape and locations at the edge of water such as lakes or the ocean.

John Zelenski and Elizabeth Nisbet (2012) from Trent University in Peterborough, Canada, found in their studies about people's nature relatedness that people who are more connected with nature report being happier than people who are less connected. They also compared outdoor walking with indoor walking and found that outdoor walks in nearby woods contributed more to personal happiness. Moreover, they found that people tend to systematically underestimate how happy short walks in nature will make them (Nisbet and Zelenski, 2011).

Taylor and Kuo (2009) found that children with attention deficits concentrate better after walking in a park, most likely due to the restorative effect of nature. Berman *et al.* (2008)

found that a relaxing 3-mile walk in an arbo-retum refreshed people and they showed more of an increased cognitive ability than people who had taken a 3-mile walk in an urban industrial area. In a different study, Berman *et al.* (2012) found cognitive and affective benefits for people with major depressive disorder, indicating that time in nature could be a clinically viable supple-mentary treatment for this disorder. Others have found contact with nature to decrease depression and mental illness and increase feel-ings of self-efficacy, self-worth, self-confidence, and personal contentment (Pretty *et al.*, 2006; Van den Berg *et al.*, 2007).

In her book *The Ecology of Imagination in Childhood*, Cobb (1977) summarized years of observation and research that showed a cor-relation between deep experiences in the nat-ural world during childhood and healthy development, adult cognition, and psycho-logical well-being. She found that a strong, loving bond between children and nature is indicative of adult creativity. Chawla (1990) concluded that early experiences in nature lead to adult creativity and indicated that this link is related to ecstatic childhood moments, or moments of intense emotion during forma-tive years, most often occurring while in nature. Chawla concluded that these trans-cendent experiences or enchantment can be found in a spectrum of natural areas, includ-ing a weedy patch in an apartment building's parking lot.

Other researchers in Sweden, Australia, Canada, and the US have found that children tend to show more curiosity and participate in more creative play, including participating in more fantasy and make-believe, in natu-rally green playgrounds as opposed to manu-factured playgrounds. These studies showed a difference in the social standing of children and the social distinctions between girls and boys depending on setting. In nature-based play areas, the social hierarchy among chil-dren tended to be based on language skills, creativity, and inventiveness. In manufac-tured areas, the social hierarchy tended to be established through physical competence and the social distinctions between girls and boys were more pronounced (Taylor *et al.*, 2001; Bell and Dyment, 2006).

Environmental stewardship or strong environmental protection feelings have been correlated with time spent in wild or semi-wild places with an adult who taught respect for nature (Chawla and Hart 1988; Sobel, 2008). The theory is that people protect what they are emotionally attached to; in this case, children become emotionally attached to nature and therefore want to protect it.

Animals, such as dolphins, have been shown to produce healing or beneficial changes in humans, perhaps because of an emotional connection between human and animal. In a 2011 book published by Yale Press, Frohoff and Dudzinski reported on over 20 years of research about human and dolphin interactions. A neuroscientist, Lily, initiated human–dolphin interaction inter-ventions in the 1950s, followed by Betsy Smith, an educational anthropologist, and then David Nathanson, a psychologist, both at Florida International University. Nathanson (1998) and MdYusof and Chia (2012) demon-strated that when children swim with dol-phins they become less anxious and more teachable, and their ability to pay attention increases by 500%. Language, speech, gross and fine motor functioning improve for chil-dren more effectively than when conven-tional speech or physical therapy is used.

One of several theories about why heal-ing occurs with dolphins is that the uncon-ditional acceptance stimulates the immune system, enhances self-worth, and gives hope for the future. This fits with the theory put forth by Bernie Siegal, author of *Love, Medicine, and Miracles* (2011), who said that disease comes from a lack of unconditional love, causing the immune system to be vul-nerable. Swimming with the dolphins may increase immunoglobin or I-killer cells, thus stimulating the immune system.

Finally, there are numerous examples of animals warning or saving people from fires, floods and other disasters through a seem-ingly emotional connection between animals and humans. One typical headline might be: 'Dog sounds alarm in Magnolia house blaze', followed by the story about Joey, a normally quiet dog, running through the house bark-ing until the five people woke up and escaped safely from the home (*Delaware State News*, 2013).

Other examples include a pod of dolphins protecting an injured surfer from sharks (Celizic, 2007), and a female gorilla protecting and caring for a toddler who fell 24 feet into the gorilla exhibit at a zoo (King, 2008). Many indigenous people have taken cues from animal or plant behavior over the course of time for seasonal clues or for storm, earthquake, and tidal wave warnings.

This emotional and spiritual link to nature hugely impacts our well-being and ability to live in a compassionate and peaceful world through the immunizing effect for the human psyche, opportunities to recover from mental stress and gain protection from future potential stress, and opportunities for psychological restoration. Humans' emotional connection to the natural world is profound; when humans do not understand this connection their mental well-being can suffer.

Our connection to the natural environment influences our physical, emotional, social, spiritual, and intellectual well-being. There are a number of intricacies to our connection with nature. Some aspects of humans' connection with nature are absolute; we simply have to have air, water, and food and these are sustained by the natural world. We may never be able to make air, water, and food without nature and if we discover a way to do so, we may not want to because of the importance of the more than physical ways we are attached and interrelated to nature.

This complex and interconnected system of life illustrates that human-made changes in the natural environment can have negative consequences for health and well-being. Next the impact of worldviews is discussed.

What are Worldviews?

Worldviews are comprised of collections of images and stories that people use to help make sense of the complex world around them (Marten, 2001). In other words, worldviews are ontological in that they refer to a way of being and knowing about the world. In this case, the word 'world' is not used in a geographical sense, but rather to refer to the entire perceptual content of an individual (Aerts *et al.*, 1994).

Our worldview is personal, though it is conceptualized by social, cultural, and environmental interactions. These interactions dictate to a large degree what we value and how we view the world around us. In this case, the world we view is the totality of our lives, including all to which we relate from our spiritual, physical, affective or emotional, intellectual or cognitive, and social areas of life. On a practical level, having a worldview increases our understanding of our place and relationship to everyone and everything within our sphere, and therefore can offer us security and comfort. Given that humans live in groups, we tend to share a paradigm (some groups of people call this a story) or a collection of paradigms (stories) which can be seen as a collective worldview that we use to explain life, the world, and how we should be in the world. These shared views become dominant worldviews. Paradigms or worldviews are held for particular areas, such as religion, philosophy, science, political movements, and how we perceive our connection with nature. These societal worldviews shape behavior, politics, relationships, institutions, and the totality of our perceptions, and therefore how we act as a group. As different worldviews become dominant they influence so much of our thought or perspective about a subject that this influences the overall direction of development of that area. For example, large-scale archeological evidence of warfare dates back to less than 7000 years ago, but a dominant paradigm today is that defense and warfare are inevitable and part of human life (and death). The rest of this chapter focuses on the development of dominant worldviews as they pertain to the natural environment and, consequently, how they impact our relationship to the natural world.

Even with dominant worldviews, there are people within that society and people and groups outside the dominant culture who have had and have worldviews differing from the predominant paradigm. Sometimes within a culture there are clashes with worldviews, as displayed in the 2012 US elections concerning control over women's reproductive rights.

Throughout history we can identify dominant worldviews about nature as well as alternative worldviews that have been present. As we present the various collective worldviews in this chapter, remember to think about the voices and therefore alternative perspectives of worldviews not often talked about in history or in the present, or even known about by many other people. Also remember that worldviews do not have to be fixed. Given our cognitive abilities as human beings, we can consciously work toward a worldview that provides sustainability for the human species.

The History of Worldviews/Evolving Worldviews about Nature

Perceptions of nature are driven by worldviews and impacted by both culture and environment, and have varied depending on the historical time period. While throughout history there have been multiple worldviews circulating at the same time, humans seem to have established periods where dominant worldviews can be identified. The moral philosopher Denis Kenny (2001, in Eckersley, 2004) found that over human history there have been four substantially different cosmological stories. These include: (i) the *enchanted universe* in which the world is alive with forces, powers, and influences, often personified as gods; (ii) the *sacred universe* of Abrahamic religions in which the world is created by an all powerful, singular God; (iii) the *mechanical universe* of Newtonian physics, embodying a world that runs like clockwork according to a set of physical laws; and (iv) the *organic universe* of Einstein, relativity, and quantum physics in which the distinction between the material and spiritual no longer holds. As we examine major shifts in dominant worldviews these four stories and others will be referenced.

For the purpose of this book, six major historical shifts have occurred in dominant society's perceptions of nature or the collective stories about humans' relationship to nature. These include: (i) first humans; (ii) sacred cycles; (iii) agricultural; (iv) early modernity; (v) industrial; and (vi) technological. During both the first humans and the sacred cycles stages, humans were primarily hunters and gatherers, shifting to an agrarian society about 10,000 years ago. The last four major shifts have occurred relatively recently compared with the notion that humans have been on Earth beginning sometime between 400,000 and 250,000 years ago, with each shift coming significantly faster than the previous shift (see Table 2.1). These stages are referenced with other geologic and archeological time periods. Different fields of study name time periods in regard to their particular frame of reference, which is confusing because the start and ending times vary, and even within disciplines there is often disagreement about start and ending times. Therefore, readers are encouraged to learn the relationships between the time periods and to place less emphasis on the exact dates.

First humans stage

The first human stage dates from the early *Homo sapiens* to about 50,000 years ago and overlaps with parts of the Stone Age and parts of the Early and Middle Paleolithic. Referred to as prehistory because there is no written record, archeologists primarily identify activity through their cultural artifacts of stone tools. The use of the technology of the period to describe the culture continues to the present in part because these artifacts are durable through time and thus preserved. Over time more archeological evidence continues to be found, causing updates in our perceptions of past cultures (see Box 2.1). The current dominant paradigm of equating more technology to more civilized or advanced cultures is, however, overlain on history and continues in the next stages. Because signs of warfare do not show up until later, these people were thought to be hunters and gatherers who lived in small equalitarian societies. They lived in the natural environment and possibly did not differentiate between the earth and the cosmos. Without electric lights the stars appeared closer and the stars were used in the cycles of their lives.

Little to nothing is known about the worldview of pre-modernity humans; however, our understanding of brain evolution

Table 2.1. Six major shifts have occurred in the known history of humanity each with its dominant worldview as gleaned through artifacts and written history.

Stage	Beginning date	Other geologic and archeological period interfaces	Dominant worldview in relationship to nature
First humans	~300,000 years ago	Prehistoric Old Stone Age Early and Middle Paleolithic	Unknown; possibly the importance of rearing children and cooperation, the usefulness of learning through experimentation, a sense of connection or oneness with nature, and the values-of-belonging. Cosmological story may have been the *enchanted universe* and that nature's cycles (stars, seasons, migratory paths) generated a rhythm for life
Sacred cycles	~50,000 years ago	Behavioral modernity Old Stone Age Middle and Upper Paleolithic Reindeer Age	Awe and respect for femaleness, birth, and natural systems, importance of care and cooperation, the importance of rearing children, the usefulness of learning through experimentation, expecting change, a sense of connection or oneness with nature, and the values-of-belonging. Cosmological story of the *enchanted universe* possibly expanded to include *regenerativity* and *reciprocity with nature*
Agricultural	~10,000 years ago	New Stone Age Mesolithic Neolithic Ancient Egypt Roman Period Bronze Age Iron Age	A split in worldviews: one continued with the cooperation and care notions, extending that to the natural world, the other understood ownership and agriculture in terms of dominance and control—that nature was to be cultivated and subdued. This view expanded to dominance and control of other humans. Cosmological story for some continues to include *regenerativity* and *reciprocity with nature* while shifting to the *sacred universe* for those people who worked for dominance and control
Early modernity	~1400 AD	Renaissance Enlightenment Age of Reason	Nature was to be used, cultivated, and subdued. Nature was seen as the source of raw materials for growth. Through Western expansion and based on the Cartesian mechanistic model and Newtonian physics, people believed they could learn mechanistic functions and then control processes, including nature and people. The *mechanical universe* cosmological story evolved, embodying a world that runs like clockwork according to a set of physical laws

Table 2.1. Continued.

Stage	Beginning date	Other geologic and archeological period interfaces	Dominant worldview in relationship to nature
Industrial	~1700 AD	Classical Modernity Late Modernity Industrial Revolution	Nature continued to be seen as the source of raw materials for growth, a commodity. Capitalist economies externalized environmental costs. Western people saw themselves as separate and above nature. Nature romanticism developed in Western cultures. Many pockets of indigenous people continue their reciprocal relationship with nature. The *mechanical universe* cosmological story prevailed in Western economies
Technological	~1900 AD	Space Age	Nature continued to be seen as the source of raw materials for growth with growing numbers of people lobbying for the rights of nature. Capitalist economies externalized environmental costs. The *mechanical universe* cosmological story combined with an *organic, regenerative universe* story and perhaps back to the values-of-belonging

---

Box 2.1. Understanding worldviews of the past

Information about worldviews is gained through different types of records, including tools, other artifacts, and written material, when available. This means that history is pieced together from the remains that survived and the ways these remains are interpreted. Missing evidence or remains that may be discovered in the future can change the prevailing beliefs of the past. For example, recent evidence of the use of fire by early people dates it to 1 million years ago, much earlier than previously thought. In Germany, eight well-preserved spears found in 2011 and dated to be 300,000 years old show humans using these tools many thousands of years earlier than thought (University Tübingen, 2012).

---

helps us construct theories. We know that in this first 250,000 years of human development our brains evolved in close contact with nature. Humans were part of ecosystems, developing with the other flora and fauna in a mutualistic manner. They may have relied on instinct and intuition along with vision, smell, hearing, and touch; they were connected in every sense to the natural world. On most days humans walked about 12 miles and our brains have developed to work better with exercise rather than being sedentary. The outdoor environments by their nature have a certain amount of instability and unpredictability, thus humans evolved to be problem solvers through exploration and while we are in motion. Because the size of the human birth canal is limited, children are born needing years of parental care. Brains remember and encode what we pay attention to (Medina, 2008). Humans, especially females, had to fully pay attention to childbirth and rearing, which likely helped human brains strengthen neural networks for love and care. Humans have mirror neurons and babies, young children, and adults have the ability to mimic behavior. Mirror neurons might have helped early humans take cues from animals. For example, Sheldrake (2005) wrote about villagers in Thailand who followed stampeding buffalo up a hill, not seemingly sure why they did. Their lives were saved from the tsunami.

While the early humans may not have articulated a worldview, they may have had an innate ability to care for their young, believed in immediacy, believed in the importance of rearing their young to survival and of children and adults learning through experimentation, and understood cooperation at least at a family level. Human brains are emotionally empathetic, which may have influenced a worldview of the importance of tending to babies. Brains evolved such that people felt a need to belong, which helped children stay close to kin as they learned to navigate the world and helped family groups stay intact to help each other. This feeling of a need to belong, also helped with oxytocin, first developed with immediate family and kin and later spread to kith. Learning to cooperate and trust each other is a capacity that allowed for the expansion of the human species. While early humans may not have had a choice in traveling miles each day, they possibly incorporated it into their worldview that moving is a part of life. Returning to the concept of humans' connectedness with nature, our brains today continue to work better when we are moving our body. Current research supports that people who exercise outperform sedentary people in long-term memory, reasoning, attention, and problem-solving tasks (Medina, 2008). Implications of this impact of movement on brain functioning are discussed more in Chapter 7.

Carol Lee Flinders (2002/2003) said that the relationships the hunter–gatherers had with the natural world, with one another, and with their concept of spirit included trust, inclusion, and mutual reciprocity, complementing her understanding of their society's core values: intimate connection with the land, empathetic relationship with animals, self-restraint, balance, expressiveness, generosity, egalitarianism, playfulness, and non-violent conflict resolution. If they had a cosmological story, it may have included Flinders' description of 'values-of-belonging' and it may have

been the *enchanted universe*. It would be easy to imagine early humans believing that their world was alive with forces, powers, and heavily influenced by their constant immersion in the natural world. They allowed nature's cycles (stars, seasons, migratory paths) to generate a rhythm for their lives.

Sacred cycle stage

Depending on geographical region, with the northern regions developing more slowly than the southern regions because of the influence of Ice Age remnants, the sacred cycle stage comes about 50,000 years ago. This stage is punctuated by behavioral modernity. Behavioral modernity is marked by specific behaviors interpreted through the diversity of artifacts found in settlement remains and the occasional midden. A widely accepted trait definition in anthropology, archeology, and sociology of behavioral modernity is the point at which *Homo sapiens* demonstrated an ability to use complex symbolic thought and express cultural creativity, often thought to coincide with the origin of language. Symbolic thought includes being able to engage in symbolic thinking such as number systems, writing systems, maps, and models, and demonstrates flexibility in our brain. Developing the capacity for symbolic thought or dual representation allowed culture to be effectively communicated and shared among a larger group of people through language and other symbols; therefore settlements could evolve. It meant humans could share other information such as distances and location of hazards or food. It also meant that people had to be able to understand the symbolism in their culture to be full participants, therefore some people belonged and some did not. Now called a universal developmental task, Judy DeLoache (2010) found that children begin to be able to engage in dual representation at about 18 months and reliably at 3 years of age for some tasks and later for other areas. She defined dual representation or symbolic reasoning as the ability for a person to attribute characteristics and meaning to things that do not really have them. As children develop symbolic reasoning they learn emotional intelligence or how to understand one another's intentions and motivations. This understanding of intentions and motivations allows for cooperation and community building.

Beginning about 50,000 years ago cultural universals or key elements deduced from archeological evidence have been shared by all groups of people since. In addition to the use of complex language, these cultural elements include use of natural resources (humans' geographic range expanded, they used different hunting techniques for different species, and began to use marine resources—fish and shellfish), technology (finely made tools, including bone tools, projectile point, special purpose tools, composite tools, and tools with blades and backed scrapers, and the control of fire, including cooking and seasoning foods), social organization (having myths, spiritual practices, and/or religion, expanded exchange or barter networks, organized group hunting, settlements with living spaces and hearths, systematic burial of adults and children, care for the elderly and infirm, game playing, and music), and art (systematic use of jewelry for decoration or self-ornamentation, the use of ochre and then other pigments, and the creation of figurative art such as cave paintings, petroglyphs, and figurines). Hunting and using reindeer was more common than previously, which is why the period beginning 50,000 to 40,000 years ago is sometimes referred to as the Reindeer Age. There still is no evidence of warfare.

There was an upsurge of visual art and music during the sacred cycle stage. The earliest flutes were found during this stage and current research shows that peak experiences of music, present in all human communities, releases dopamine, which emotionally engages the reward system in the brain (Salimpoor *et al.*, 2011). Seasonal rites, initiation rituals, and other ceremonies related to the participation in the sacred ceremonies of life were reflected in cave art (Noble, 1993). The earliest figurative art found is the Venus of Schelklingen and thousands of other wood and bone carvings of female figures have been unearthed. The earliest known ceramic is Venus of Dolní Věstonice, from about 30,000 to 25,000 BCE. These artifacts serve a referential function and given the plethora of female artifacts it seems logical to conclude that they displayed awe and respect for femaleness, birth, and natural systems.

The artifacts, including burial rituals, seem to indicate an understanding of the importance of cycles and the female procreative energy. In some areas children and women were given preferential treatment to be buried inside the settlements in specific relationship to certain parts of the home (Naumov, 2007). They were buried in the fetal position thought to honor the birth and death cycle. Deduced from thousands of artifacts it seems that the female womb was sacred as well as a life-generating deity that was nature herself (Noble, 1993). As both mystery and source of power, the female body was a metaphor for nature; the womb was ever able to renew herself with the cycle of birth. These artifacts give form to the axiom that we intuitively love what is born.

Medina (2008) provides evidence that our brains are wired for flexibility and improvisation, thought to be a consequence of living in the ever-changing natural environment. The evolutionary milestone, the prefrontal cortex, housed in the frontal lobe and controlling executive functions including problem solving, maintaining attention, and inhibiting emotional impulses, allowed for the sacred cycle stage. Humans now were even better equipped to learn through experimentation and adapt to changing natural environments.

With the description of the time period there could be many interpretations of their worldview in regard to the natural environment and because the population was dispersed there may have been concurrent worldviews. Putting together the evolution of symbolic reasoning, the long childhood needed for learning, and female carvings and other artifacts in addition to tools, these people's worldview may have included awe and respect for nature and natural processes, including childbirth, as well as a shared sense of care and cooperation among humans and with natural systems. Their regard for nature might have contributed to positive outcomes when learning through experimentation, another trait we carry in human brains. The values-of-belonging combined with the *enchanted universe* cosmological story was still part of many cultures; many possibly added the *regenerative universe* and *reciprocity with nature* as foundational for cosmological stories. Cave paintings symbolize their value for natural systems, life in general, and birthing and children in particular. Like other parts of nature, human birth and regeneration were revered. Nature's cycles (stars, seasons, and migratory paths) generated a rhythm for life.

Any story made up about humans in the sacred cycle stage is theory, including that these people were the first to exhibit cultural creativity, implying that earlier *Homo sapiens* did not. Published descriptions of the sacred cycle people tell us about the worldviews of the academicians of the present perhaps more than the people of that time period. For example, of the two main theories (i) that this shift in behaviors occurred gradually over hundreds of thousands of years of human evolution as *Homo sapiens* accumulated knowledge, skills, and culture and (ii) that the language and settlements occurred more like a revolution or sudden event as the result of a genetic mutation or major biological reorganization of the brain that brought forth languages, scientists have labeled the second theory the Great Leap Forward showing a bias that the increase in technology is synonymous with cultural progress. There is a current cultural bias to see early humans in a survival mode with nature. While these people were active, faced environmental challenges, and had no technology, there is no reason to believe that they thought of themselves in a survival mode. They had the time and space for enough art that numerous artifacts survived.

Early humans and the sacred cycle stage are not usually associated with cultivating plants. However, new research sheds light on possible practices that may have occurred before what has commonly been recognized as agriculture, including tending certain plants to increase species diversity, selectively harvesting specific parts of a plant so it grows back prolifically, and replanting sensitive seeds (Turner *et al.*, 2000). The agricultural stage discussed next marks a new relationship with plants.

Agricultural stage

The third stage occurred when humans moved from a primarily nomadic hunting and gathering society to an agricultural one. Toward the end of the behavioral modernity stage people began to harvest wild grains, which may have helped to lead to farming. This development was possible because the climate became more temperate and therefore people and animals could shed their nomadic lifestyle, build permanent homes, and accumulate material goods. The evolution of symbolic reasoning helped people have the ability to form cooperative relationships, which contributed to larger settlements. Along with permanent homes came the ability to save grain for winter or as insurance in case of a drought or other weather-related challenges. During this stage the social structure changed and people could differentiate into different occupations. There was a growth in population, putting people in closer contact than in earlier stages.

The propensity toward cooperation that was the most likely characteristic of the beginning of this stage encouraged reciprocal actions, though sometime during this stage a great shift in beliefs about power and control and the natural world seemed to occur. Leaders of some settlements controlled the group activities while other settlements continued to be egalitarian. There seemed to be a split where some people continued with the cooperation and care notions that included the natural world, while another worldview evolved to be the dominant Western culture that understood agriculture in terms of power and control, and used food surpluses not only to survive but also to dominate other humans. Both of these worldviews are motivated by self-preservation even though the way to go about self-preservation is close to opposite. Flinders (2002/2003) said that people who continued with the values-of-belonging maintained their intimacy with nature and retreated from cultures based on aggressiveness, cunning, and greed.

Human brains have a great deal of diversity in how they interpret sensory information and two people experiencing the same event easily can learn different pieces of information and life lessons. This is why police detectives expect people who saw the same event to have different stories about what happened, and why they are concerned about lying when stories are exactly the same. When children are about 2 years old and again in their teens there is prolific brain activity and neuronal wiring. This is when life experiences turn into belief systems from which we operate. Therefore, for some humans the fear of scarcity led to a perceived need to dominate, while for other humans this concern led them to cooperate and share.

Those who took the opportunity to dominate also believed that nature's primary purpose was to be conquered and cultivated. Society became hierarchical with kings, merchants, farmers and slaves. People could hoard food and use it as power or a weapon and they had to protect the land and material property they had, hence the movement towards warfare. This ability to accumulate and use material wealth as power gave rise to the belief that materialism is good, and launched a preoccupation with symbols of wealth rather than actual measures of health. Neuroscience tells us that humans are herd animals and as such often do what leaders say. Herd behavior has been identified by philosophers, economists, psychologists, sociologists, and marketers. Freud called it crowd psychology and Jung called it the collective unconsciousness; Nietzsche called it a herd instinct while Kierkegaard called it the crowd or herd morality. The book *Instincts of the Herd in Peace and War* popularized the term herd behavior (Trotter, 1914). Markets depend on the theory that enough members of a group want to mimic group members of a higher status, which is why companies hire these people to be pictured using products they are selling.

There is debate about the first archeological evidence of warfare; some references say a battle of some type occurred about 14,000 years ago. Evidence of a battle at Mesopotamia was dated to about 5000 years ago, the same time period that evidence of large-scale military engagements has been found in Syria. Bows, maces, and slings were the common weapons found through the Neolithic and Bronze Ages. Warfare and the agricultural stage appear related because it was near or during the agricultural stage that warfare seemed to begin. Warfare has the same foundation of power and control that grew in many agrarian communities.

Some cultures never used agriculture and remained hunter–gatherers such as the Bushmen of the Kalahari and the Batek people living in forests of Peninsular Malaysia. They also remained peaceful and cooperative people who travel in small bands tied spiritually and materially to their land. Indigenous Australians have occupied regions of northwest Australia for at least 30,000 years. A predominantly peaceful group relying on dreams to guide their day-to-day survival, they did not cultivate crops or maintain permanent settlements. A number of other Australian indigenous people maintain a worldview that retains a close relationship with nature in practice as well as considering their dreams part of their reality. Closely connected to nature, they embrace natural phenomena and life as part of a vast and intricate system. Having dream time and a harmonious existence with nature is the foundation of their worldview. They have maintained this worldview even though the English colonizers introduced a worldview inclusive of warfare, material accumulation, and dominance over nature and other humans. While some people label this indigenous worldview as primitive or label Australian indigenous people as primitive—as an example, they did not know about metal until it was introduced to them in the late 1700s—their spiritual and physical relationship with nature maintained the biological diversity in their environment.

The Nharo Bushmen are another example of people who maintained a sustainable relationship with their environment until recent invasion by Europeans. Even today they want to keep their culture and continue to raise their children on the land as part of the land (Apelian, 2013). As Apelian said, they 'hold true connection—to self, each other, and the land around them' (p. 13). Medina's (2008) conclusion that human brains are wired to learn experientially continues to be practiced in Bushman communities to the extent allowed by law.

The way of life in these cultures, guided by their worldview of close connection with and respect for nature, is sustainable within our current knowledge of natural processes. Embedded in their cultures and language are their traditional wisdom and practices that honor the reciprocity between people and their natural surroundings.

Early modernity stage

The fourth stage began around the 14th century as humans transitioned from an agricultural, feudal, and barter economy toward an industrialized society. It is characterized by the split between science and the church, and movement toward capitalism, rationalization, and secularization. The people in the agricultural stage who turned to power, control, and violence entered the early modernity stage and basically split into two factions: one aligned with the church, and one aligned with the new science. Both factions viewed nature as a commodity.

Nature continued to be something to be cultivated and subdued, with an accompanying worldview emerging that nature's primary purpose was as the source of raw materials for growth. The 15th century saw the invention of the printing press, marking a huge difference in the way information could be distributed. Western imperialism burgeoned in the 16th century. The dominant scientific view was positivist, believing that there was one truth. Concurrent with these changes in the dominant scientific view was the rise of Western religion, though the connection between science and spirituality was severed by the time of the Renaissance.

Unlike the older spirit religions and Eastern religions, the Judeo-Christian worldview holds humankind as being separate from and above nature (White, 1967; Simkins, 1994; Marten, 2001). Part of this belief system stems from the Judeo-Christian ideal that humans are similar to God because they are made in God's image. While nature was still considered sacred by Christian religions—some aspects of the church such as Bible verses, hymns, and other writings praise nature—the worship of nature was rejected

and replaced with the ideal of humans as the stewards and keepers of nature. The biblical cosmology of the creation story that emphasizes the task of humans to subdue nature and the anthropomorphic notion that humans stand above nature helped church followers see nature as subordinate. Some Christians interpret Genesis 1:28 passage of 'Be fruitful and multiply, and fill the earth and subdue it; and have domination over the fishes of the sea, over the birds of the air, and over every living thing that moves on the earth' to mean use and pollute. Pagans, druids, and other nature-affiliated groups of people were persecuted. Some factions of the church maligned nature as evil, wild, and something needing taming.

Copernicus went to jail for presenting a new model of the solar system with the Sun instead of the Earth at the center, symbolizing some of the tensions of the period. In the 17th century Descartes helped cement a mechanized view of the world, including nature, by describing the universe as a giant 'clockwork' with individual mechanical parts. This Cartesian view of the world as mechanistic influenced science to change from observing and experientially learning about nature intact as a system, as humans had done previously, to a mechanistic and reductionist enterprise with the primary purpose to control nature. Newtonian physics reinforced this separation by visualizing the universe as the interaction of billiard ball-like objects. The belief prevailed that matter is dead and inert—and that humans can rearrange it. Reductionism, along with binary and dichotomous thinking, became part of the dominant worldview leaving us with the Newtonian–Cartesian mechanical model of reality, which championed rational objectivity instead of sympathetic intuitive understanding of nature and spirit. Westernized humans now had a firm perception of dominance, believing that if we can understand it mechanistically then we can control it.

During the agriculture stage people started living in larger settlements and therefore no longer developed their identity in small, close-knit, nomadic or semi-nomadic, tribal living conditions in which healthful attachment bonds to the mother and the symbolic mother—the Earth—were formed. Shepard (1995) and Chalquist (2013) argued that without these nourishing bonds humans remain in childish and adolescent confusion that includes not taking responsibility for the health of the earth or for each other. This immaturity, coupled with the inventions of potentially destructive technology, led to power and dominance as a base for many cultures' worldviews by the beginning of the industrial stage.

Life became more sedentary and wars more frequent. Medina (2008) talked about the human brain thinking better when people are physically moving. New research at Princeton shows that physical exercise reorganizes the brain to reduce the stress response (Schoenfeld *et al.*, 2013). Exercise reduces anxiety, allowing the executive functions in the frontal lobe to work more reliably. Gould, who supervised Schoenfeld's research, hypothesized that those early humans who were temporarily sedentary would have benefited by being more anxious; it would increase their avoidant behavior and possibly keep them out of harm's way. However, as Western society generally became sedentary the lack of movement leading to anxiety has had more negative side-effects than benefits, possibly including more violence.

Wilderness has become a special aspect of nature. Nash (2001) describes the development of the Western view of wilderness as having its roots in a concern for survival and a desire and felt need for humans to control their environment. He postulates that for much of the history of civilization 'wilderness' was a place to be feared, as contrasted with paradise—a gentle, easily controlled pastoral environment where food could easily be grown and humans could feel safe from predators and environmental dangers. This view, informed to an extent by Christian ideals and extended by seemingly endless supplies of natural resources, guided the treatment of the New World by the European settlers. It wasn't until the romantics in the late 18th century and the transcendentalists in the 19th century—both reactions against the rationalization of nature of the 17th century onward—that nature was again perceived by some Western people as aesthetically, culturally, and spiritually important.

The dominant worldview about nature of the early modernity stage can be summarized as nature was to be used, cultivated, and subdued. Nature was seen as the source of raw materials for growth. Through Western expansion, and based on the Cartesian mechanistic model and Newtonian physics, people believed they could learn mechanistic functions and then control processes, including nature and people. The *mechanical universe* cosmological story evolved, embodying a world that runs like clockwork according to a set of physical laws, leading to the industrial age and ever-increasing pressure on natural resources. This pressure not only included increased use, but also reliance on the environment to absorb the polluting effects of industrialization.

Once again there were groups of people who continued the traditions of the sacred cycle stage including earthen spiritualities, animistic, and shamanistic traditions in which humans believe that they are integral with nature, sharing the same life essence.

Industrial stage

The pursuit of power and dominance helped usher in an industrial society. In the 1700s, when global population was a tenth of its current size and the world's resources seemed infinite, the concept of growth seemed like a road to a better life. Nature was integral to growth, providing raw materials and a platform on which to externalize manufacturing costs. Catalyzed by the invention of the first fossil-fuel power in the form of the steam engine, bringing with it unfathomable changes in the magnitude and speed of travel, the industrial age propelled people into a time of rapid change. Worldviews driving the Industrial Revolution included an economy dependent on growth and education based on the assembly model, resulting in a culture of consumption accompanied with materialism and consumerism held in esteem. Nature was an endless means to an endless end.

The belief in economic growth in the late 1700s was the catalyst for an economic revolution characterized by devotion to modern commerce and efficiency (McKibben, 2007) and dependent on consumer spending. Since World War II, growth has been integral in Western policy and aggressively exported to the world at large. The steam engine fueled growth, playing a key role in removing water from mines, transporting coal to the cities using locomotives, ships and roadways, and enhancing the manufacturing process in factories and mills. Even today, 90% of US electricity is from steam turbines.

This modern market economy does not reflect the full costs of goods and services, place any value on ecosystem services, or respect nature and basic ecological principles, resulting in a market that provides misleading information to economic decision makers. The incredible growth manifested in the US during a few short decades (100% growth in years, rather than millennia) led to a turbulent decade in the 1970s in which rivers caught fire, cities became thick with smog, and dependence on foreign oil shocked our economy (McKibben, 2007). Nature was reaching carrying capacity for many pollutants, motivating the first calls for limits to growth including E.F. Schumacher's (2009) *Small is Beautiful*. These clarion calls were quickly eclipsed by the political will of the 1980s and 1990s during which economic growth once again took center stage.

During the 19th and 20th centuries education was adjusted to align with the philosophy and perceived needs of the economic market. With the Industrial Revolution came the need to train the population en masse 'to perform as parts of machines – with precise, repetitive, mind-numbing action' (Ackoff and Greenberg, 2008). Achieving success with the pursuit of industrialization required that children receive compulsory mass schooling intended to prepare them to be obedient parts of the new economic engine. The resulting mechanistic approach to education paralleled the increasingly mechanistic view of the world (Sterling, 2001).

In large part resulting from the above economic and educational paradigms, the dominant Western cultural narrative or worldview has become one in which everyone could and should pursue the acquisition of material wealth as the pathway to happiness and freedom. A pervasive narrative has been created that people are separate from one another and the rest of the natural world and that the purpose of life is to primarily serve individual interests. Playing into the humans-are-separate-from-nature narrative was the fact that a transition from rural to urban environments meant that fewer people in the US lived in rural areas, therefore fewer people had nature as a large part of their everyday lives.

In summary, nature continued to be seen as the source of raw materials for growth, a commodity. Capitalist economies externalized environmental costs. Western people saw themselves as separate and above nature. Beginning in the late 1800s, a subculture of nature romanticism developed in Western cultures including authors Henry David Thoreau and John Muir. By the 1900s scientists and writers like Ellen Swallow Richards, Florence Nightingale, and Rachel Carson were expressing the importance of the interface of nature and human health, encouraging people in politics and at home to clean up the environmental pollution as well as take time to experience the beauty and awe of nature. There were people who believed it was important for children to be outside, resulting in the recreation movement to help children have safe outdoor play environments. The *mechanical universe* cosmological story prevailed in Western economies and was exported to as many cultures as were receptive. Many pockets of indigenous people continued their reciprocal relationship with nature; however, many were simply overrun by the industrial machine.

Technological stage

In the US the last 100 years have seen the beginning of the technology age and changes in society that hugely diminish the time people spend in contact with nature. By 1900 only 40% of US households lived on farms and by 1990, 1.9%. By the late 19th century technology was in place to transmit electrical current on a widespread basis. In 1882 the first central power plant was built in Manhattan providing light to about 500 homes. Technology for light spread rapidly and by 1895 a large-scale power plant was in place at Niagara Falls, replacing Edison's direct current system with alternating current more efficient for long-distance transmission. The ubiquitous availability of electric light reduced people's need to synchronize their activities with natural day and night rhythms, creating another physical separation from nature.

The evolution of manufacturing and technology has allowed us to control things that were once deemed uncontrollable. In the not-too-distant past, a society's livelihood may have depended on local rainfall. Now, not only do we have the ability to ship foods and goods all around the world, effectively making up for poor weather conditions in certain areas, but we even have the technology to actually make it rain through cloud seeding and other techniques. While much of these technologies have enabled us to live healthier, longer lives, they have also had the important side-effect of radically changing our interaction with the world around us.

In 1992 the average US household made 2.3 trips to the grocery store weekly averaging 35–40 minutes each, which equates to 4.3 days per year spent food shopping; in contrast our ancestors in the Paleolithic Age are thought to have spent about 20 hours per week or 85 days per year securing food—while leaving significant time for shelter building and perhaps leisure. This lack of attention on food gathering takes our consciousness away from nature and with it our connection with nature. Many people feel alienated from nature and part of that comes from our separation or lack of knowledge about where our food comes from. While we have more food choices than ever in the US and other developed countries, we have less understanding and physical, emotional, and spiritual connection to food, including food gathering, food cultivation, animal killing, and processing.

This sixth shift, from the Industrial Revolution to a technical revolution, has occurred in a relatively short period of time. Now humans are able to live entire lifetimes seemingly without having to encounter nature. Exceptions to this isolation often occur only in the midst of natural disasters such as earthquakes, fires, tornadoes, hurricanes, flooding, mudslides, rock fall, and tidal waves and tsunamis. Considering the impact such encounters imprint on those who experience them, it is not surprising that many people have a reaction of pervasive fear and mistrust of the natural world (Mitten and Woodruff, 2009). Another effect of the distance from the natural environment that technology exacerbates is that humans are further removed than ever before in history from the consequences of their actions on natural systems. This leads to a vicious cycle of dramatic human-caused changes to the natural systems, more natural disasters, and a buildup of negative health consequences because of our self-polluted living environment. The outcome is a web of crises fueled by our near-pathological pursuit of technological fixes that has brought human civilization to the brink of collapse.

From the first binary programmable computer in the late 1930s to the iPhone today and who knows what tomorrow, technology has altered culture. The transition in birthing practices over time is indicative of a larger cultural shift in which science and technology have become revered above natural processes. By the mid-1900s most hospitals had become sterile, often large complexes more concerned with efficiency and cost rather than patient satisfaction. Outdoor terraces and balconies disappeared and parking lots replaced natural areas (Malkin, 1992; Ulrich, 1992; Ulrich and Parsons, 1992; Horsburgh, 1995).

In June 2013 the American Medical Association labeled obesity a disease, and in doing so went against the recommendation of its own council on Science and Public Health. The human body is adapted to storing extra calories as fat; eating more calories than needed and then gaining weight is normal. Labeling obesity a disease is a signal of how far we have distanced ourselves from natural processes and from taking responsibility for lifestyle choices. We are now in a place where we have to consciously try to connect with the natural environment rather than it being something that happens on a daily basis.

The Wilderness Act of 1964 and the wilderness movement illustrated that even when humans wanted to go into the wilderness they thought of themselves as separate. Wilderness is defined by the Wilderness Act of 1964 as a place where 'man' [sic] is separate: 'A wilderness, in contrast with those areas where man and his own works dominate the landscape, is hereby recognized as an area where the earth and its community of life are untrammeled by man, where man himself is a visitor who does not remain' (Anon., 1964, section 2(c)). The principal author Howard Zahniser, along with John Muir and other prominent preservationists of the time were working within the dominant worldview of nature being separate from humans, and thus acting to preserve and protect the natural environment by keeping it separate from human civilization. This view of nature came from the same dominant and separate perspective of people who believed that nature was intended for human use and wanted to consume it as natural resources for growth. Aspects of the current environmental movement retain this separate from nature view. Rather than conceiving ourselves as part of the ecosystem, we see ourselves as living outside it and modifying our behavior—recycling, turning down thermostats, eating local, and installing compact florescent light bulbs—to 'save' the environment.

Our current worldview or paradigm, characterized by anthropocentrism, materialism, and alienation from nature, has resulted in humans leaving a clear and unique record in the Earth's geologic history, causing some geologists to label this historic period the Anthropocene (Crutzen and Stoermer, 2000). While most epochs have lasted millions of years (the Mesozoic lasted for hundreds of millions of years; the Eocene and others lasted more than 20 million years), the most recent period, the Holocene, has lasted 11,000 years, approximately since the end of the last Ice Age. The significant human-driven processes—including nitrogen pollution (humans now synthetically fix more nitrogen than is fixed by all the world's ocean and land plants, including ocean acidification), overfishing, patterns of consumption, and population growth, that are likely to have lasting effects for tens of millions of years—have influenced the jump to a possible new epoch, the Anthropocene (Zalasiewicz et al., 2008).

Zalasiewicz et al. (2011) note that George Perkins Marsh addressed the anthropogenic global change in Man and Nature published in 1864 and in the 1870s Italian geologist, Antonio Stoppani, used 'Anthropozoic' to label the transformation caused by humans. While geologists are slow to embrace change, the term Anthropocene was adopted relatively quickly after Paul Crutzen (one of three chemists who shared the 1995 Nobel Prize for discovering the effects of ozone-depleting compounds) and Eugene Stoermer (2000) used the term Anthropocene in an article published by the International Geosphere–Biosphere Programme (IGBP).

Geologists will not be quick to pin down an Anthropocene epoch dates. This entry into a new geologic period, demonstrated by humans' record left on Earth and in particular the stratigraphic record, is especially evident since the onset of the Industrial Revolution (Connor, 2010) and analyses of air trapped in polar ice showed the beginning of global changes in concentrations of carbon dioxide and methane from the late 1800s. Other scientists say the Anthropocene might not begin for up to 50 years from now because future changes may dwarf our current changes. This dialogue is useful in terms of bringing to light the deep impact humans have on Earth, leaving a unique and clear record, though the actual reality of the change and delineation of epochs will be more precise in retrospect, perhaps 100 years in the future or more.

The impact of the last three of these six major shifts in the dominant worldview of nature has been to reinforce a perceived disconnection of people from the natural world, as well as to reinforce a value of manufactured and non-nature-based goods, including pharmaceuticals and foods that have replaced nature-based goods. Humans' dependency on fossil fuels comes to light as we extract oil from more difficult sources and transport it over greater distances, risking and causing greater and greater environmental disasters.

As shown by imaging, what humans do in their physical life impacts the neuronal configuration of their brains. The change to an agrarian lifestyle, then to an industrial lifestyle and then to a technology lifestyle, and the associated increased sedentary nature has changed the brain; however, it may not be in our best interests. Neural mechanisms guide our social relationships, including motivation and drive, reward and prediction, perception and memory, impulse control and decision making. As beliefs get inculcated into the culture, newborn members grow up learning the new behaviors and beliefs, not knowing other options. Thus, the new worldviews become ingrained. In this manner the *mechanical universe* cosmological story continues to prevail in the technological stage, combined with the *regenerative universe* story believed by some.

The subculture in the industrial stage, which emphasized the need to clean and preserve the environment for human health and wellness continued into the technology stage. There are many small groups of people now working for more connection with the Earth and a sustainable future. Early in the technology stage, Aldo Leopold (1966) questioned the sentiment of humans as conquerors of nature and wrote in his famous Land Ethic that the land and all of its parts, including humankind, should be considered as members of the same community. Leopold did not oppose the wilderness concept, but rather saw it as a means of preserving the art and skill of travel and wilderness recreation, as a scientific laboratory, and as a reserve for wildlife. This was another indication of a movement in the US seeking to reunite humans with the natural environment and recognizing the importance of nature to the health and well-being of humankind.

Even today many indigenous people live in harmony with nature and natural landforms and have kept the knowledge of this mutual dependency between nature and humans at the forefront of their culture. Some Native Americans, flying in the face of a materialistic culture, continue the tradition of potlach or gift-giving festival, though this practice has been made illegal in Canada and the US. A number of aboriginal populations continue to use initiation or rite-of-passage ceremonies, used for thousands of years, to aid in healthy development and maturation. A study about health promotion and illness prevention in Chinese elders revealed that the elders today continue to believe 'conformity with nature' is the key to health and wellness (Yeou-Lan, 1996). In Scandinavian history the importance of nature, popularized by Ibsen with the concept of friluftsliv, Naess with the concept of deep ecology, and others, continues to be strong. More recently, beginning in the mid-20th century many practitioners of outdoor and environmental education, worldwide, have understood the value of being outdoors and have educated people about the natural environment and environmental ethics. Even within dominant cultures, like the US where many people live most of their lives disassociated from nature, numerous folks still spend time outside learning wilderness living and traveling skills (see Jon Young and Wilderness Awareness school; for an historical account of women learning and traveling in the outdoors see Mitten and Woodruff, 2009).

The Resultant Perception of Disconnection from Nature in Western Cultures: Where Are We Today?

As we can see from the first part of this chapter, for over 95% of the time that *Homo sapiens* has been on Earth humans have lived in direct interrelationship with the rest of the natural world and probably understood that they depended on nature (Oelschlaeger, 1991; Glendinning, 1995; Suzuki, 2007; Young *et al.*, 2008). Early people may have been guided by Flinders' (2002/2003) values-of-belonging, perceiving that everything in nature is connected. This perception of connectedness aligns with the worldview of many past and present indigenous populations who follow the seasons, time activities to the blooming of certain flowers and the rising and setting of the sun, and eat foods that are grown, caught, or hunted locally. In the sacred cycles stage there literally was on-the-ground evidence illustrating a worldview where humans considered themselves integral with nature and sharing the same life essence.

However, for some humans during the agricultural stage as they defined accumulation of material goods, including food, and the physical symbolic representations of their spiritual beliefs as wealth, they significantly changed their view of nature as 'other'. During this major shift in their paradigm humans also started engaging in warfare. Fortunately, there was enough space that many cultures continued to live close to the earth guided by the values-of-belonging, and there is evidence that not everyone reacts with a scarcity consciousness that leads to war (see Box 2.2).

Over time the paradigm of competition and dominance was used by many people in the Western world and was starkly manifested when the Judeo-Christian religions preached nature as other and as beneath humans and god. Like a runaway train, this worldview of superiority of humanity over nature continued during the industrial and technological revolutions. Nature became a commodity and over-indulgences a theme. Beginning with the Industrial Revolution at the end of the 18th century, people became the dominant force of change to the Earth's systems. Meanwhile, under-consumption is a crisis for the billions of people around the world whose lives are oppressed by other peoples' material consumption, which has another set of complex social, economic, and ecological impacts.

Swan (1992) suggests that the word dominant is inaccurate and believes that the disconnection from or superior feelings over nature is a result of fear. As humans lost the value-of-belonging (Flinders, 2002/2003) they experienced inner doubts and fears; they felt a lack of self-worth; they did not feel loveable and capable, which are the two ingredients necessary in order to sustain healthy relationships (Clarke, 1998). Roszak (1995) agrees that a deep despair underlies Western culture. Many humans seem lost, unaware of their indigenous roots, and unaware of being part of something larger. We are in a vicious cycle of consumerism as the foundation of our economics, and economics have degraded the global environment. As personal identity becomes further entangled with consumer behavior, it becomes harder to challenge existing patterns of consumption.

The impact of a sense of landlessness

According to Bowers (2006), the wonders inherent in nature are often usurped by modern temptations and economic greed, causing relationships with nature to become insignificant. The cosmology story of *enchantment*, in which the world is alive with forces, powers, and influences, used to come alive for many US children and others as they played or worked day after day in the outdoors, contributing positively to human development of creativity and values (Cobb, 1977). As the modern way of life in industrialized societies has removed many humans from the natural world, people have forgotten the enchantment or have never known it. Many humans, as individuals or societies, no longer have a personal relationship with the land in which they reside. This makes it easy to disregard humans' place in the biotic community. Baker (2008), echoing Leopold, calls this almost

---

Box 2.2. Women may react to stress with tend and befriend behavior

This perception of connectedness also manifests today as relic and relevant behavior around stress. Suzanne Braun Levine (2004) reported that women identified their biggest work challenge as plugging into a source of collective professional energy; they want to be a part of supportive collaborative networks, now described as a tend and befriend response. Sparked by noticing in her stressful work experience that women formed supportive groups, Shelly Taylor began exploring women and stress and labeled the behavior she found tend and befriend. The tend aspect involves nurturing activities that help protect self and others and the befriend aspect involves creating social networks to aid in tending. She and her colleagues hypothesized and later gained evidence for the idea that under stress women are devoted to their offspring (possibly matching the energy to the attachment that babies need for survival). This response, underpinned by the hormone oxytocin, by opioids, and by dopaminergic pathways, suggests that oxytocin may provide an impetus for affiliation. She suggested that females create, maintain, and use social groups, especially with other females, to manage stressful conditions (Taylor, 2006).

complete detachment from nature landlessness. She describes landlessness as the constant absence of awareness, respect, and kinship with the land, advancing the loss of wild, untouched places as well as the overdevelopment of urban areas. Leopold (1966) said that landlessness is significant because when people do not love, value, or associate with nature, there is little desire to protect it and explained, 'The problem, then, is how to bring about a striving for harmony with the land among a people, many of whom have forgotten there is any such thing as land, among whom education and culture have become almost synonymous with landlessness' (1966, p. 210). Landlessness therefore perpetuates more landlessness. Landlessness also negatively impacts spiritual, mental, emotional, and social health and development.

Landlessness is a loss of a culture's sense of place. A sense of place is rooted in the concept that people used to, can, and do form emotional, spiritual, and meaningful bonds with natural areas, making the welfare of the land personally significant (Williams and Stewart, 1998). Developing a strong sense of place provides the foundation from which caring relationships with nature are built (Russell and Bell, 1996). As explained later in the book, the paradigm of place-based education is a means to help students develop a fascination and enchantment with nature and regain a personal connection to the land. A sense of place deepens the sense of community with the biotic world, including strengthening relationships among people and societies (Noddings, 2005). Sobel (2008) and others suggest that building this personal connection to the land helps students comprehensively explore the social, political, cultural, and natural components that define their community. With a sense of place it is hard for people to be passive, indifferent observers. The welfare of one's community, including the natural elements, becomes personally significant and people usually then choose to actively integrate into their personal community (Haluza-Delay, 1999; Lane-Zucker, 2004; Gruenewald, 2008). Elaborated later in this book, spiritual, mental, emotional, and social health all improve with grounding in a sense of place.

Emerging consciousness

At the level of absolute truth, there is no reason to suffer. But at the relative level, we're all in considerable pain. The cause of our discontent is our mistaken feeling of separateness. This isn't based on anything tangible. It's based on beliefs and concepts. The duality of subject and object, self and other, is an illusion imputed by the mind.
Pema Chodron from *No Time to Lose*

Yet humans are not separate from nature even though a substantial portion of people around the world have changed their relationship with nature in ways that negatively impact health and well-being. Laws of science now recognize that all human activity is inherently embedded in the natural world, which sustains us. We are still connected through co-evolution, biological, psychological, and spiritual needs, emotions, and probably intuition. While we often are physically separate from nature, our disconnection is a perception of disconnection. As Chodron said, humans have a mistaken feeling of separateness; but this is not true of all humans.

Some indigenous people risk their lives and their culture daily in order to stay connected with nature. They have retained a worldview that they are a part of nature and that nature is a part of them, and have maintained that perspective through to the present time. Their culture exhibits values-of-belonging and their members typically feel like they belong. For tens of thousands of years these indigenous cultures have taught a story about inherent goodness and the connectedness of a living universe. Following this ancient wisdom, science is now discovering evidence that humanity is hardwired for connection and compassion: from the vagus nerve, which releases oxytocin at simply witnessing a compassionate act, to the mirror neurons, which cause us to literally feel another person's pain. Darwin emphasized that humankind's real power comes in the ability to perform complex tasks together, to sympathize and cooperate; he did not say that human survival depended on competition (Shadyac, 2010).

These indigenous societies have been largely marginalized (Apelian, 2013). Cultures that have a close connection to the natural world have been mistreated and maligned, yet they exhibit the moral resolve to maintain interwoveness with nature. Those groups that are in remote areas and who have a deep positive relationship with nature have been less affected by the dominant society, but with the intense use of certain natural resources, including trees, oil, and gas, few people and areas are untouched by the dominant paradigm of greed and consumption. As we learn more about cultures that are rooted in their land and exhibit values-of-belonging—for example, Apelian's (2013) work with the Nharo Bushman—we can continue to learn about living in harmony and connection with nature. For too long, the notion that power and material gain equals civilized has dominated globally, because until the deer have their historians, tales of the hunt shall glorify the hunter.

At the same time many people have recognized that we need to integrate and interweave our lives more with nature. Even within the dominant culture during the age of industrialism and growth people have championed a closer relationship with nature. As we moved into the 20th century, the leaders of environmental movements, including Rachel Carson, Dian Fossey, Jane Goodall, Aldo Leopold, Dana Meadows, John Muir, Ellen Swallow, and Howard Zahniser, were motivated to educate people that the world was facing an emerging environmental crisis. The environmental movement, sometimes criticized because nature can still be treated as other, has gained support and made progress in educating people about our dependency on the natural systems. Many individuals such as Julia Butterfly Hill performed radical acts to call attention to our need to connect with nature. In Hill's case she climbed a 1500-year-old Californian redwood tree to make a point and ended up staying in the tree for 738 days. Her vigil ended with an agreement to leave a 200 foot buffer of trees in the logging operation and a US$50,000 donation to Humboldt State University to research sustainable forestry.

Religion remains an essential institution in shaping worldviews, and many religious people work within the church to connect church ideology with environmental problem-solving and citizen action. Examples include Mary Evelyn Tucker and John Grim from the Yale School of Forestry and Environmental Studies creating in 2006 the Forum on Religion and Ecology; the United Church of Christ commissioning a study in 1987 on environmental justice in the US; the Earth Ministry hosting a web page featuring creation care sermons; and Warren Wilson College holding an 'eco-sermon challenge' in 2008 to highlight the role of the church in the environmental movement.

Humans have the capacity for moral thinking and some people do not follow the herd, as exemplified by postconventional (Rest *et al.*, 1999) thinkers and doers such as Mother Teresa and the four students from the all-black North Carolina Agricultural and Technical College, Clarence Henderson, Frank McCain, Josepha McNeil, and Billy Smith, who sat at the all-white Woolworth lunch counter (Woolworth changed its segregation policy to serve everyone at the counter after 6 months and a one-third drop in total overall sales), beginning a lunch counter movement that lasted years. The notion of independent moral thinking and therefore behavior is supported by neural research demonstrating that there is a great deal of diversity in how individual brains are wired and supporting the concept of multiple intelligences (Medina, 2008). Interfacing with philosophy, Hume (1739) cautioned people not to confuse *ought* with what we see as *is*. Just because warfare has been part of humanity for the past 7000 years or so does not mean that it ought to be part of the human experience or that it is natural or inevitable. It means that humans have to use the strength in their frontal cortex and executive functions to consciously make other choices of behavior that lead to a sustainable paradigm. Rooted in the circuitry of the brain is the ability to care for the well-being of self, offspring, mates, kin, and kith (Medina, 2008). In fact, recent research flies in the face of the flight or fight stress response in humans being the primary option.

As we moved from being hunter–gatherers to farmers, to city-dwellers, our relationship with the natural world changed. However, we need to reconsider what type of relationship with the natural environment would be beneficial to human health, development, and well-being. From the past we see the psychological benefits of kinship with nature and connection with our ancestral heritage. Because humans have denied this close relationship with nature, we have alienation and anger with the accompanying psychosomatic illness and aggression.

Korten (2006) says we have been living in the *mechanical universe* cosmological story that some people term the 'Empire story'; in this story people are hardwired for greed, competition, and violence. He says we need a different story; such a story seems to be emerging and it is supported by recent scientific findings, the ageless teachings of the great religious prophets and wisdom traditions, as well as our daily experiences. It is that human beings are born to connect, care, learn, and serve (Korten, 2006). Research has shown that since the mid-20th century, there has begun to be a cultural shift embodied by both changes in individual values and behaviors, and the emergence of social movements that focus on the creation of a progressive and sustainable human culture. Yearning for Balance, a 2004 study by The Harwood Group, found many possible postconventional moral thinkers not following the crowd. They found that 'people from all walks of life share similar concerns about a culture of materialism and excess, and the consequences for future generations'. Many expressed excitement when learning that others shared their concerns about misplaced values, children, and the environment and their longing for a more balanced life (Harwood Group, 2004). Supporting research confirms that high material wealth does not correlate with happiness (Speth, 2009) and that economic growth no longer contributes to the happiness of most people (McKibben, 2009). Validating the idea of cultural change, a study of people in the US and several European nations between 1970 and 2001 found a pronounced shift from materialist to postmaterialist values (Inglehart *et al.*, 2004). Additional studies reveal that 35% of Americans are 'cultural creatives', people who have a made

a comprehensive shift in their worldview, values, and way of life to emphasize relationships, communities, spirituality, and ecological sustainability (Ray, 2008). This notion of the ability to choose a course of action, and research about current choices gives great hope for new stories and worldviews.

To create a new worldview in which both nature and people are valued and sustained and our profound interdependence is recognized, we need to create sustainable education that extends beyond formal schooling toward 'a vision of continuous re-creation or coevolution where both education and society are engaged in a relationship of mutual transformation' (Sterling, 2001). Through community building and education we need to connect with our ancestral heritages to awaken and value our indigenous consciousness and our multiple intelligences, including naturalistic intelligence. Buddhist philosopher, environmental activist, and systems theorist Joanna Macy (2006) has an optimistic view of humans' ability to reconnect with the values-of-belonging. According to Macy, on one hand we face 'the great unraveling', in which our environment, social structures, and species are collapsing under the weight of current stresses, and on the other is 'the great turning', where humankind will learn how to reconnect strongly with nature because we can—and must—live in harmony with all life. While the unraveling is expressed through destabilization and degeneration, the turning is made visible by the largest social and ecological justice movement the planet has ever known (Hawken, 2007), in which numerous small, medium, and large organizations address various dimensions of human activity, revealing ever more clearly the interrelationship between diverse areas of concern (Marques, 2011).

Natural systems are providing a rich source of ideas and wisdom for how we can approach the challenge of creating schools that have the capacity to successfully transition to a new living systems-based paradigm. Natural systems and processes work everywhere, no matter the culture, group, or person, because they are basic dynamics shared by all living beings (Wheatley, 2005). The underpinning philosophy and approach of the ecological paradigm is whole systems thinking in which the dominant analytic, linear, and reductionist forms of thinking are replaced by holistic thinking that reference natural systems to be understandable, accessible, and practicable. Supporting Macy and Hawken, Wheatley and Frieze (2007) assert that the way in which all large-scale change happens in the natural world is the process of separate, local efforts connecting and strengthening their interactions and interdependencies, which they refer to as 'emergence'. In our effort to transition to the new, whole systems thinking-based ecological paradigm for both our society and our education system, this natural concept of change through emergence is motivating.

As Duane Elgin, founder of the Great Transition Stories project, emphasizes, 'it is vital that the human community come together and consciously co-create visions and stories of a sustainable and thriving relationship with the earth and one another'. The 'great transition' is a vision of how humanity could turn the planetary phase of civilization into an opportunity to create a global society that reflects egalitarian social and ecological values, affirms diversity, and defeats poverty, war, and environmental destruction (Raskin et al., 2002). The defining feature of the great transition is the ascendancy of a new suite of values—human solidarity, quality of life, respect for nature, and interconnectedness.

Sterling (2001) argues that 'the fundamental tension in our current age is between the mechanistic and organicist way of viewing the world'. In harmony with the hope of a great transition, Kenny (2001, in Eckersley, 2004) posits that we are on the threshold of a fifth cosmology: the *creative universe* in which the universe is understood to be a self-organizing and creative process in which 'the human species is given the opportunity to take full control of our future'. Duane Elgin (2009) states that such a cosmology must understand the universe to be fundamentally alive in order to radically transform how humanity relates to life. Under this cosmology, rather than searching for meaning, people will create it by taking responsibility for the design of our personal, social, and planetary future.

Humans have a unique and extraordinary capacity to choose our destiny by conscious collective action, which is driven by their worldview: a cultural system of customary beliefs, values, and perceptions that encode our shared learning. Human brains are flexible and learn about what we pay attention to; we can train our brains to understand systems thinking and pay attention to the land and our place in the land. We know that positive affiliation with nature is good for human health, development, and well-being and we can use our cognitive abilities to put this into practice and into our worldview. Chapter 3 begins to capture the rich history of the ways that humans have interfaced with the natural environment for health benefits.

References

Ackoff, R. and Greenberg, D. (2008) *Turning Learning Right Side Up: Putting Education Back on Track.* Prentice Hall, Upper Saddle River, New Jersey.

Aerts, D., Apostel, L., De Moor, B., Hellemans, S., Maex, E., et al. (1994) *Worldviews: From Fragmentation to Integration.* VUBPRESS, Brussels.

Anderson, A. (2004) *Real Medicine Real Health.* Holographic Health Press, Waynesville, North Carolina.

Anon. (1964) *Wilderness Act of 1964. 16 U.S.C. 1131–1136, 78 Stat. 890.* United States of America.

Apelian, N.M. (2013) Restorative ecotourism as a solution to intergenerational knowledge retention: an exploratory study with two communities of San Bushmen in Botswana. PhD thesis, Prescott College, Prescott, Arizona.

Backster, C. (2003) *Primary Perception: Biocommunication with Plants, Living Foods, and Human Cells.* White Rose Millennium Press, Anza, California.

Baker, M. (2008) Landfullness in adventure-based programming: promoting reconnection to the land. In: Warren, K., Mitten, D. and Loeffler, T.A. (eds) *Theory and Practice of Experiential Education.* Association of Experiential Education, Boulder, Colorado, pp. 359–367.

BBC News Africa (2011) Somalia famine: UN warns of 750,000 deaths. *BBC News*, 5 September 2011. Available at: http://www.bbc.co.uk/news/world-africa-14785304 (accessed 15 April 2013).

Bell, A.C. and Dyment, J.E. (2006) *Grounds for Action: Promoting Physical Activity through School Ground Greening in Canada*. Evergreen, Toronto, Ontario.

Berman, M.G., Jonides, J. and Kaplan, S. (2008) The cognitive benefits of interacting with nature. *Psychological Science* 19(12), 1207–1212.

Berman, M.G., Kross, E., Krpan, K.M., Askren, M.K., Burson, A., *et al.* (2012) Interacting with nature improves cognition and affect for individuals with depression. *Journal of Affective Disorders* 140(3), 300–305.

Bowers, C.A. (2006) *Transforming Environmental Education: Making Renewal of the Cultural and Environmental Commons the Focus of Educational Reform*. Ecojustice Press, Eugene, Oregon.

Celizic, M. (2007) Dolphins save surfer from becoming shark bait. *NBC News*, 8 November 2007. Available at: http://www.today.com/id/21689083#.UWf48b8jlFt (accessed 12 April 2013).

Chalquist, C. (2013) Mind and environment: a psychological survey of perspectives literal, wide, and deep. Keynote speaker, *Fifth Annual Sustainability Education Symposium*, Prescott College, Prescott, Arizona, 16–18 May 2013.

Chawla, L. (1990) Ecstatic places. *Children's Environments Quarterly* 7(4), 18–23.

Chawla, L. and Hart, R. (1988) Roots of environmental concern. In: Lawrence, D., Habe, R., Hacker, A. and Sherrod, D. (eds) *People's Needs/Planet Management: Paths to Coexistence*. Environmental Design Research Association, Pomona, California, pp. 15–18.

Clarke, J.I. (1998) *Self-Esteem: A Family Affair*. Hazelden Foundation, Center City, Minnesota.

Cobb, E. (1977) *The Ecology of Imagination in Childhood*. Routledge & Kegan Paul Publishers, London.

Connor, S. (2010) Mankind leaves mark on the planet with the end of the 12,000-year Holocene age. Landmark in the Earth's 4.7bn-year history as geologists hail dawn of the 'human epoch'. *The Independent*, 6 April 2010. Available at: http://www.independent.co.uk/news/science/mankind-leaves-mark-on-the-planet-with-the-end-of-the-12000year-holocene-age-1936725.html (accessed 17 September 2013).

Crutzen, P.J. (2002) Geology of mankind. *Nature* 415(6867), 23.

Crutzen, P.J. and Stoermer, E.F. (2000) The 'Anthropocene'. *Global Change Newsletter* 41(May), 17–18.

Delaware State News (2013) Dog sounds alarm in Magnolia house blaze. *Delaware State News*, Friday 4 January 2013.

DeLoache, J.S. (2010) Early development of the understanding and use of symbolic artifacts. In: Goswami, U. (ed.) *The Wiley-Blackwell Handbook of Childhood Cognitive Development*, 2nd edn. John Wiley & Sons, Inc., Hoboken, New Jersey, pp. 312–336.

Earhart, B.H. (ed.) (2001) *Religious Traditions of the World: A Journey through Africa, Mesoamerica, North America, Judaism, Christianity, Islam, Hinduism, Buddhism, China, and Japan*. HarperCollins, New York.

Earp, S.E. and Maney, D.L. (2012) Birdsong: is it music to their ears? *Frontiers in Evolutionary Neuroscience* 4(14), 1–10.

Earth Ministry (2013) Creation Care Sermons. Available at: http://earthministry.org/resources/worship-aids/sermons/creation-care/Creation%20Care/?searchterm=creation care sermon (accessed 21 September 2013).

Eckersley, R. (2004) A new worldview struggles to emerge. *The Futurist* 38(5), 20–24.

Elgin, D. (2009) *The Living Universe*. Berrett-Koehler Publishers, San Francisco, California.

Flinders, C. (2003) *Rebalancing the World: Why Women Belong and Men Compete and How to Restore the Ancient Equilibrium*. HarperCollins, San Francisco, California. Originally published as: Flinders, C. (2002) *The Values of Belonging: Rediscovering Balance, Mutuality, Intuition, and Wholeness in a Competitive World*. HarperOne, New York.

Francis, C. and Cooper-Marcus, C. (1991) Places people take their problems. In: *Proceedings of 22nd Annual Conference of the Environmental Design Research Association*. Environmental Design Research Association, Oklahoma City, Oklahoma, pp. 178–184.

Frohoff, T. and Dudzinski, K. (2011) *Dolphin Mysteries: Unlocking the Secrets of Communication*. Yale University Press, New Haven, Connecticut.

Glendinning, C. (1995) Technology, trauma, and the wild. In: Roszak, T., Gomes, M.E. and Kanner, A.D. (eds) *Ecopsychology: Restoring the Earth, Healing the Mind*. Sierra Club Books, San Francisco, California, pp. 41–54.

Gruenewald, D. (2008) The best of both worlds: a critical pedagogy of place. *Environmental Education Research* 14(3), 308–324.

Haluza-Delay, R. (1999) Navigating the terrain: helping care for the Earth. In: Miles, J.C. and Priest, S. (eds) *Adventure Education*. Venture Publishing, Inc., State College, Pennsylvania, pp. 445–454.

Harwood Group (1995) Yearning for Balance: Views of Americans on Consumption, Materialism, and the Environment. Prepared for the Merck Family Fund by The Harwood Group. Available at: http://www.iisd. ca/consume/harwood.html (accessed 7 September 2013).

Hawken, P. (2007) Blessed Unrest: How the Largest Social Movement in History is Restoring Grace, Justice and Beauty to the World. Penguin Books, London.

Horsburgh, C.R. (1995) Healing by design. New England Journal of Medicine 333(11), 735–740.

Hume, D. (1739) A Treatise of Human Nature. John Noon, London.

Inglehart, R., Basáñez, M., Díez-Medrano, J., Halman, L. and Luijkx, R. (eds) (2004) Human Beliefs and Values: A Cross-cultural Sourcebook based on the 1999–2002 Values Surveys. Siglo XXI Editores, Coyoacan, Mexico.

Ki-moon, B. (2012) Secretary-General stresses negative impact of climate change in drylands to International Conference on Food Security in Drylands, Qatar, 14 November 2012. Available at: http://www.un.org/ News/Press/docs/2012/sgsm14631.doc.htm (accessed 7 September 2013).

King, B.K. (2008) What Binti Jua knew. The Washington Post, 15 August 2008. Available at: http://www. washingtonpost.com/wp-dyn/content/article/2008/08/14/AR2008081403049.html (accessed 12 April 2013).

Klokov, K. (2007) Reindeer husbandry in Russia. International Journal of Entrepreneurship and Small Business 4(6), 726–784.

Koelsch, S. (2010) Toward a neural basis of music-evoked emotions. Trends in Cognitive Sciences 14(3), 131–137.

Korten, D. (2006) The Great Turning: From Empire to Earth Community. Kumarian Press, Bloomfield, Connecticut.

Lane-Zucker, L. (2005) Forward. In: Sobel, D. (ed.) Place-Based Education: Connecting Classrooms and Communities. The Orion Society, Great Barrington, Massachusetts.

Laski, M. (1961) Ecstasy: A Study of Some Secular and Religious Experiences. The Cressett Press, London.

Leopold, A. (1966) A Sand County Almanac. Oxford University Press, New York.

Levine, S.B. (2004) Inventing the Rest of Our Lives: Women in Second Adulthood. Penguin, New York.

Macy, J. (2006) The great turning as compass and lens. Yes! Magazine, 10 May 2006. Available at: http:// www.yesmagazine.org/issues/5000-years-of-empire/the-great-turning-as-compass-and-lens (accessed 7 September 2013).

Malkin, J. (1992) Hospital Interior Architecture. Van Nostrand Reinhold, New York.

Marapana, R.A.U.J., Hewamanage, D.S., Seresinhe, R.T. and Senaratne, R. (2012) Study on behavioral changes of animals prior to a tsunami natural disaster. In: Senaratne, R., Filson, G.C. and Janakiram, J. (eds) Rebuilding of Tsunami Affected Areas in the Southern and Eastern Provinces of Sri Lanka: Workshop Proceedings. Tharanjee Prints, Maharagama, Sri Lanka, pp. 65–72.

Marques, R. (2011) Individuation within the context of ecological identity: outdoor education and the passage rite into adulthood. MSc dissertation, University of Edinburgh, Edinburgh, UK.

Marten, G.G. (2001) Human Ecology: Basic Concepts for Sustainable Development. Earthscan/James & James, London.

Martusewicz, R. (2005) Eros in the commons: educating for eco-ethical consciousness in a poetics of place. Ethics, Place & Environment: A Journal of Philosophy & Geography 8(3), 331–348.

McGregor, C. (2010) Partnering with Nature: The Wild Path to Reconnecting with the Earth. Atria Books/ Beyond Words, Hillsboro, Oregon.

McKibben, B. (2007) Deep Economy: The Wealth of Communities and the Durable Future. Macmillan, London.

McKibben, B. (2009) Human nature, community, and 'deep economy'. In: Buzzell, L. and Chalquist, C. (eds) Ecotherapy: Healing with Nature in Mind. Sierra Club Books, San Francisco, California, pp. 186–191.

MdYusof, M.S.B. and Chia, N.K.H. (2012) Dolphin encounter for special children (DESC) program: effectiveness of dolphin-assisted therapy for children with autism. International Journal of Special Education 27(3), 1–14.

Medina, J. (2008) Brain Rules. Pear Press, Seattle, Washington.

Miller, J. (2012) Smarter than the average bear: bears use nightfall to avoid hunters. Yale Environment Review, 10 December 2012. Available at: http://environment.yale.edu/yer/article/smarter-than-the-average-bear (accessed 6 April 2013).

Mitten, D. and Woodruff, S.L. (2009) Women's adventure history and education programming in the United States favors friluftsliv. In: Henrik Ibsen: The Birth of 'Friluftsliv'. A 150 Year International Dialogue Conference Jubilee Celebration. North Troendelag University College, Levanger, Norway, Mountains of Norwegian/Swedish Border, 14–19 September 2009. Available at: http://norwegianjournaloffriluftsliv. com/doc/212010.pdf (accessed 7 September 2013).

Nash, R. (2001) *Wilderness and the American Mind*, 4th edn. Yale University Press, New Haven, Connecticut.

Nathanson, D.E. (1998) Long-term effectiveness of dolphin-assisted therapy for children with severe disabilities. *Anthrozoös* 11(1), 22–32.

Naumov, G. (2007) Housing the dead: burials inside houses and vessels in the Neolithic Balkans. In: Barrowclough, D.A. and Malone, C. (eds) *Cult in Context: Reconsidering Ritual in Archaeology*. Oxbow Books Limited, Oxford, UK, pp. 257–268.

Nisbet, E.K. and Zelenski, J.M. (2011) Underestimating nearby nature: affective forecasting errors obscure the happy path to sustainability. *Psychological Science* 22(9), 1101–1106.

Noble, V. (ed.) (1993) *Uncoiling the Snake: Ancient Patterns in Contemporary Women's Lives*. HarperCollins, San Francisco, California.

Noddings, N. (2005) Place-based education to preserve the Earth and its people. In: Noddings, N. (ed.) *Educating Citizens for Global Awareness*. Teachers College Press, New York, pp. 57–68.

Oelschlaeger, M. (1991) *The Idea of Wilderness*. Yale University Press, New Haven, Connecticut.

Ordiz, A., Støen, O.-G., Sæbø, S., Kindberg, J., Delibes, M., *et al.* (2012) Do bears know they are being hunted? *Biological Conservation* 152, 21–28.

Philo, C. (2009) Medical geography. In: Gregory, D., Johnston, R., Pratt, G., Watts, M. and Whatmore, S. (eds) *The Dictionary of Human Geography*, 5th edn. Blackwell, Oxford, UK, pp. 451–453.

Pollan, M. (2002) *The Botany of Desire: A Plant's-Eye View of the World*. Random House Trade Paperbacks, New York.

Pretty, J., Hine, R. and Peacock, J. (2006) Green exercise: the benefits of activities in green places. *Biologist* 53(3), 143–148.

Raskin, P., Banuri, T., Gallopín, G., Gutman, P., Hammond, A., *et al.* (2002) *The Great Transition: The Promise and the Lure of the Times Ahead*. Tellus Institute, Boston, Massachusetts.

Ray, P. (2008) *The Potential for a New, Emerging Culture in the US: Report on the 2008 American Values Survey*. Wisdom University, Mill Valley, California.

Reiter, P. (2001) Climate change and mosquito-borne disease. *Environmental Health Perspectives* 109(Suppl. 1), 141–161.

Rest, J., Narvaez, D., Bebeau, M. and Thoma, S. (1999) *Postconventional Moral Thinking: A Neo-Kohlbergian Approach*. Lawrence Erlbaum Associates, Mahwah, New Jersey.

Roszak, T. (1995) Where psyche meets Gaia. In: Roszak, T., Gomes, M. and Kanner, A. (eds) *Ecopsychology: Healing the Earth, Restoring the Mind*. Sierra Club Books, San Francisco California, pp. 1–20.

Russell, C.L. and Bell, A.C. (1996) A politicized ethic of care: environmental education from an ecofeminist perspective. In: Warren, K. (ed.) W*omen's Voices in Experiential Education*. Kendall/Hunt Publishing Co., Dubuque, Iowa, pp. 172–181.

Salimpoor, V.N., Benovoy, M., Larcher, K., Dagher, A. and Zatorre, R.J. (2011) Anatomically distinct dopamine release during anticipation and experience of peak emotion to music. *Nature Neuroscience* 14(2), 257–262.

Santmire, H. (1985) *The Travails of Nature: The Ambiguous Ecological Promise of Christian Theology*. Fortress Press, Minneapolis, Minnesota.

Schoenfeld, T.J., Rada, P., Pieruzzini, P.R., Hsueh, B. and Gould, E. (2013) Physical exercise prevents stress-induced activation of granule neurons and enhances local inhibitory mechanisms in the dentate gyrus. *The Journal of Neuroscience* 33(18), 7770–7777.

Schumacher, E.F. (2009) Small is beautiful. *The Top 50 Sustainability Books* 1(116), 38–41.

Shadyac, T. (2010) *I Am* [film]. Los Angeles, California.

Sheldrake, R. (2005) Listen to the animals: why did so many animals escape December's tsunami? *The Ecologist* 35(2), 18–20.

Shepard, P. (1995) Nature and madness. In: Roszak, T., Gomes, M. and Kanner, A. (eds) *Ecopsychology: Healing the Earth, Restoring the Mind*. Sierra Club Books, San Francisco California, pp. 21–40.

Siegel, B.S. (2011) *Love, Medicine and Miracles: Lessons Learned about Self-Healing from a Surgeon's Experience with Exceptional Patients*. HarperCollins, New York.

Simkins, R. (1994) *Creator & Creation: Nature in the Worldview of Ancient Israel*. Hendrickson Publishers, Peabody, Massachusetts.

Skutnabb-Kangas, T., Maffi, L. and Harmon, D. (2003) *Sharing a World of Difference: The Earth's Linguistic, Cultural and Biological Diversity*. UNESCO–Tarralingua–WorldWide Fund for Nature, Paris.

Sobel, D. (2008) *Childhood and Nature: Design Principles for Educators*. Stenhouse Publishers, Portland, Maine.

Speth, G. (2009) *The Bridge at the Edge of the World: Capitalism, the Environment, and Crossing from Crisis to Sustainability*. Yale University Press New Haven, Connecticut.

Sterling, S.R. (2001) *Sustainable Education*. Green Books for the Schumacher Society, Great Barrington, Massachusetts.

Suzuki, D.T., McConnell, A. and Mason, A. (2007) *The Sacred Balance: Rediscovering our Place in Nature*. Greystone/David Suzuki Foundation, Vancouver, Canada.

Swan, J. (1992) *Nature as Teacher and Healer: How to Awaken Your Connection with Nature*. Random House, New York.

Taylor, A.F. and Kuo, F.E. (2009) Children with attention deficits concentrate better after walk in the park. *Journal of Attention Disorders* 12(5), 402–409.

Taylor, A.F., Kuo, F.E. and Sullivan, W.C. (2001) Coping with ADD: the surprising connection to green play settings. *Environment and Behavior* 33(1), 54–77.

Taylor, S.E. (2006) Tend and befriend biobehavioral bases of affiliation under stress. *Current Directions in Psychological Science* 15(6), 273–277.

Tompkins, P. and Bird, C. (1974) *The Secret Life of Plants*. Harper & Row, New York.

Trotter, W. (1914) *Instincts of the Herd in Peace and War*. University of Michigan Press, Ann Arbor, Michigan.

Turner, N.J., Ignace, M.B. and Ignace, R. (2000) Traditional ecological knowledge and wisdom of aboriginal peoples in British Columbia. *Ecological Applications* 10(5), 1275–1287.

Ulrich, R.S. (1992) Effects of interior design on wellness: theory and recent scientific research. *Journal of Healthcare Design* 3, 97–109.

Ulrich, R.S. and Parsons, R. (1992) Influences of passive experiences with plants on individual well-being and health. In: Relf, D. (ed.) *The Role of Horticulture in Human Well-being and Social Development*. Timber Press, Portland, Oregon, pp. 93–105.

United Church of Christ Commission for Racial Justice (1987) Toxic Wastes and Race in The United States: A National Report on the Racial and Socio-economic Characteristics of Communities with Hazardous Waste Sites. Available at: http://www.ucc.org/about-us/archives/pdfs/toxwrace87.pdf (accessed 7 September 2013).

University Tübingen (2012) Skilled hunters 300,000 years ago. *Science Daily*, 17 September 2012. Available at: http://www.sciencedaily.com/releases/2012/09/120917085535.htm (accessed 5 July 2013).

Van den Berg, A.E., Hartig, T. and Staats, H. (2007) Preference for nature in urbanized societies: stress, restoration, and the pursuit of sustainability. *Journal of Social Issues* 63(1), 79–96.

Vitebsky, P. (2005) *Reindeer People: Living with Animals and Spirits in Siberia*. HarperCollins UK, London.

Warren Wilson College (2008) Eco-Sermon Challenge. Available at: http://www.warren-wilson.edu/~advancement/Eco-Sermons.php (accessed 16 December 2012).

Wheatley, M.J. (2005) *Finding Our Way – Leadership for an Uncertain Time*. Berret-Koehler Publishers, San Francisco, California.

Wheatley, M.J. and Frieze, D. (2007) How large-scale change really happens – working with emergence. *The School Administrator*, Spring 2007 issue. Available at: http://www.margaretwheatley.com/articles/largescalechange.html (accessed 7 September 2013).

White, L. (1967) The historical roots of our ecological crisis. *Science* 155(3767), 1203–1207.

Williams, D. and Stewart, S. (1998) Sense of place: an elusive concept that is finding a home in ecosystem management. *Journal of Forestry* 96(5), 18–23.

Yale Forum on Religion and Ecology (2013) The Forum on Religion and Ecology at Yale. Available at: http://fore.research.yale.edu/ (accessed 21 September 2013).

Yeou-Lan, D.C. (1996) Conformity with nature: a theory of Chinese American elders' health promotion and illness prevention processes. *Advanced Nursing Science* 19(2), 17–26.

Young, J., Haas, E. and McGown, E. (2008) *Coyote's Guide to Connecting with Nature*. OWLink Media, Shelton, Washington.

Zalasiewicz, J., Williams, M., Smith, A., Barry, T.L., Coe, A.L., *et al.* (2008) Are we now living in the Anthropocene? *GSA Today* 18(2), 4–8.

Zalasiewicz, J., Williams, M., Haywood, A. and Ellis, M. (2011) The Anthropocene: a new epoch of geological time? *Philosophical Transactions of the Royal Society A: Mathematical, Physical and Engineering Sciences* 369(1938), 835–841.

Zelenski, J.M. and Nisbet, E.K. (2012) Happiness and feeling connected: the distinct role of nature relatedness. *Environment and Behavior* (in press).

3

The Historical Connection between Natural Environments and Health

The quality of life, depends upon the ability of society to teach its members how to live in harmony with their environment—defined first as family, then the community, then the world and its resources.

Ellen Swallow Richards, founder of
ecology (1842–1911)

Since the dawn of humanity human health has been explicitly tied to the natural environment. Our ancestors lived with the daily reality that nature influences health and well-being, and early systems of *medicine* (medicine as derived from Latin *ars medicina* or the art of healing) were constructed using this concept of interrelatedness and a systems approach to healing. Even today, it is estimated that about 80% of the world's population relies on nature medicine, largely in the form of plants, for their healthcare. In regions and cultures around the world the deep reliance of people on nature for healing is a recognized and respected part of daily life. For example, Amchis, the Tibetan Medicine healers in Nepal, currently use up to 1800 plants in their work (Adhikari, 2009). However, as societies around the world industrialize, modernize, and urbanize, a frequent side-effect is the perception that nature is separate from humans—perhaps something to be observed, preserved, conserved, or consumed—but not something with which we

are integrally connected and therefore is necessary for our health and well-being. In this chapter, we explore facets of the history of humanity that focus on the specific use of nature to maintain or increase humans' health and well-being. As we trace this history, a strong thread of connection or integration with nature is found. Recent research that supports these nature-based healing practices is also presented.

Our connection and interrelatedness with nature is such that we receive healing and medicine from nature and the next section explores that history. See if you can interpret possible paradigms or worldviews that may have been present as you read this chapter.

Early Indigenous Conceptions of Health and Nature as Connection to the Earth's Rhythms and Attachment to Mother Earth

Part of our current drive as humans is to understand our needs and recognize what behaviors will lead to more health and well-being. Being aware of some of the knowledge and practices of the past helps shed light on current directions. The further back in human history we venture, the stronger the perceived and lived connection and mutuality we see between humans and nature. For most of human history, humans were deeply embedded in the land and natural world in interrelatedness and a lived appreciation of the dependency on nature for daily needs and survival, including healing.

In the time period beginning 40,000 years ago, sometimes referred to as the Reindeer Age, it is thought that Shamans (healers) recognized that everything was interrelated and that healing occurs with harmony with the natural world. In these early times most likely people did not think about themselves in discrete domains of health (physical, intellectual or cognitive, affective or emotional, social, and spiritual); they considered themselves as a whole and these aspects wove into their lives in a manner that defied separation. Therefore when people thought about healing it was automatically in a holistic manner. The medicine or healing was primarily intuitive, perhaps genetically inculcated in humans as discussed in Chapter 4. We suspect that they lived what Doreen Martinez (2008) refers to today as indigenous consciousness or the internal sense or psychological knowledge about the spiritual relationship with all other sentient beings as well as the Earth's processes. Some authors believe that all humans have an intra-indigenous consciousness that can be awakened or activated through time in the outdoors and direct experiences with nature (Mitten, 2010).

At least 5000 years ago and up until at least 3500 years ago medicine and healing continued to emphasize coming back into harmony with natural cycles and rhythms, often symbolized by the medicine wheel. During this time a number of medicine traditions crystallized, including Ayurvedic Medicine (the Indigenous Indian medical system), Native American Medicine, Tibetan Medicine, Traditional Chinese Medicine, and others. In all of these traditions nature was seen as integral to healing, health, and well-being. Indigenous people seem to have an innate wisdom about how to interact, bond, and benefit from a close relationship with nature. Because disease was seen as disconnection from the natural rhythms of the Earth, healing was, in part, to reconnect with nature and the natural rhythms within us that related to natural rhythms in our environment. These medical systems continue to be practiced today; the US Department of Health and Human Services, National Institutes of Health, National Center for Complementary and Alternative Medicine (NCCAM) classifies these medical traditions as an alternative medicine in the category of whole medical systems, defined as complete systems of theory and practice that have evolved over time in different cultures and apart from conventional or Western medicine.

These systems of healing were holistic by their nature—incorporating the human domains of physical, spiritual, social, emotional, and intellectual into the healing. An integrative and interdisciplinary systems approach (remember Chapter 1) was the norm, tending to start from the least invasive approach, such as a change in diet or social and physical activities, which continues to drive these medicine systems today. However, invasive procedures also had begun to be developed. For example, evidence of cataract and other surgical procedures was found in the Susrutasamhita of Susruta, an Ayurveda text dating back approximately 3000 years (circa 6th century BCE).

Both historical and modern Chinese medicine highlights the interconnection between macrocosm and microcosm, teaching that medical knowledge is from a cosmological source (a macrocosmic reality) and influences the microcosmic reality in a detailed and multi-layered manner, supporting the belief in the influence of celestial bodies on human life. This results in a complex system of diagnosis and therapy requiring a systems view of the world. At the same time having its roots in what might be called folklore and embedded in the mythic-poetic mode of observing and describing nature, classical Chinese medicine provides a link to folk biology (see Chapter 6) and many, though not all, practitioners continue to use mythical poetic language with little or no reference to Western scientific language.

These early traditional medicines often incorporated breathing through yoga and meditation which helped maintain health partly through keeping the body in a well-oxygenated state. In recent years scientists have come to show the value of bringing oxygen into the body through meditation and physical exercise for maintaining health. Having a well-oxygenated body (our physical bodies are about 65% oxygen) is preventive for cancer (Edwards, 2008). Understood by the 1931 winner of the Nobel Prize for Medicine, Otto Warburg, and confirmed by scientists since, many types of cancer cells thrive in anaerobic or oxygen-depleted environments and cannot survive in oxygen-rich environments (Warburg, 1966).

Another way to conceptualize health is to understand and live the importance of interdependence and attachment. As discussed in Chapter 5, the attachment of a child to the mother is crucial for healthy development, especially mental health. In the same manner, humans' attachment to the Earth may be necessary for mental health. According to archeological evidence, from the Reindeer Age up until about 3000 BCE, and through recent times for some cultures, people may have worshiped the Earth as a living, female being. Female figures discovered in Europe date this belief back at least to the Paleolithic period, before lunar worship changed to solar worship throughout Neolithic Europe. This female connection symbolized the belief in attachment, dependence, and interrelationship with the Earth, as in mother and child. This same phenomenon is seen in Vedic culture (about 1700 BCE), who worshipped a mother deity called Rigveda Maimata, translating to Mother Earth. Names for the Earth include Mother Nature, Terra, Changing Woman, Ala, Nerthus, and others. These names symbolize a belief in attachment as a child to a mother. It is a dependence that grows to interdependence for health and well-being much in the same way child development specialists talk about healthy human maturation today. Lynn Margulis (1998), a molecular biologist and scientist who shifted thinking in biological science to understanding cooperation and symbiosis as crucial drivers in evolution, was a 20th century Western scientist who confirmed this attachment to the earth theory (Margulis and Sagan, 2007).

Evidence of nature worship or integration is found in Greek culture with evidence in the time period of 750–146 BCE of worshipping the earth goddess Gaia. The worshipping of an earth goddess is an important indicator of these humans' tie to nature and the health they derived from being attuned to the Earth's rhythms. It symbolizes that, as a species, humans knew either by intuition or cognition that we are connected to the 'whole' and that our attachment to nature was life-giving and therefore necessary for health and well-being. In fact as people sat in temples to honor Gaia, they apparently spoke of genius loci or the spirit of place, which relates to the sense of place or being place-based that environmental educators and others believe so important now in terms of helping people connect to the Earth, discussed more in Chapter 8. Even Aristotle weighed in, saying that our minds are linked with nature and that we have intuitive knowledge of the flows, cycles, systems, and creatures of the natural world (Swan, 1992), though he believed the Earth and universe were fixed in time—not evolving.

Understanding the Spiritual Health Connection with Nature

Nature and spirituality interface both in practices and in symbolism. Spiritual health, described in Chapter 1, can be enhanced through nature and spiritual practices and religions reference nature in their teachings. Many people are aware of the ancient polytheist religions including Norse (Northern Europe) and Celtic traditions, commonly called Pagan, that were nature-centered, honoring the Earth's rhythms and nature connections. Druidism—recorded by early Greek and Romans—was a pagan religion that revered oak trees and used mistletoe growing in Velonia oaks to cure infertility. According to spiritual texts, approximately 2000 years ago a number of spiritual teachers recognized the interrelatedness between spiritual, physical, mental, and behavioral health and nature. These actual and symbolic incorporations of nature into spiritual life illustrate ancient practices that demonstrate some of the current findings about the importance of nature for spiritual well-being (Burton-Christie, 1999;

Lodewyk *et al.*, 2009; McCormick and Gerlitz, 2009; Reese and Myers, 2012). How this connection is talked about varies according to the spiritual belief or religion; however, in most cases nature is used to enhance spirituality or the sense of something larger than oneself. In Christianity this connection is exemplified by Jesus' suggestion to 'contemplate the flower and learn how to live' (Tolle, 2005, p. 2). Contemplation, supposedly originating with contemplation of a flower, is the origin of Zen in the 14th century. As the tenth child, Hildegard of Bingen's parents dedicated her to the church at birth and she became a 12th century mystic and healer who used nature, art, and music. Her visionary writings and theology included natural history and the medicinal uses of plants, animals, trees, and stones (Flanagan, 1989). Some of her recorded visions speak to the psychological connection between humans and the natural world very similarly to the bases for ecopsychology (see Chapter 6).

The use of a white bird, the dove, as a symbol of peace or the Holy Spirit illustrates another use of nature. Jesus spent 40 days and 40 nights in the wilderness as a pathway to purification and transformation. Buddha used time in nature as a bridge between the physical and spirit domains and he gained his enlightenment while sitting under a fig tree. In some Eastern religions, the lotus flower is considered a window to the spirit and the beauty of the flower is symbolic of the beauty of a person's essence. The turkey vulture (Latin name *Cathartes aura*, the golden purifier) has had a strong relationship with humans for over 10,000 years. Tibetan people buried their dead in an air burial. The deceased person was set out (sometimes cut into smaller pieces) and the vultures ate the carrion, symbolically taking the soul from Earth. Today in India the Parsis people, descendants from Persians, carry their dead to a large stone amphitheater at the top of a hill and set the corpses out for vultures. This group of people believes that the dead are released from their spirit and purified (McDonald, 1993; Siegel, 2012). There is a distinct health benefit to this practice because many of the dead are often diseased and possibly contagious. The vultures are able to digest the contaminated bodies because

of enzymes in their gut that purify the diseased flesh, in a manner of speaking. Humans may have used vultures to locate fresh kills that may be edible. One might draw a symbolic relationship to angels who have large wings, and in the Christian religion carry the spirit to heaven; vultures have the largest wing span of any living bird. Other cultures including Greeks and Zawi Chemi believed in the healing power of the vulture spiritually and physically.

The practice of Taoism is dated as having originated sometime between the 6th and the 4th century BCE. Taoists understood that the natural environment was important to health and well-being and therefore intertwined time in nature with encouraging their populations to cultivate optimism, passivity, and inner calm. These health benefits from being in nature have been confirmed by modern research. Followers were (and are) encouraged act in harmony with nature because illness was thought to be a result of being out of harmony with nature. Taoist tradition embraces nature as equivalent to the Western religious concept of 'heaven'. As acceptance of and acknowledgement of the necessity and interconnectedness of all, and a description of systems thinking, Lao-tze wrote: 'The real is originally there in things, and the sufficient is originally there in things. There's nothing that is not real and nothing that is insufficient. Hence, the blade of grass and the pillar, the leper and the ravishing beauty, the noble, the sniveling, the disingenuous, the strange—in Tao they all move as one and the same'. Taoism today continues to encourage followers to act in harmony with nature, viewing life as a series of transformations.

While the current general writings of Christian religion and Confucianism can be interpreted as an indictment of nature as evil, Taoism and Buddhism continue to espouse that we are nature and have interdependency with nature from a survival and health perspective as well as a natural aspect of the order of life (His Holiness the 14th Dalai Lama of Tibet, 1992). The Dalai Lama, current leader of Tibetan Buddhism, is quick to point out that Buddhist teaching tells us that care of the environment is pragmatically self-care and that we should avoid drastic changes to the environment in order to preserve our lives. From a practice aspect he advises us that humans are gentle and that we identify with gentler mammals such as deer. He says that we have a non-violent inner core and can and should be non-violent towards each other and nature (His Holiness the 14th Dalai Lama of Tibet, 1992).

Trees, spirituality, mythology, and metaphor

Folklore and mytho-poetic stories continue to influence moral and ethical regard for nature (Hulmes, 2009). However, as science and the church divided, nature was more or less 'forgotten' by the Christian church. There were attempts to incorporate some nature-related customs into this religion, such as churches in the 12th century built incorporating animals and nature symbols into their architecture (Kellert *et al.*, 2008; Barnett, 2009; HRH The Prince of Wales, 2010) and eggs associated with the Christian festival of Easter. The use of eggs may, to some extent, be an adaptation of ancient pagan practices of such folks as Druids, related to equinox and spring rites. The egg symbolizes fertility and rebirth. In Egyptian mythology, the phoenix burns its nest to be reborn later from the egg that is left; Hindu scriptures relate that the world developed from an egg.

Altman (1994) writes about a number of groups for whom trees are prominent in their creation stories. He says that Fiji islanders believe they descended from trees and other sacred plants; ancient Greeks saw humanity coming from ash trees; Scandinavian legend has the first woman coming from an elm tree and the first man from an ash. Christmas trees may have originated with German tradition in the 16th century where evergreens were used to symbolize life and rebirth at the time of the winter solstice, although bringing trees into homes goes back to ancient Celtic times when the fir tree was revered as a sacred tree. Trees figure prominently in cultures and myths such as the Tree of Knowledge in the Garden of Eden or the Tree of Life, still popular in children's books. Doug Hulmes (2009) has researched sacred trees in Nordic cultures, finding that the tradition of sacred trees in Scandinavia goes back to at least the pre-Christian Viking Age. Called a 'Vårdträd' in Swedish and a 'Tuntre' in Norwegian, these trees are planted in the center of a yard and, according to the knowledge passed down, provide a direct connection with nature spirits. Hulmes says, 'The caring for the tree demonstrates respect for ancestors' spirits that were/are believed to reside in the tree, and is a moral reminder of caring for the farm or place where one lives' (p. 2).

One could wonder if humans continue—some in their consciousness and some in their subconscious—to value trees' healing and spiritual qualities. Releafing of trees in the spring or after a storm can symbolize hope for people. Some psychologists use tree drawings as a diagnostic complement, finding that severely ill people often draw extreme tree depictions such as stumps or abused people draw wounds in trees reflecting their wounding (Torem *et al.*, 1990). Writer Toni Morrison draws parallel between tree form and women claiming their power. People are familiar with the concept of a family tree. Many tree metaphors sprinkle our language including the concept grounding or rooting our memories, a tangle of roots, old trees sometimes being described as patriarchs, losing energy described as feeling sapped, a feeling of loss after a move sometimes is described as feeling uprooted, and some say the idea took root. A sense of place or attachment to a place may be described using tree examples such as having deep roots in a place or roots as deep as a tree, feeling uprooted, pining to be home, losing grounding, finding roots, and rooted to the spot. People can be described as feeling rotten inside, taking a stand, lazy as a bump on a log, tall as a tree, a hollow person, a hollow leg, branching out, and feeling fruitless. A common expression is the apple doesn't fall far from the tree. While other nature metaphors also are common, some theorists believe this attachment to trees is due to the spiritual connection between them and humans (Perlman, 1994).

Living in an area with trees increases health according to a number of research reports. Current research in Japan is reviving a practice 'Shinrin-yoku' (taking in the forest atmosphere or forest bathing) with research looking at the calming and stress-reduction effect on the five senses of being in a forest atmosphere (Tsunetsugu *et al.*, 2010). A US Forest Service study showed that the loss of 100 million trees in the eastern and midwestern US influenced death rates. Data analyzed from 1296 counties in 15 states between 1990 and 2007 found that people living in areas infested by the emerald ash borer, a beetle that kills ash trees, suffered an additional 15,000 deaths from cardiovascular disease and 6000 more deaths from lower respiratory disease when compared with uninfected areas. While no causal link was proven, this association between loss of trees and human mortality is troubling; the magnitude of the effect was greater in counties with above-average median household income (Donovan *et al.*, 2013).

The Advent of Modern Medicine and New Therapies

From our earliest days, humans relied on observation, experience, intuition, and indigenous consciousness to inform knowledge and understanding of the natural world, including the healing effect of the natural environment. Transitioning to the 15th century, the cure of the body and the care of the soul were separated and hospitals were now built as separate structures established by individual merchants or guilds (Elmer, 2004).

The term *science* (from Latin *scientia*, meaning 'knowledge') as a body of knowledge about the natural world encompassing empirical, theoretical, and practical knowledge has been around since the 17th century. The term scientist is relatively new; before the 19th century people investigating nature were called natural philosophers. This is because they were trying 'to find natural, rather than supernatural explanations for natural processes' (Gaarder, 1995, p. 27). Today, most science historians consider nature inquiries to be rigorous and adequate science, even those done before the scientific revolution of the 16th and 17th centuries (Hendrix, 2011). The importance of the scientific revolution is that science began to be

shaped as an alternative to the Western religion, with religion having the moral realm and science the factual realm for questions in life. Over time the use of nature in healing became less and less.

Science, in terms of facts, grew to be known as a marker for the progress of human civilization. The more facts we know the more progress we have made. Natural philosophers or scientists used observation, explanation, and prediction to help explain the natural world and human interactions with it. Using the definition of science to be 'the ordered knowledge of natural phenomena and the relations between them', it is easy to see science as a thread in humans' existence well before the 17th century. People have used some sense of ordered knowledge of the Earth and universe, including the Earth's cycles, for millennia as they derived health benefits from nature. The positivist phase (mid-1800s)—where the philosophy of science was positivism, or a sense that the only truth is the truth in scientific knowledge with narrowly defined methodologies—helped science become even more reductionist and removed from common life. Positivists believed that metaphysical knowledge from theology, in addition to lesser known forms of knowledge generation, were not valid. Adding another layer of complication, science was seen as the mechanism by which nature and natural processes could be predicted and controlled. This gross generalization of the epistemology of truth, or way of knowing, perpetuated a separation from nature. The fact is, however, that many modern-day negative stereotypes of science as being cut-and-dried knowledge of only what we can measure were not valid for early scientists (pre-1800s) and are not valid for many scientists today.

In the Greek civilization the natural philosophers within the church who used the natural sciences to determine how to cure illnesses were called temple healers, while those moving towards what is now called a profession might be called physicians. This split between spirituality and science may symbolize and embody the split between healing and curing, another misfortunate consequence of using science to predict and control. As science with the purpose to control and the domination of nature continued, a prevailing thought was that we could find cures for illnesses using nature or natural material. The concept of cure comes from a mechanistic view certainly put forth by Aristotle, including deductive reasoning and empiricism or the notion that truths can be arrived at through observation and induction. For example, understanding that amber is fossilized resin from pine trees can be deduced from seeing samples with trapped insects within them. This building block conceptualization was fixed during the Cartesian period and is with Western societies today. As we conceive of a more holistic view of science using systems theory and realize the ubiquitous nature of natural philosophy, we see science as a useful part of a paradigm.

Paracelsus was a Swiss-German physician and botanist in the early 16th century who incorporated cosmos, spiritual influences, and chemistry in his medical practice. He was a practicing astrologer as many physicians of the time were, though he rejected the popular so-called magical beliefs of the time. His medical views were different from his peers' because he felt that a person's health depended on the harmony of the person (microcosm) and nature (macrocosm). In this manner his views were more like Chinese medicine; however, they were also different in that he was only concerned with the physical body. He used material from nature for cures but mainly relied on what he referred to as balancing minerals in a body.

Healing, hospitals, and herbs

The deliberate integration of natural elements, most often plants, into healthcare environments and practices has a long and rich history, and represents a continued recognition of the healing power of nature. Women healers, sometimes called wise women, were accomplished natural scientists and often buried in history (see Box 3.1). Their use and knowledge of natural remedies and bodily functions laid the foundation for later physicians.

Indicators can be found that more than 1000 years ago people in both Asian and Western cultures believed that plants and gardens are beneficial for patients in healthcare environments (Ulrich and Parsons, 1992). There is evidence of Chinese Taoists creating gardens and greenhouses believed to be beneficial for health as early as 500 BCE. In the Middle Ages court physicians would prescribe walks in the garden to help disturbed royalty calm themselves. Throughout the struggle for the control of medicine, gardens remained popular through the evolving history of hospitals. Hospital comes from a Latin root meaning host (also the root for hostel, hotel, and hospitality). Throughout the Middle Ages, hospitals were often almshouses and resting places for pilgrims. Hospitals often were situated near monasteries so that guests and patients could receive relief through walking in the gardens growing medicinal and culinary herbs and sitting in the outdoor courtyards of the monasteries (Irvine and Warber, 2002). Other hospitals were attached to a church as part of their charity mission as they embraced a charity doctrine. A binding force during the feudal period, a medieval monastery by the 12th century was a farm, an inn, a school, a library, and a hospital. Some historians note that monasteries created elaborate gardens to bring pleasant, soothing distraction to the more well-to-do ill (Gierlach-Spriggs et al., 1998). There was an understanding that this contact with nature increased wellness. The church allowed certain women to be midwives, though by the Middle Ages that was their only legal access to healing.

Gardens and plants remained as prominent features in European and American hospitals into the 1800s (Nightingale, 1996). The hospital gardens provided fresh air, restful physical activity, and a place to socialize as they had in past centuries. In the US during tuberculosis epidemics there was a shortage of hospital beds and the concern of infection. Therefore some hospitals decided to move tuberculosis and some mentally ill patients outside. It turned out to be fortuitous as these patients tended to recover faster and more fully. By the late 1800s the Quakers' Friends Hospital purposefully used nature in the

---

Box. 3.1. Women healers were caught in the cross-fire between nature-based healing and a physician system based on patriarchal values

---

Women in many cultures traditionally have been healers and used nature in their healing most significantly through the use of herbs, many of which are still used today. These women, called wise women, used centuries of empirical evidence gained though practice-based research in prescribing their medicinal cures and healing tonics. Humans' connection to nature in terms of benefiting from the healing use of plants and other materials was evident. This history continues to be uncovered: 'It was witches who developed an extensive understanding of bones and muscles, herbs and drugs, while physicians were still deriving their prognoses from astrology and alchemists were trying to turn lead into gold. So great was the witches' knowledge that in 1527, Paracelsus, considered the father of modern medicine, burned his text on pharmaceuticals, confessing that he had learned from the Sorceress all he knew' (Ehrenreich and English, 1973). We know that by the 14th century, though, many women, possibly in the millions, were persecuted and put to death for practicing their healing arts (Murray, 1921). A sample case follows.

The establishment of medicine as a profession, requiring university training, made it easy to bar women legally from practice. With few exceptions, the universities were closed to women (even to upper-class women who could afford them) and licensing laws were established to prohibit all but university-trained doctors from practice. It was impossible to enforce the licensing laws consistently since there were only a handful of university-trained doctors compared with the great mass of lay healers. But the laws could be used selectively. Their first target was not the peasant healer, but the better-off, literate woman healer who competed for the same urban clientele as the university-trained doctors. Take, for example, the case of Jacoba Felicie, brought to trial in 1322 by the Faculty of Medicine at the University of Paris, on charges of illegal practice. Jacoba was literate and had received some unspecified 'special training' in medicine. That her patients were well off is evident from the fact that (as they testified in court) they had consulted well-known university-trained physicians before turning to her. The primary accusations brought against her were that she would cure her patient of internal illness and wounds or of external abscesses. She would visit the sick assiduously and continue to examine the urine in the manner of physicians, feel the pulse, and touch the body and limbs. Six witnesses affirmed that Jacoba had cured them, even after numerous doctors had given up, and one patient declared that she was wiser in the art of surgery and medicine than any master physician or surgeon in Paris. But these testimonials were used against her, for the charge was not that she was incompetent, but that— as a woman—she dared to cure at all (Ehrenreich and English, 1973).

---

treatment of mental illness and tuberculosis through the use of a greenhouse as well as outdoor sanitariums where people lived in order to recover. It is debated if the canvas and wood 'tents' used by the Jewish Consumptives' Relief Society, Colorado (opened in 1904) or the Firland Sanatorium in Seattle (opened in 1911), combined with the state-of-the-art 'heliotrope' treatment of fresh air, sunshine, and good food, was a successful treatment for tuberculosis. Nevertheless, Colorado and other western states advertised heavily for ill people to recover in the dry air and sun so successfully that by 1920 it is estimated that 60% of Colorado's population were ill people and their families who migrated there from the east coast (Anon., 2009).

The concept of gardening for health extends past the monastery and hospital grounds to gardening by lay people. The 1699 book, *English Gardner*, tells readers that there is no better way to preserve your health than spending time gardening. In the 16th and 17th century gardening became popular, especially by the more well-to-do class as a way to relate to nature (see Thomas Hill's *The Gardeners Labyrinth* (1577) or Hugh Platt's *Floraes Paradise* (1608)). In her book, *Green Desire*, Rebecca Bushnell (2003) writes about the pleasures of garden labor, which she also notes was strictly divided by gender. The gardening fad may have been a way to reclaim some relationship with nature and it seemed to stimulate a social outlet. Gardening continues in popularity as a means to interact with nature and current research shows that people being together in nature leads to stronger social bonds. Many manuals were about the artistry of plant propagation to get a certain color or shape. Gardens were mostly cultivated by upper-class people and they used the garden as an artistic and poetic outlet. While there were books and manuals that described the asymmetrical gardens of China, most English gardens were symmetrical which Bushnell believes was a sign of a need to control the plants or a need to put order in an unpredictable world. However, the irregular placement from the Chinese style was eye-catching and beautiful, causing some gardeners to plant an asymmetrical arrangement. In a sense these early books helped define humans' relationship with nature by creating a worldview that included relating this gardening relationship to nature with health.

Homeopathy, naturopathy, and Bach flower remedies: newer medical systems relying on nature

In 1796 another medical system relying on nature was developed. Using theory from the 16th century, Samuel Hanhnemann, a German physician, developed principles for a medical system in which remedies from plants, animals, and minerals in non-material doses stimulate the body to heal itself and cure symptoms in ill people, while these same substances often would cause ill symptoms in healthy people. This medical practice used observation and an understanding of the characteristics of the plants, animals, and minerals to connect the natural item to healing, demonstrating a belief and practice in humans' connection to the natural world. Hahnemann, followed by others, developed over 3500 remedies. Homeopathy was brought to the US in 1825. A number of hospitals in the US were homeopathic hospitals, such as the Hahnemann University Hospital in Philadelphia affiliated with Drexel University College of Medicine, the Massachusetts Homeopathic Hospital that combined with the Female Medical College of Boston in 1873 to become the Boston University School of Medicine, and the Minnesota Homeopathic Medical College established in 1886. By 1862, 110 hospitals were homeopathic in practice as well as over 30 nursing homes, 62 orphan asylums and retirement homes, and 16 insane asylums.

By 1888 there were over 2400 homeopathic physicians in the US. The American Institute of Homeopathy was founded in 1844 as the first national medical organization in the US and many medical schools in the US taught homeopathy well into the 20th century. However, homeopathy was almost eradicated in the US after the American Medical Association (AMA), founded in 1847, created a charter containing specific language against what they labeled alternative practices and forbid consultation with homeopathic physicians. While homeopathy struggled with the AMA in the US, it has been practiced beside allopathic medicine in many countries, including the UK, and is seeing a revival since the late 20th century in the US. It has come back into some medical schools and alternative schools, including the University of Minnesota.

Both homeopathy and naturopathy are modern whole medical systems classified as alternative medicine by NCCAM. The 2007 National Health Interview Survey asked about the use of Ayurveda, homeopathy, and naturopathy. Although relatively few respondents said they had used Ayurveda or naturopathy, homeopathy ranked tenth in usage among adults (1.8%) and fifth among children (1.3%) (Barnes et al., 2007).

Naturopathic medicine at its core has the principle of the healing power of nature (Vis Medicatrix Naturae). For naturopathic medicine this refers to the belief in an inherent self-healing process in individuals that is ordered and intelligent. In this systems approach an inherent self-organizing and healing process establishes, maintains, and restores health when the physician and patient work together to create a healthy internal and external environment. The origin that the American Association of Naturopathic Physicians claims for first formulating the concept of 'the healing power of nature' is Hippocrates, a Greek physician who lived 2400 years ago. This concept is at the core of many medical systems. In this integrative system, natural healing methods such as botanical medicine, homeopathy, nutritional therapy, hydrotherapy, manipulative therapy, acupuncture, and lifestyle coaching are employed. Naturopathy has its roots in the 19th century Nature Cure movement of Europe exemplified by

Thomas Allinson of Scotland who in the 1880s advocated his hygienic medicine, which promoted a natural diet and exercise while avoiding coffee, alcohol, tobacco, and overwork. In the US Dr Benedict Lust founded The American School of Naturopathy, graduating the first class in 1902. After enjoying popularity, advances in surgical techniques, the discovery of antibiotics, and the growth of the pharmaceutical industries replaced the more low-tech traditional healing processes. This period of deep scientific reductionism lasted through the 1950s and now naturopathy is becoming more popular again.

In the 1930s Dr Edward Bach, a British physician, homeopath, bacteriologist/pathologist, and spiritual writer, inspired by classical homeopathy, developed Bach flower remedies based on his philosophy that disease was caused by a disharmony between body and mind. He believed that a new system of medicine could be derived from nature and he focused on plants. Intuitively derived, using the scientific method, Bach discerned the energy of the flower. He believed the dew or morning water on the leaves contained healing powers and bottled this water with brandy. The remedy is prescribed based on the patient's personality rather than the specific aliment. Today Bach flower remedies can be found in most health shops in the US and other countries.

Vacation, parks, recreation, and well-being

In the US, dating back to the 1800s, romantic notions about nature often fueled spending time outside when it was not for work. Frequently these romantic notions were about being refreshed by spending time in nature, akin to the current restoration theories. Thoreau (1817–1862) wrote about our connections with nature and the importance of being in nature in order to know oneself. This time period saw Henrik Ibsen's writing about friluftsliv (1864) or a way that the tonic of nature gets under one's skin, solidifying in one's being. A unifying concept for outdoor enthusiasts and educators in the Scandinavian countries, friluftsliv is a principal tradition

for outdoor education with the goal to seek to seep nature into one's bones. John Muir's (1838–1914) writings speak to this concept: 'Climb the mountains and get their good tidings, nature's peace will flow into you as the sunshine into the trees. The winds will blow their freshness into you, and the storms their energy, while cares will drop off like autumn leaves'.

The concept of a vacation (a time away for rest and regaining of health) in Europe was most likely a result of the influences of industrialization and urbanization combined with a wish for the romantic. In the early 1800s the concept of going to the mountains for health was becoming in vogue. For example, an English woman, Isabella Bird Bishop (1831–1904), traveled at the advice of her physician. When she was at home in Britain she often was ill with the vapors and other non-descript ailments. When she traveled in Japan, Malaya, Canada, Scotland, and the Rocky Mountains and California in the US while pursuing her natural history interests in nature, she was not ill. By the 1820s mountaineering and adventure pursuits as vacations were ensconced in Europe and by the 1930s the British Mountaineering Leadership Scheme was developed. This beginning of exploration and adventure excursions formed the basis for adventure travel and challenge course programs in Australia, Canada, Great Britain, New Zealand, and the US (Priest, 1986). The concept of vacationing for health and renewal has been better maintained in Europe, New Zealand, and Australia where six weeks of vacation per year are the norm with no less show in work productivity.

The idea of the natural environment being instrumental in attention restoration in addition to being used by hospitals was also used in intentional design to incorporate nature into parks and planned communities. In the 1860s Frederick Law Olmsted (1822–1903), one of the landscape architects responsible for the planning and design of Central Park in 1857, was certainly aware of this phenomenon, as evidenced in the following quote: 'The enjoyment of scenery employs the mind without fatigue and yet exercises it; tranquilizes it and yet enlivens it; and thus, though the influence of the mind over the body, gives the effect of refreshing rest and reinvigoration to the whole system' (Olmsted, 1865 as cited in Nash, 2001).

When Olmsted designed Mount Royal Park, Montreal he combined what we now might describe as a socio-ecological approach to health and well-being. He believed what has now been confirmed by research—that providing pleasantly wooded open spaces would encourage city dwellers to enjoy fresh air and take walks, providing them with healthful exercise. He believed in both a therapeutic and a mystical effect of the natural landscape upon people. While he may have had the prevailing Anglo-Saxton attitudes toward nature of using it for the good of people, he also found in nature, according to Murray (1967), 'the emotional intensity and moral qualities which had hitherto been reserved for formal Christianity' (p. 163). Olmsted considered beautiful scenery to be an effective therapy against mental disease. In his report he stressed the: 'power of scenery to eliminate conditions which tend to nervous depression or irritability. It is thus in a medical phrase, a prophylactic and therapeutic agent of vital value.... And for the mass of the people it is practically available only through such means as are provided through parks' (Olmsted, 1967).

Olmsted developed sites in line with their intrinsic qualities and many sites have stood the test of time. He understood the importance for stress relief as people became city dwellers rather than working with the land. Olmsted was involved with Yosemite and the oldest US state park, the Niagara Reservation in Niagara, New York, as well as one of the country's first planned communities in Riverside, Illinois. There he designed the drive to and from the planned community to Chicago to be lined with trees and green to help alleviate the stress after work. This same concept was used in his design of a boulevard ring as part of the 'emerald necklace' in south Chicago. Perhaps a lesser known park, Presque Isle Park in Marquette, Michigan is a compact version of Olmsted's values of getting people into the environments to relax and recover from stress as well as fortify themselves to return to city life. He strove to have geniality between nature and people; he wanted the

parks to welcome people and he wanted people of poor health and from all classes to have access.

The late 1800s and early 1900s also saw the rise of the camping movement, which was a return to the outdoors for health and romantic ideals. In 1861 Frederick and Abigail Gunn, who ran the Gunnery Camp in Washington, Connecticut for 12 years, decided to take the boys on a 2-week hike to the beach as part of their curriculum in order to keep the boys fit. A focus of the boys' camps was physical fitness and recapturing their rugged individualism reminiscent of the US frontier life (Miranda and Yerkes, 1987), as well as competition, challenge, and conquering the wilderness (Mitten and Woodruff, 2009). Dr Joseph Trimble Rothrock founded the North Mountain School of Physical Culture in 1876 in Wilkes-Barre, Pennsylvania devoted to 'weakly' boys. Boys stayed at the camp for 4 months combining the pursuit of health with practical knowledge. The Boy's Club began in 1900 in Salem, Massachusetts with 76 boys and in 30 years grew to 26,088 campers. The Boy Scouts began in Britain in 1907 and in the US in 1910. The first Boy Scout Camp was on Brownsea Island off the coast of England. The Boy Scouts continue to epitomize what Miranda (1987) indentifies as a form of resistance to a changing world in their (and other boys' camps) conservative view of using the outdoors to preserve a rugged manliness and an allegiance to certain ideals. The boys achieved health gains from the time outdoors; however, the political framing was one of back to the days of the pioneering spirit.

Girls' camps were not extensions of boys' camps for at least two reasons. The women who started girls' camps were caught in the conundrum of feeling the need to be socially appropriate for the time period, and the rugged individualism was not considered feminine. Secondly, though conservative, these women leaders were embracing change and the new roles women would play alongside men in the urbanized US. Using a pedagogy born of feminine and feminist ideals (Miranda, 1987), women camping leaders created educational conditions with a focus on relationships and community. They worked towards cooperative government where the girls took an active role in organizing and leadership, which continues to be a vital theme in Girl Scouts today. They also framed camp as providing a time for networking, relaxation, skills acquisition, and civic engagement. Women camp leaders wanted their programs to emphasize 'the aesthetic and spiritual kinship of girls to nature and to one another'; the pedagogy was for women to have tools to thrive in the changes caused by urbanization, therefore women leaders made the girls' camps 'into excellent social incubators for what would become a new type of woman and the politically active citizen' (Miranda and Yerkes, 1996). By 1874 the first YWCA camp in the Philadelphia chapter of the YWCA, called the 'vacation project', was designed to provide a relaxing environment for young women who worked at tedious factory jobs with little free time. Luther Halsey Gulick in 1890 opened a private camp so his daughter could attend camp and then formed the Camp Fire Girls in 1914, soon reaching 500 participants. However, women's camps started by men tended to be socially normative, creating spaces for girls to learn to cook and be homemakers. In 1902 Laura Mattoon started Camp Kehonka for girls in Wolfeboro, New Hampshire. In 1912 Girl Scouts entered the camping movement. There were enough camps that by 1910 the Camp Directors Association of America (CDAA) was formed to help standardize camp curriculum, and Mattoon was the secretary-treasurer. By 1916 the National Association of Directors of Girl's Camps formed and by 1924 it merged with CDAA, which in 1935 became the American Camping Association, now the American Camp Association (ACA). In 1948 the ACA adopted its first set of camp standards, which eventually became the basis for accreditation of camps across the US. Both genders received the physical and mental health benefits from being outdoors at camps that current research highlights; however, these examples illustrate the influence educational pedagogy has on the health benefits in the social and developmental realms gained in groups in the outdoors. ACA has completed and recorded research about the health benefits of camps.

The founder of the Appalachian Trail, Benton MacKaye, had similar ideas about the importance of nature for health and recuperation. In 1921 he published an article envisioning the trail as a resource for citizens to access recreation, health, recuperation from illness, and as a means of creating jobs (MacKaye, 1921). The trail was developed to provide access to mountains and nature for those who could not afford to take a vacation to the west. MacKaye recognized that larger plots of undeveloped land such as those being developed into western national parks no longer existed in the east and saw the possibility of a linear span of wild lands connecting Maine to Georgia as a practical alternative. According to the Appalachian Trail Conservancy (2013), to date over 13,500 people are recorded as hiking the entire 2000 miles of the Appalachian Trail since records began to be kept in the 1930s, which showed five completed thru-hikers.

Lloyd Burgess Sharp (1895–1963), a pioneer and leader in modern outdoor education, incorporated Dewey's experiential education philosophies into youth camping. Sharp believed that educational lessons and principles could be incorporated into camp settings, thus beginning the school camping movement, which eventually became what we know today as outdoor education. He believed that outdoor and experiential education techniques were essential to the learning process, for if certain things were not experienced, they could not be fully understood: 'That which can best be taught inside the schoolrooms should there be taught, and that which can best be learned through experience dealing directly with native materials and life situations outside the school should there be learned' (Sharp, 1943).

In addition to organized youth camps, the US was seeing the increased popularity of camping as a pastime for individuals and families. This was fueled by the industrialization and prosperity of the 1920s and by the recent establishment of the National Parks. Yellowstone was named the first national park in 1872, followed by Yosemite in 1890, and the establishment of the National Park Service in 1916. Americans began to travel to see these natural wonders, and the culture of camping took off. The rugged individualism of the times was embodied by Theodore Roosevelt, who was president from 1901 to 1909. Roosevelt was an avid hunter and outdoorsman who founded the Boone and Crockett Club with George Bird Grinell, and who was responsible for signing the legislation behind much of the conservation movement of the era. Camping's popularity in the US was evidenced by John Steinbeck's (1937/1994) *Of Mice and Men* as well as the National Park System making Recreation Demonstration Areas a part of the federal government's work relief program. Out of this program, 34 of the areas were organized camp facilities to be used by organizations that did not have their own camping areas. Most of these areas became state parks after the depression.

Health through horticultural therapy

An important development during the early 1900s was the academic world believing that nature is therapeutic. One of the first examples in modern time is the blossoming of the academic field of horticultural therapy. In 1955 the University of Michigan developed a combination horticultural and occupational therapy master's degree and others followed. The groundwork was laid earlier. In the 1920s horticulture therapy began to be practiced in occupational therapy and then showed up in occupational therapy textbooks. The Association for Occupational Therapists in England formally acknowledged the use of horticulture therapy as a treatment for certain psychiatric and physical disorders in 1936. In 1942 Milwaukee Downer College offered a course in horticulture therapy within its occupational therapy degree program; the first such course known to be offered in a US institution of higher learning. During World War II garden club volunteers partnered with occupational therapists to use plants and gardening activities for rehabilitation. McDonald (1995) reported that use of this therapeutic practice for World War II veterans provided important testing and showed the validity of its use in physical and mental rehabilitation, including reduced hospital stays. This

involvement by garden club members continues today. Louis Lipp, a propagator for the Arnold Arboretum of Harvard University, developed a horticultural therapy program as an outreach program at a nearby veterans' hospital in 1953. In 1951, Alice Burlingame, a trained psychiatric social worker at the Pontiac, Michigan State Hospital, started a horticulture program in the hospital's geriatric ward (Lewis, 1976). Burlingame became convinced of the validity of this method with geriatric populations, and she persuaded Dr Donald Watson to jointly convene a week-long workshop in horticulture therapy at Michigan State University, which then led to the eventual degree. The two teamed to write the first textbook, *Therapy Through Horticulture*, in 1960. The first degree recipient, Genevieve Jones, who worked as an occupational therapist at the Hines Veterans Administration Hospital in Chicago, went on to write the *Handbook of Horticulture Therapy* published by the Federation of Garden Clubs in Michigan, the organization that provided Jones a scholarship to pursue her degree. New York University Medical Center's Rusk Institute for Rehabilitative Medicine's horticultural therapy program was so effective that the horticultural therapist was made part of the treatment team with doctors and psychologists, using horticulture both diagnostically and rehabilitatively—a first for people with disabilities. It was called the 'glass garden program' because it operated in a greenhouse.

In 1971–1972, a curriculum was developed at Kansas State University providing students with training in horticulture and psychology leading to a bachelor's degree in Horticultural Therapy followed by a 7-month clinical internship at the Menninger Foundation (Lewis, 1976). Clemson University was the second institution of higher education to offer a horticultural therapy degree.

In England two organizations, the Society for Horticultural Therapy and Rural Training and the Federation to Promote Horticulture for Disabled People, were formed with more of a focus on practice rather than the US focus of professionalizing the practice. The mission of the Society for Horticultural Therapy and Rural Training was 'To relieve persons who are physically or mentally ill, disabled or handicapped, or who are in necessitous circumstances, by the advancement of education in the use of land through horticulture, agriculture, farming and gardening in all their forms'.

Today horticultural therapy is described as: 'a professionally conducted client-centered treatment modality that utilizes horticulture activities to meet specific therapeutic or rehabilitative goals of its participants. The focus is to maximize social, cognitive, physical and/or psychological functioning and/or to enhance general health and wellness' (Haller and Kramer, 2006).

The success of horticultural therapy may be its non-threatening, familiar modes of therapy and rehabilitation. It has been shown to be successful with people who have physical, mental, psychological, or developmental disabilities, as well as those who are victims of abuse, prisoners, young children, and older adults. Horticultural therapy has been shown to help people recover from illness as well as help spiritually heal those who will not recover and who seek quality of life in their final days. Horticultural therapy seems to appeal to gardeners at all levels of involvement.

Horticultural therapy has grown to occupy a useful position within the healthcare system. The People–Plant Council formed in 1990 to document existing research and to encourage new research uses an interdisciplinary symposium, 'The Role of Horticulture in Human Well-Being and Social Development', as one venue to distribute horticultural therapy trends and information.

The Beginning of the Modern Environmental Movement: Ecology

Ecology has become a household word because Ellen Swallow Richards (1842–1911) was a visionary environmental leader who coined the word in about 1892. Swallow Richards understood the connection between the natural environment and human health. As a sickly child growing up on a farm, she experienced firsthand that being in the clean outdoor air made her healthier and stronger. She connected people to their environment

and educated many people at venues such as the 1893 World Fair exhibit, Rumford Kitchen, where she concentrated on preparing nutritious meals from natural foods, published *Air, Water, and Food from a Sanitary Standpoint* (1900), and helped to start Woods Hole Marine Biological Laboratory. It was at the Woods Hole Marine Biological Laboratory that Rachel Carson later studied and had experiences that profoundly changed her, leading to her environmental work. Swallow Richards graduated and worked at Massachusetts Institute of Technology, making headway in sanitation and in teaching children environmental education. She symbolizes a culture's coming to terms with the healthful interface between humans and our environment even as we industrialize.

The modern environmental movement, recreation, and understanding the Earth as an organism

By the mid-20th century, while many people in the US were moving off farms and into suburbs and cities, a host of people began to engage in an environmental movement. Perhaps people intuitively missed nature. From the mid-20th century on many practitioners of outdoor and environmental education, worldwide, have understood the value of being outdoors from both a scientific and intuitive level and have educated people about the natural environment and environmental ethics. Aldo Leopold's (1949) land ethic in *A Sand County Almanac*, published posthumously, created an awareness of interconnectedness with nature and the expansion of 'community' from only humans to the more than human world. The 1964 Wilderness Act presumed the need for people to have time in space untrammeled by man [sic]. Many nature and health connections were publicized during this time, including Rachel Carson's (1962) *Silent Spring*, which talked about our connection with the Earth and how we need to avoid poisoning ourselves with the toxic chemicals used in industrial societies. Her book is often credited with awakening public attention to the negative impacts that humans were wreaking on the natural environment. This time period also saw Robert Greenway's coining of the term ecopsychology in 1963, now a discipline that explores how our psychological health relates to the ecological health of planet Earth. A college professor at Sonoma State University, Greenway annually took students on wilderness trips and then surveyed the students to see if and how this experience impacted their mental health. In the data he found that 90% of the students described an increased sense of aliveness, well-being, and energy during and after their trips and 90% said that the experience allowed them to break an addiction (defined broadly) including nicotine and chocolate (Greenway, 1996). He noted a significant gender difference in that 60% of the men and 20% of the women stated that a major goal of the trip was to conquer fear, challenge themselves, and expand limits and that 57% of the women and 27% of the men stated a major goal was to come 'home to nature', a gender difference noted by other authors as well (Mitten 1985, 1992).

The 1970s saw a continuation of the environmental movement of the 1960s. April 22, 1970 was celebrated as the first Earth Day, and also considered to be the birth of the Environmental Protection Agency (EPA) (Lewis, 1985). This time period saw the advent of legislation to protect the environment, including the Clean Air Act of 1970 and the Endangered Species Act of 1973. Earth Day is still celebrated in 2013. In 1968 we received the first photo of the Earth from space, now labeled the Earthrise photo, which has become a symbol of the holistic nature of the Earth's processes. Many astronauts claim to have been spiritually and emotionally changed forever after seeing the Earth from space (Poole, 2008).

In 1977 Edith Cobb published *The Ecology of Imagination in Childhood*, demonstrating from her research that childhood natural places are critical for healthy human development. She found that adults who spent time in nature during childhood formative years that included establishing a relationship and bond with a particular tree, section of a brook, or the like were more likely to mature into creative and optimistic adults (Cobb and Mead, 1977). In 1979 Drs Lovelock and Margulis,

a chemist and a biologist, respectively, published the Gaia hypothesis, which talks about the Earth as an organism, personifying the primal Greek goddess, Gaia. This understanding of connectedness is reminiscent of the kind of attachment and interrelatedness described by early people until about 10,000 years ago, and for some populations through today. Margulis understood rhythm and heartbeat from a biology perspective and Lovelock, from an atmospheric sciences perspective, found a similar pulse-like effect. This pulse of sorts has been further identified and the mutualistic and systems concept in relationship to the Earth and its processes confirmed. The theory is that the various systems on and in the Earth work together to create an environment optimal for life (see Chapter 6 for a more complete discussion of systems theory). Understanding the Gaia hypothesis and being able to apply systems theory will positively influence health and well-being. Knowing that humans are meant to be part of this complex system rather than apart and that humans cannot define and change the system using reductionist or mechanistic thinking without negative effects propels humans into systems thinking about life at the most basic level as well as throughout all aspects of health care and wellness.

The leisure industry boomed in the US starting in the 1970s. Academic programs in leisure grew at many higher education institutions. Academics and professionals have defined and researched leisure, which has been instrumental in helping to understand why people engage in leisure and if and how people are affected by time in nature.

In their summaries of the Recreation Experience Preference scales, Driver *et al.* (1991) indicated that people are motivated to spend time in nature in order to escape daily pressures, enjoy nature, experience tranquility, and be with other people. Montes (1996) found that people select natural environments for leisure because of perceived restorative or re-creative benefits. Iso-Ahola (1999) says that testing of leisure theory from a psychological perspective shows that people engage in leisure activities, including activities in nature, when they perceive freedom and are intrinsically motivated. These results indicate the importance of choice in leisure and the individual nature and large range of potential benefits.

Whether through informal or formal channels people seek natural areas for health and well-being. Many people write about finding themselves in nature or going to nature in times of turmoil. For example, Joan Anderson, author of two non-fiction *New York Times* bestsellers, when talking about her personal growth in *The Second Journey: The Road Back to Yourself* (2008), references the importance of being in nature to her well-being. This theme of being in nature to feel better or to gain psychological health is common. People seem to know they need to be in nature or notice that they feel better in nature. The anecdotal evidence seen in a number of books is supported by a number of studies including one that asked a sample of university students in the San Francisco area what settings they sought when they felt stressed or depressed. Seventy-five per cent of the sampled population cited outdoor places such as wooded urban parks, places offering scenic views of natural landscape, and locations at the edge of water such as lakes or the ocean (Francis and Cooper-Marcus, 1991). Dr Dawn Yankou (2002), School of Nursing, University of Western Ontario found that walking outside helps alleviate depression and generally improves mood states and well-being.

The 1980s saw landmark research about nature and health benefits. Roger Ulrich (1984) demonstrated that hospital patients who viewed natural scenes from their rooms recovered faster, required fewer painkillers, and had fewer post-operative complications. Research from Ulrich and others about the medical effect of nature views or time in nature now includes dozens of studies and hundreds of publications. E.O. Wilson (1984) published the biophilia hypothesis (the innately emotional affiliation of human beings to other living organisms that is genetically inculcated) and inspired research which suggests that our relationships with nature are a fundamental component of building and sustaining good health. This was followed by Stephen and

Rachel Kaplan's (1989) development of the attention restoration theory (ART), demonstrating the restorative effect of nature and the ability for people to gain mental clarity and to focus through spending time in nature.

At the same time that momentum was building about the benefits of being in nature, the 1980s showed the beginning of a downward trend in people in the US spending time in nature. For the first time since data had been recorded in 1939, numbers of visits to National Parks stopped trending upward and began to drop steadily (Pergams and Zaradic, 2008). This same trend was found at other recreation areas and the decrease in visits did not correlate with causal variables such as videophilia, gas prices, foreign travel, extreme outdoor recreation, family incomes, government funding, and park capacity (overcrowding). The authors were surprised at the feedback they received from a number of readers claiming that people were not fundamentally shifting away from nature-based recreation and that there must be another explanation. It seems that as a general population the US is shifting away from nature-based recreation, but that those for whom recreating in nature is important, it is very important. Like other matters in the US, people may be moving to the extremes.

Camps also began to see decreases in camper numbers. In the late 1990s ACA shifted its focus to positive youth development and began a new research program that included work on developmental outcomes, program quality, environmental leadership, health and safety as well as key business operations. Currently it is estimated that 5,068,600 youth campers and over 300,360 staff participate in approximately 2700 ACA-accredited camps around the country that generate over US$2.8 billion a year in revenues. ACA's 20/20 organizational vision is a commitment to serve 20 million children by the year 2020.

By 1992, with the work of ecofeminists and others, ecopsychology cemented as an interdisciplinary field with the work of Theodore Roszak, Mary Gomes, and Allen Kanner in their 1995 edited volume *Ecopsychology: Restoring the Earth, Healing the Mind.*

Bringing us into the current millennium we have seen a huge jump in research about nature and health benefits, made public by a number of authors including Richard Louv in his books, *Last Child in the Woods* (2005) and *The Nature Principle* (2011). Timing is important; Louv published a similar book in 1998, *The Web of Life: Weaving the Values That Sustain Us*, that received little notice. As Malcolm Gladwell in *The Tipping Point* (2002) and *Outliers: The Story of Success* (2008), comments, timing and being in the right place at the right time make a difference in the beginning of movements as well as who is credited. More sampling of the many books published in the last decade can be found in Chapter 11 in the resources.

In September 2006, the US Fish and Wildlife Service (FWS) and the Conservation Fund hosted the National Dialogue on Children and Nature at the National Conservation Training Center in Shepherdstown, West Virginia, attended by leaders in education, healthcare, outdoor recreation industry, residential development, urban planning, conservation, and academia. Subsequently the Conservation Fund launched a Children and Nature Forum to incubate and grow new and existing programs to foster childhood nature experiences. One of these was the National Environmental Education Foundation's Children and Nature project, which partners healthcare providers with 'nature providers' such as the FWS, the National Park Service, and the National Audubon Society. Many government agencies, using the growing research about the importance of being outside in nature for health and well-being, started programs to reinforce getting people outside, such as the FWS Connecting People with Nature, with slogans such as the 2007 FWS 'Let's go outside!' and the US Forest Service 'More Kids in the Woods' and the 2008 Bureau of Land Management 'Take it outside!' (Kruger *et al.*, 2010). This is reinforced with outdoor activity books now including specific information about the health benefits such as *Hiking and Backpacking* edited by Goldenberg and Martin (2008). Public service announcements on television and the internet, and billboards along the highways now bear the 'Unplug' campaign of the US Forest Service. The National Park Service (2013) began a Healthy Parks, Healthy People program and is currently working to identify best practices of nature programs that serve to reconnect children with nature.

The medical community has begun to weigh in on the specific health benefits of time in nature. Howard Frumkin, then director of the National Center for Environmental Health at the Centers for Disease Control and Prevention, described the clear benefits of nature experiences to healthy child development and to adult well-being: 'In the same way that protecting water and protecting air are strategies for promoting public health, protecting natural landscapes can be seen as a powerful form of preventive medicine' (Frumkin and Louv, 2007, p. 4). The American Academy of Pediatrics (Ginsburg, 2007) stated that free and unstructured play is essential for child development and Louv (2010) offered a list of suggestions to the Academy for fostering outdoor play in nature.

Urban planners have also begun to connect health and nature as exemplified with Tim Beatly's work with biophilic cities. Realizing that more people will continue to be in urban environments, Beatly at the University of Virginia School of Architecture began a project to connect policy and research in order to help city planners and citizens understand biophilic design on an urban-wide scale. The goal is to creatively and effectively incorporate natural spaces into cities (Beatley, 2013). These biophilic cities contain abundant nature, and promote connections and daily contact with the natural world. Their work so far has centered on seven cities in four countries, including Perth in Western Australia, and Phoenix, Portland, and San Francisco in the US.

Chapter Summary

This chapter explored how people through the ages have related to nature within the context of health and healing. Throughout our history people have relied on nature-based medicine, from the Shaman of the

Reindeer Age to the system of homeopathy that was developed in the 19th century. The benefit of using a historical perspective is that we can see how a systems approach to medicine can benefit health care. We see how respecting and living with the Earth's rhythms, understanding the role of attachment to caregivers and the Earth, knowing the specific effect of various herbs and other substances, and understanding the role of being in nature for our mental health and rest all combine to increase health and well-being.

Recent research continues to find support for the ways early humans understood our interrelatedness with nature. Whole medicine systems from Tibetan Medicine to Native American Medicine are increasingly being acknowledged by Western medical systems for the knowledge and practices that are being empirically shown to improve human health. As such, health-enhancing methods and practices that are more intertwined with nature are being popularized and developed. More than any specific medicine or healing system, the basic practice of spending direct time in nature is essential for human health and well-being. Nature is known to be healing; as humans we need nature and are worse without it. Understanding some of the knowledge and practices of the past helps shed light on current directions.

References

Adhikari, S. (2009) Commonly used ethno-medicinal plants by the indigenous people of Nepalese Himalaya. Presented at *Ecological Society of America 94th Annual Meeting*, Albuquerque, New Mexico, 2–7 August 2009.

Altman, N. (1994) *Sacred Trees*. Sierra Club Books, San Francisco, California.

Anderson, J. (2008) *The Second Journey: The Road Back to Yourself*. Hyperion, New York.

Anon. (2009) Video: Sanatorium Scenes. *Harvard Magazine*, March–April 2009. Available at: http://harvardmagazine.com/2009/03/video-sanatorium-scenes (accessed 8 September 2013).

Appalachian Trail Conservancy (2013) 2000 milers. Available at: http://www.appalachiantrail.org/about-the-trail/2000-milers (accessed 8 September 2013).

Barnes, P.M., Bloom, B. and Nahin, R.L. (2007) *Complementary and Alternative Medicine Use Among Adults and Children: United States, 2007. National Health Statistics Report Number 12*. National Center for Health Statistics, Hyattsville, Maryland.

Barnett, R. (2009) Serpent of pleasure: emergence and difference in the medieval garden of love. *Landscape Journal* 28(2), 137–150.

Beatley, T. (2013) Biophilic Cities. Available at: http://biophiliccities.org/ (accessed 8 September 2013).

Burton-Christie, D. (1999) Into the body of another: Eros, embodiment and intimacy with the natural world. *Anglican Theological Review* 81(1), 13–37.

Bushnell, R. (2003) *Green Desire: Imagining Early Modern English Gardens*. Cornell University Press, Ithaca, New York.

Carson, R. (1962) *Silent Spring*. Houghton-Mifflin, Boston, Massachusetts.

Cobb, E. and Mead, M. (1977) *The Ecology of Imagination in Childhood*. Columbia University Press, New York.

Donovan, G.H., Butry, D.T., Michael, Y.L., Prestemon, J.P., Liebhold, A.M., *et al.* (2013) The relationship between trees and human health: evidence from the spread of the emerald ash borer. *American Journal of Preventive Medicine* 44(2), 139–145.

Driver, B., Tinsley, E. and Manfredo, M. (1991) Leisure and recreation experience preference scales: two inventories designed to assess the breadth of perceived psychological benefits of leisure. In: Driver, B., Brown, P. and Peterson, G. (eds) *Benefits of Leisure*. Venture Publishing Inc., State College, Pennsylvania, pp. 263–286.

Edwards, T. (2008) Cancer's Achilles' heel. *Ode Magazine*, June 2008 issue. Available at: http://odewire.com/61867/cancers-achilles-heel.html (accessed 8 September 2013).

Ehrenreich, B. and English, D. (1973) *Witches, Midwives, and Nurses: A History of Women Healers*. The Feminist Press at the City University of New York, New York.

Elmer, P. (ed.) (2004) *The Healing Arts: Health, Disease and Society in Europe 1500–1800*. Manchester University Press, Manchester, UK.

Flanagan, S. (1989) *Hildegard of Bingen, a Visionary Life*. Routledge, London.

Francis, C. and Cooper-Marcus, C. (1991) Places people take their problems. In: *Proceedings of 22nd Annual Conference of the Environmental Design Research Association*. Environmental Design Research Association, Oklahoma City, Oklahoma, pp. 178–184.

Frumkin, H. and Louv, R. (2007) *The Powerful Link between Conserving Land and Preserving Health*. Land Trust Alliance, Washington, DC.

Gaarder, J. (1995) *Sophie's World: A Novel about the History of Philosophy*. Phoenix/Orion Books Ltd, London.

Gierlach-Spriggs, N., Kaufman, R.E. and Warner, S.B. Jr (1998) *Restorative Garden: The Healing Landscape*. Yale University Press, New Haven, Connecticut.

Gladwell, M. (2002) *The Tipping Point: How Little Things Can Make a Big Difference*. Back Bay Books, Boston, Massachusetts.

Gladwell, M. (2008) *Outliers: The Story of Success*. Little, Brown and Company, New York.

Ginsburg, K.R. and American Academy of Pediatrics Committee on Communications and Committee on Psychosocial Aspects of Child and Family Health (2007) The importance of play in promoting healthy child development and maintaining strong parent–child bonds. *Pediatrics* 119(1), 182–191.

Goldenberg, M. and Martin, B. (eds) (2008) *Outdoor Adventures: Hiking and Backpacking*. Wilderness Education Association and Human Kinetics, Champaign, Illinois.

Greenway, R. (1996) Wilderness experience and ecopsychology. *International Journal of Wilderness* 2(1), 26–30.

Haller, R. and Kramer, C. (eds) (2006) *Horticultural Therapy Methods: Making Connections in Health Care, Human Service, and Community Programs*. The Haworth Press, Binghamton, New York.

Hendrix, S.E. (2011) Natural philosophy or science in premodern epistemic regimes? The case of the astrology of Albert the Great and Galileo Galilei. *Teorie vědy/Theory of Science* 33(1), 111–132.

Hill, T. (1577) *The Gardeners Labyrinth; or, A New Art of Gardening*. Reprint: 1652, Jane Bell, London.

His Holiness the 14th Dalai Lama of Tibet (1992) A Buddhist Concept of Nature. Transcript of an address on February 4, 1992, at New Delhi, India. Available at: http://www.dalailama.com/messages/environment/buddhist-concept-of-nature (accessed 17 February 2013).

HRH The Prince of Wales (2010) *Harmony: A New Way of Looking at our World*. Blue Door, London.

Hulmes, D.F. (2009) Sacred trees of Norway and Sweden: a friluftsliv voyage. Presented at *Henrik Ibsen: The Birth of 'Friluftsliv'. A 150 Year International Dialogue Conference Jubilee Celebration*. North Troendelag University College, Levanger, Norway, Mountains of Norwegian/Swedish Border, 14–19 September 2009.

Irvine, K.N. and Warber, S.L. (2002) Greening healthcare: practicing as if the natural environment really mattered. *Alternative Therapies in Health and Medicine* 8(5), 76–83.

Iso-Ahola, S.E. (1999) Motivational foundations of leisure. In: Jackson, E.L. and Burton, T.L. (eds) *Leisure Studies: Prospects for the Twenty-first Century*. Venture Publishing, Inc., State College, Pennsylvania, pp. 35–51.

Kaplan, R. and Kaplan, S. (1989) *The Experience of Nature: A Psychological Perspective*. Cambridge University Press, Cambridge, UK.

Kellert, S.R., Heerwagen, J.H. and Mador, M.L. (eds) (2008) *Biophilic Design: The Theory, Science, and Practice of Bringing Buildings to Life*. John Wiley & Sons, Hoboken, New Jersey.

Kruger, J., Nelson, K., Klein, P., McCurdy, L.E., Pride, P., *et al.* (2010) Building on partnerships: reconnecting kids with nature for health benefits. *Health Promotion Practice* 11(3), 340–346.

Leopold, A. (1949) *A Sand County Almanac*. Oxford University Press, London.

Lewis, C.A. (1976) The evolution of horticulture therapy in the US. Presented at *Fourth Annual Meeting of the National Council for Therapy and Rehabilitation through Horticulture*, Philadelphia, Pennsylvania, 6 September 1976.

Lewis, J. (1985) The birth of EPA. *EPA Journal* 11, 6.

Lodewyk, K., Chunlei, L. and Kentel, J. (2009) Enacting the spiritual dimension in physical education. *Physical Educator* 66(4), 170–179.

Louv, R. (1998) *The Web of Life: Weaving the Values That Sustain Us*. Red Wheel Weiser & Conari Press, Newburyport, Massachusetts.

Louv, R. (2005) *Last Child in the Woods: Saving our Children from Nature-Deficit Disorder*. Algonquin Books of Chapel Hill, Chapel Hill, North Carolina.

Louv, R. (2010) Grow Outside! Keynote Address to the American Academy of Pediatrics National Conference, 4 October 2010. Available at: http://blog.childrenandnature.org/2010/10/04/grow-outside-keynote-address-to-the-american-academy-of-pediatrics-national-conference/ (accessed 8 September 2013).

Louv, R. (2011) *The Nature Principle: Human Restoration and the End of Nature-deficit Disorder*. Algonquin Books of Chapel Hill, Chapel Hill, North Carolina.

MacKaye, B. (1921) An Appalachian Trail: a project in regional planning. *Journal of the American Institute of Architects* 9, 325–330.

Margulis, L. (1998) *Symbiotic Planet: A New Look at Evolution*. Basic Books, New York.

Margulis, L. and Sagan, D. (eds) (2007) *Dazzle Gradually: Reflections on the Nature of Nature*. Chelsea Green, White River Junction, Vermont.

Martinez, D.E. (2008) Indigenous consciousness and the production of knowingness. Presented at the *American Sociological Association Annual Meeting*, Boston, Massachusetts, 31 July 2008.

McCormick, R. and Gerlitz, J. (2009) Nature as healer: Aboriginal ways of healing through nature. *Counseling and Spirituality/Counseling et Spiritualité* 28(1), 55–72.

McDonald, H. (1993) India: the Parsi dilemma. *Far Eastern Economic Review* 156(40), 36.

McDonald, J. (1995) A comparative study of horticulture therapy profession in the United Kingdom and the United States of America. MSc thesis, University of Reading, Reading, UK.

Miranda, W. (1987) The genteel radicals. *Camping Magazine* 59(4), 12–15, 31.

Miranda, W. and Yerkes, R. (1987) Women's outdoor adventure programming. In: Meier, J., Morash, T. and Welton, G. (eds) *High Adventure Outdoor Pursuits*. Publishing Horizons, Inc., Columbus, Ohio, pp. 259–267.

Miranda, W. and Yerkes, R. (1996) The history of camping women in the professionalization of experiential education. In: Warren, K. (ed.) *Women's Voices in Experiential Education*. Kendall/Hunt Publishing Company, Dubuque, Iowa, pp. 24–32.

Mitten, D. (1985) A philosophical basis for a women's outdoor adventure program. *Journal of Experiential Education* 8(2), 20–24.

Mitten, D. (1992) Empowering girls and women in the outdoors. *Journal of Physical Education, Recreation & Dance* 63(2), 56–60.

Mitten, D. (2010) Friluftsliv and the healing power of nature: the need for nature for human health, development, and wellbeing. In: *Henrik Ibsen: The Birth of 'Friluftsliv'. A 150 Year International Dialogue Conference Jubilee Celebration*. North Troendelag University College, Levanger, Norway, Mountains of Norwegian/Swedish Border, 14–19 September 2009. Available at: http://norwegianjournaloffriluftsliv.com/doc/122010.pdf (accessed 8 September 2013).

Mitten, D. and Woodruff, S.L. (2009) Women's adventure history and education programming in the United States favors friluftsliv. In: *Henrik Ibsen: The Birth of 'Friluftsliv'. A 150 Year International Dialogue Conference Jubilee Celebration*. North Troendelag University College, Levanger, Norway, Mountains of Norwegian/Swedish Border, 14–19 September 2009. Available at: http://norwegianjournaloffriluftsliv.com/doc/212010.pdf (accessed 8 September 2013).

Montes, S. (1996) Uses of natural settings to promote, maintain and restore human health. In: Driver, B.L., Dustin, D., Baltic, T., Elsner, G. and Peterson, G. (eds) *Nature and the Human Spirit: Toward and Expanded Land Management Ethic*. Venture Publishing Inc., State College, Pennsylvania, pp. 105–115.

Murray, A.L. (1967) Frederick Law Olmsted and the design of Mount Royal Park, Montreal. *Journal of the Society of Architectural Historians* 26(3), 163–171.

Murray, M.A. (1921) *The Witch-cult in Western Europe: A Study in Anthropology*. Clarendon Press, Oxford, UK.

Nash, R. (2001) *Wilderness and the American Mind*, 4th edn. Yale University Press, New Haven, Connecticut.

National Park Service (2013) *The National Parks & Public Health: A NPS Healthy Parks, Healthy People Science Plan*. US Department of the Interior, Washington, DC.

Nightingale, F. (1996) *Notes on Nursing (Revised with additions)*. Ballière Tindall, London.

Olmsted, F.L. (1967) Frederick Law Olmsted and the design of Mount Royal Park, Montreal. *Journal of the Society of Architectural Historians* 26(3), 163–171.

Pergams, O.R.W. and Zaradic, P.A. (2008) Evidence for a fundamental and pervasive shift away from nature-based recreation. *Proceedings of the National Academy of Sciences USA* 105(7), 2295–2300.

Perlman, M. (1994) *The Power of Trees*. Spring Publications, Dallas, Texas.

Platt, S.H. (1608) *Floraes Paradise*. London.

Poole, R. (2008) *Earthrise: How Man First Saw the Earth*. Yale University Press, New Haven, Connecticut.

Priest, S. (1986) Outdoor leadership preparation in five nations. PhD thesis, University of Oregon, Eugene, Oregon.

Reese, R.F. and Myers, J.E. (2012) EcoWellness: the missing factor in holistic wellness models. *Journal of Counseling & Development* 90(4), 400–406.

Roszak, T., Gomes, M.E. and Kanner, A.D. (eds) (1995) *Ecopsychology: Restoring the Earth, Healing the Mind*. Sierra Club Books, San Francisco, California.

Sharp, L.B. (1943) Outside the classroom. *The Educational Forum* 7(4), 361–368.

Siegel, R. (producer) (2012) Vanishing vultures: a grave matter for India's Parsis. *National Public Radio,* 12 September 2012. Available at: http://keranews.org/post/vanishing-vultures-grave-matter-indias-parsis (accessed 24 September 2013).

Steinbeck, J. (1937/1994) *Of Mice and Men.* Penguin, New York.

Swan, J. (1992) *Nature as Teacher and Healer: How to Awaken Your Connection with Nature.* Random House, New York.

Tolle, E. (2005) *A New Earth: Awakening to Your Life's Purpose.* Penguin Books, New York.

Torem, M.S., Gilbertson, A. and Light, V. (1990) Indications of physical, sexual, and verbal victimization in projective tree drawings. *Journal of Clinical Psychology* 46(6), 900–906.

Tsunetsugu, Y., Park, B. and Miyazaki, Y. (2010) Trends in research related to 'Shinrin-yoku' (taking in the forest atmosphere or forest bathing) in Japan. *Environmental Health and Preventative Medicine* 15(1), 27–37.

Ulrich, R.S. (1984) View through a window may influence recovery from surgery. *Science* 224(4647), 420–421.

Ulrich, R.S. and Parsons, R. (1992) Influences of passive experiences with plants on individual well-being and health. In: Relf, D. (ed.) *The Role of Horticulture in Human Well-being and Social Development.* Timber Press, Portland, Oregon, pp. 93–105.

Warburg, O. (1966) The prime cause and prevention of cancer with two prefaces on prevention. Revised lecture at the Meeting of the Nobel-Laureates on June 30, 1966, Lindau, Lake Constance, Germany. Available at: http://www.whale.to/a/warburg.html (accessed 8 September 2013).

Wilson, E.O. (1984) *Biophilia: The Human Bond with Other Species.* Harvard University Press, Cambridge, Massachusetts.

Yankou, D. (2002) MorningStar: A place of restoration. *MorningStar Adventures* 18(2), 3.

4

Concepts and Theories

How strange that nature does not knock,
And yet does not intrude.

Emily Dickinson

Many, if not most, people have an inherent belief that there is a connection between human health and natural environments. While it is true that some individuals or groups have experienced negative events in natural settings (Davidson, 2012), most would report a positive experience (Hartig *et al.*, 2011). Moreover, this positivity has a long history and involves virtually countless individuals, worldwide, and in a broad spectrum of natural settings (Garst *et al.*, 2010). But do we really know there is a connection between health and natural environments which goes beyond anecdotal memories and intuition? Moreover, what explanations currently exist that inform us as to how natural environments actually affect human health? This chapter explores these issues by addressing three underlying questions: (i) Do natural environments actually affect human health? (ii) How do natural environments affect human health? (iii) What are some of the influencing variables that contribute to this relationship, such as dosage or length of exposure to natural environments?

We begin with a description of several salient theories related to genetics and inherent disposition as well as psychologically based theories that speak to humans' seemingly intuitive and positive affiliation with nature, possibly

based on evolutionary history or, alternatively that humans learn to enjoy and want to be in natural environments through repeated experience (Fenton, 2008). The chapter concludes with a summary of spiritual affiliations, natural environments, and health and an explanation of intentionally designed experiences.

Underlying this chapter, we define theory as a set of statements or principles that seek to explain a group of facts or phenomena. Theories and concepts are used to construct knowledge. We have highlighted theories that have been repeatedly tested or widely accepted and that can be used to make predictions about natural phenomena. Clarifying these relevant concepts and theories helps advance the understanding and contextualization of the connection between natural environments and human health and well-being. Readers will notice overlap among the concepts and theories presented in this section. That is because people from different fields have put forth theories or explanations about humans' connection with nature. As people come at this question or phenomenon from different angles their results vary, though there are similar explanations for the same phenomenon. When these concepts are examined and tied together, the resulting overarching theory, which some people believe as fact, is that time in nature as well as appropriate use of natural landscapes is

© CAB International 2014. *Natural Environments and Human Health*
(A.W. Ewert, D.S. Mitten and J.R. Overholt)

essential for healthy growth, development, and maintenance of human beings.

Evolutionary-Based Theories

In Chapters 2 and 3 readers learned about the long and close history humans have had with nature. This may have been through necessity because until the industrial age humans were not able to live physically separate from nature; however, humans may still need that physical connection with nature and nature may also need for humans to have that connection. In the first chapters readers also learned that many cultures consciously chose not to engage in the industrial revolution, preferring to remain mutualistic or symbiotic with the land. People from industrial cultures are now looking at the nature-based cultures to understand more about the benefits received from their nature-based lifestyle.

People have approached the concept of humans' connectedness to nature from a variety of perspectives. A number of theories have been proposed asserting that there is a basic need for humans to affiliate with nature. Some say our connection is genetic, others claim it is learned; the nature–nurture debate is discussed in Chapter 5. Many propose their theories without attributing the behavior to either genes or environment because that debate continually evolves. In this book, there is an attempt to organize theories that have arisen and been described by many people from different fields who conclude that humans have a natural affiliation with nature. First, theories commonly thought of as genetically based are described, followed by those thought of as psychologically based. However, these two categories are related.

For example, theories such as naturalistic intelligence, biophilia, and intra-indigenous consciousness may be related to genetics and be realized and further developed through nurture or simply more exposure to nature. Theories such as attention restoration theory and psycho-evolutionary theory may have originated from environmental conditions and eventually became installed in genes. These theories relate to our interactions with our environments and genetics, indicating the flexibility of genes and the challenge of drawing lines between nature and nurture. A common aspect of many of these theories is that they appear to be evolutionarily based. However, caution is necessary when ascribing an evolutionary basis to human affiliation with nature; it might be assumed that all humans have developed to have the same or similar preferences, yet this may not be so. At least one research project showed that preferences for natural spaces differed among groups of adults (Virden and Walker, 1999). This research may reflect that learned behavior more than evolutionary tendencies are present and it may support Mitten's (2010) contention that humans are born with an inclination to affiliate with nature, but that tendency often needs to be nurtured in order to fully materialize.

Ecopsychology (see Chapter 6) is a theory about humans' oneness on a psychological level with the environment and it also refers to a mental health treatment or application, showing the need for clarity and the opportunity for confusion. This section describes these theories and begins a conversation, though readers are encouraged to continue the conversation by building on information discovered almost daily.

The field of evolutionary psychology seeks to connect human psychological traits, such as community formation, language, and behavior, as manifestations representing adaptations that our ancestors developed to solve recurring problems in the environment related to survival. Whether they related to food, shelter, safety, or other compelling issues, the evolutionary psychologist would often argue that many of the behaviors, attitudes, and other psychological phenomena observed in various societies can be linked to these ancestral psychological adaptations. Therefore, through adaptation to the environment new traits are constructed. One of the most widely held arguments within this line of reasoning suggests that, as a general rule, many humans are attracted to natural environments simply because that is where our ancestors grew and developed. Grinde and Patil (2009) demonstrated that this attraction has a long recorded history beginning with the tomb writings of ancient Egypt, the development of zoological gardens, and the first hospitals in Europe. Ulrich (1999) postulates that there are four reasons why humans retain connection between human health and natural settings, including: (i) being in nature tends to involve activities that promote health (e.g. physical activity); (ii) activities done in natural environments often involve socializing, which in itself can be healthful; (iii) experiences in natural environments often facilitate a sense of escape from everyday routines or demands; and (iv) natural environments can have a salutatory health effect by itself, either psychologically or physically.

Naturalistic intelligence

An early theory that links natural environments and human health from an evolutionary perspective comes from Howard Gardner's work in multiple intelligences, which is grounded in the work of Marie Montessori (Swiderski, 2011). Gardner (2000) conceptualizes human intelligence as more than a person's cognitive ability. Naturalistic intelligence is one of the nine intelligences identified in Gardner's multiple intelligence theory, first published in 1983. Naturalistic intelligence is signaled by an affinity to be outdoors and connect with nature, sensing patterns in nature, and being aware of subtle changes in the outdoor environment. Naturalistic intelligence is the ability to focus on the large picture—including seeing how natural systems are connected. It makes sense that humans would have naturalistic intelligence through genetics, innate brain development, or other avenues because being in tune with nature would help individual and species survival (Mitten, 2010). Survival is the basis of health, followed by knowing what plants and animals provide adequate nutrition as food and to use for shelter. While all humans may have naturalistic intelligence, it may be that this intelligence, like all intelligences (e.g. emotional, social), grows with use or may be dormant or unrealized without use. There is a great deal of diversity in the strengths and functioning of individual brains, causing some people to appear naturally proficient at music or math, while others seem predisposed to high naturalistic intelligence, and still others are competent in many types of intelligences. Multiple intelligence theory states that all individuals possess the nine intelligences; however, individuals vary in how innately strong or how well developed these intelligences are depending on individual genetics and upbringing. Individuals with strong naturalistic intelligence have a highly developed level of sensory perception, may be able to categorize or catalogue things easily, and may benefit from moving around while learning. Gardner suggests that people living close to the land, such as Aboriginal peoples, may both possess and value naturalistic intelligence.

If this theory holds then people such as Charles Darwin, Rachael Carson, and Jane Goodall would have high naturalistic intelligence. Rachel Carson's (1907–1964) life exemplified naturalist intelligence. She understood the human–nature connection and was a pioneer in calling for a change in the path of industrialization. A biologist, trained at Johns Hopkins and Woods Hole, Carson understood the healing power of nature from both biological and developmental aspects. In 1952 she retired from the Fish and Wildlife Service to focus on her writing. While she is best known for her 1962 book, *Silent Spring*, in her 1954 article 'The real world around us' Carson wrote, 'the more clearly we can focus our attention on the wonders and realities of the universe about us, the less taste we shall have for destruction' (Lear, 1998). In her article 'Help your child to wonder', first published in 1956, Carson encouraged parents to take children outside and help them enjoy nature (Carson, 1998). Her gift was her ability to understand the interconnectedness of all parts of the ecosystem, including humans, and to be able to write about these connections in an accessible manner to the general public. She asked people to tend to the earth so the earth could continue to nourish us. She made connections between human industrial practices and environmental damage that would then lead to human harm. Her work and collaboration with Dr George Wallace from Michigan State University showed the connection between pesticides and environmental damage. Once a breeding ground for mosquitoes and malaria, after World War II the Michigan State University's campus was regularly dosed with DDT in high doses. Wallace, another person showing high naturalist intelligence, was concerned with the multitude of dying robins and worked to find the cause of their deaths. He found that the robins were dying because the DDT had accumulated in the worms they were eating. This research set the groundwork for the ecological concept of bioaccumulation, introduced in *Silent Spring* and now an elementary concept in ecology, which states that toxicity accumulates and intensifies in species as one goes higher in the food chain. Therefore swordfish, which feed on other fish, accumulate more toxins than sardines, which are lower in the food chain. Both Carson and Wallace underwent heavy criticism for their work. Industry lobbied to have Wallace fired and Carson received life threats.

People like Carson seem compelled to study nature and to help others understand the importance of our connections to natural systems. Carson's drive was so strong that she continued her work even when faced with an early death due to cancer. Carson made accurate connections among human mental health, human development, biology, ecology, and other areas.

Some people's naturalistic intelligence drives them to be outside or to interact with the natural environment. For others, this drive can be awakened through family, school, or community outings. Growing naturalistic intelligence through the practice of being outside and interacting with natural environments helps people develop an ethic of care through being sensitive to their role in nature. They are then drawn to nurturing other people and the environment. Naturalistic intelligence helps us know to be outdoors where health increases. This theory that humans have naturalistic intelligence is supported by research discussed throughout this book showing our affinity to nature.

Biophilia hypothesis

A second widely recognized theory that links natural environments and human health from an evolutionary perspective is the biophilia hypothesis. First discussed by Fromm (1964) and later popularized by E.O. Wilson (1984), biophilia refers to the attraction that humans often feel toward other forms of life. That is, humans have an affinity to life forms such as plants, animals, and other aspects of natural environments that stems from evolutionary and subsequent genetic processes. The underlying assumption of the biophilia hypothesis is that the human species has evolved for millions of years placed within a variety of natural environments and as a result 'prefers' or seeks natural landscapes. Conversely, not being in close contact with these types of environments can create feelings of dysfunction as well as ill health. Moreover, aspects of natural landscapes such as water, trees, and animals all have an evolutionary history that has and can provide adaptive significance for human development. The theory contends that human identity and personal fulfillment depend on our relationship to nature, as does humans' positive emotional, cognitive, aesthetic, and spiritual development. Therefore, people's success in their search for a coherent and fulfilling existence depends on their relationship to nature. Criticisms surrounding the biophilia hypothesis are numerous and focus around the degree to which genetics actually account for behaviors, as Wilson asserts, and to what extent phenomena such as past experience, culture, and social learning serve to offset any effects of evolution or genetics. In addition, the fact that, for some, natural environments represent settings that are feared, disliked, or otherwise viewed negatively (Bixler and Floyd, 1997) argues against an accretion of positive attributes and attitudes toward nature. However, part of the biophilia hypothesis is that a measured or healthy fear of some parts of the natural world is essential for humans' survival. Considerable evidence from clinical psychology and psychiatry, including evolutionary psychology, supports that the majority of phobic occurrences in humans involve strong fears about things or situations that have threatened humans throughout evolution, such as snakes, spiders, heights, closed spaces, and blood. Therefore a balance between respect or healthy fear of some aspects of nature and the human capacity to experience a positive affiliation with nature increases health and survival (Mitten, 2010).

Kellert (2003) developed a typology of a set of nine responses each person has to the natural world, each having adaptive significance: (i) aesthetic; (ii) negativistic; (iii) humanistic; (iv) naturalistic; (v) symbolic; (vi) scientific (or ecologistic-scientific); (vii) utilitarian; (viii) dominionistic; and (ix) moralistic. These biophilic responses are influenced by direct (e.g. a hike in a pine forest) and indirect (e.g. pictures or stories) experiences and can change over time (Shorb and Schnoeker-Shorb, 2010). It would be expected, for example, after the 2013 fire season where 19 young Prescott, Arizona hotshot firefighters were killed, the town of Yarnell essentially burned, and many acres were lost in the Dosie fire on Granite Mountain that people in that area would have a higher score on the negativistic sale and a lower score on the aesthetic scale. Over time, with more positive experiences with nature, these same people might score higher on the aesthetic scale and lower on the negativistic scale.

Intra-indigenous or indigenous consciousness

Indigenous consciousness, or intra-indigenous consciousness (IIC), relates to human behaviors that have not necessarily been taught but that people exhibit. The concept is that humans are born understanding or knowing certain things and that through sharing beliefs, rituals, and ceremonies this information that is already in the person's unconscious mind is unlocked and becomes conscious. This consciousness includes spirituality, responsibility, reciprocity, the earth, animals, and many other facets of life. According to Martinez (2008), early medical systems such as Tibetan Medicine and Native American Medicine encompassed and have been sustained through indigenous consciousness. Today Martinez's research with seven nations has shown that medicine and health and healing paradigms are a central manner in which indigenous consciousness is both used and maintained, including helping to temper the effects of colonization.

One can theorize that everybody is indigenous and therefore that indigenous consciousness can be unlocked through reconnection with nature, perhaps by using experiences that were significant to one's ancestors, such as an activity in the outdoors. The existence of an IIC may be why many people seek outdoor recreation experiences; these experiences may awaken their positive memories of living in nature. It is unclear if IIC is genetically stored or linked to psycho-evolutionary theory. However, like naturalistic intelligence and perhaps biophila, this unconsciousness is awake at birth and expressed by some people, while other people need a stimulus, such as time in nature, to make the knowledge conscious.

Other evolutionary-grounded theories

Other lesser known theories of human attractions to natural environments that have a basis in evolution are the savannah theory and prospect-refuge theory (Hartig et al., 2011). The savannah theory, originally introduced by Orians (1980), uses findings from a number of studies that indicated that people selected savannah-like terrain features as more or most desirable compared with other landscape types. The assumption underlying this conceptualization is the belief that savannah-type landscapes offered our ancestors areas that provided suitable sites for survival. As a result of this long-standing relationship between humans and these types of habitats, savannah theory posits that there exists an emotional connection between people and savannahs, and that this emotional connection presents a positive health ambiance.

Similar to the savannah theory, prospect-refuge theory looks at habitat suitability as a mechanism to explain the attractiveness of certain habitats and landscapes to humans. As articulated by Appleton (1975), certain landscapes afforded our ancestors and present inhabitants with opportunities for accessibility, concealment, and ease of travel. Appleton maintains that these types of settings are still 'attractive' to humans although culture, historical background, and social influences cannot be discounted in ascertaining the efficacy of this theory.

The synergistic relationship between personal well-being and planetary well-being may have an evolutionary history as described in the biophilia hypothesis, naturalistic intelligence, and indigenous consciousness. These theories are very similar and may describe the same phenomena or different aspects of the same phenomena. These evolutionary theories support each other and it is a strength that these related theories—representing a number of different fields including biology, psychology, outdoor education, and others—come to comparable conclusions. The example of the common fear of snakes (even Indiana Jones is afraid of snakes!) can be described as naturalistic intelligence, biophilia, or indigenous consciousness. Darwin, curious if fears were hardwired into our brain, designed an experiment to test this theory. He went up to the edge of a snake cage where he cognitively knew that the snake could not strike him. He tried to remain still when the snake struck at him. He found that no matter how hard he tried he was unable to overcome his fear, causing him to conclude that this fear was hardwired. Today scientists believe that the reaction originates in the amygdala or emotional center of the brain. Research has shown that specific emotional reactions to certain stimuli (angry faces, happy faces, snakes, and spiders) are readily detectable within 400 ms or less following the presentation of the stimuli; far too quickly for the response to be cognitive (Ulrich et al., 1991). Emotions can be learned, so Darwin did not conclusively prove that his reaction was evolutionary; regardless we have trouble categorizing the response as biophilic, naturalistic intelligence, or indigenous consciousness because any of those three theories could explain the phenomenon.

In another experiment LoBue and DeLoache (2008) asked adults, preschoolers, and infants to look at pictures that included snakes. They found that the adults and preschool-aged children all detected the snake more rapidly than three types of non-threatening stimuli (flowers, frogs, and caterpillars). The researchers believe that this is evidence of an evolutionary response, yet we would have the same problem choosing which of the discussed theories to ascribe to this behavior. In another study they found

that 9- to 10-month-old infants showed no differential spontaneous response to films with or without snakes in them. However, infants aged 7 to 18 months looked longer at snakes in films when listening to a frightened human voice than while listening to a happy voice. When pictures instead of film were used there was no difference in response, indicating that movement may be needed to trigger fear in this case (DeLoache and LoBue, 2009). The results in the DeLoache and LoBue experiments indicate that there may be a learning component to associating fear with snakes along with natural tendencies.

Linking more connections, perhaps naturalistic intelligence, IIC, or biophilia cause friluftsliv, traditional or indigenous ecological knowledge, ecopsychology, and outdoor recreation to continue to be practiced (these applications are discussed in Chapter 6). Enough of the human population may want to strongly affiliate with nature and they may encourage others to do so through these applications. Much of the research about the healing power of nature supports the biophilia hypothesis, naturalistic intelligence, and indigenous consciousness theories, namely that humans affiliate with nature and do so because the positive health benefits overwhelmingly outweigh the negative events.

Evolutionary-based theories as described above present a picture as to why people may be attracted to natural environments and landscapes. Doubtless, there are numerous other variables and influences that serve to create these connections, including cultural and iconic symbols, places of emotional and historical significance, and idiosyncratic events such as a budding relationship, beautiful weather, or a feeling of aesthetic awe. Moreover, merely being attracted to a setting does not necessarily implicate a positive health outcome. In this case, we will turn to one of the more powerful effects of natural environments associated with humans, namely restoration and restorative environments, later in the chapter.

Psychological and Sociological Theories

There are several psychological and sociological theories that may account for humans desiring to affiliate with nature including ecopsychology, constructing an environmental identity, psychologically deep experiences, psychophysiological stress reduction, and attention restoration.

Identity and the natural environment

Humans both materially transform the environment and symbolically construct it. Thus, an understanding of human–environment interactions necessarily entails an understanding of how the environment is constructed and how individuals have been socialized to understand and relate to the environment over time. These constructs may best be operationalized within the context of an environmental identity. Of course, it should be noted that humans have had a relationship with the natural environment, and thus an environmental identity, since the beginning of time, but a recent sense of disconnection from the natural world has led to the study of and popularization of the term environmental identity.

Environmental identity, or how individuals see themselves in relation to the natural environment, both arises within and gives rise to worldviews. Therefore, a discussion of one is incomplete without the other. For this reason, the reader is encouraged to reference Chapter 2 for a more in-depth treatment of worldviews. Thomashow's (1996) work on ecological identity highlights this relationship, where having an ecological identity is premised on an ecological worldview, or the ability to perceive ecological relationships and to sense a connection to the earth. Thomashow defines ecological identity as 'all the different ways people construe themselves in relationship to the earth as manifested in personality, values, actions and sense of self' (p. 3). The study of environmental identity has focused primarily on the life of self-proclaimed environmentalists in order to understand what it is that compels this type of lifestyle. Part of this process is the examination of four components of what Thomashow terms ecological identity work: 'how people learn about ecology, how people perceive themselves in relationship to ecosystems, how an understanding of ecology changes the way people learn about themselves, and how an ecological worldview promotes personal change' (p. 5).

Other authors have also attempted to characterize this sense of connection to the natural environment, and have utilized a variety of terms in an attempt to better understand this process. For example, researchers have used the terms place identity (e.g. Proshansky et al., 1983; Korpela, 1989), ecological identity (e.g. Thomashow, 1996), and environmental identity (e.g. Clayton, 2003; Weigert, 2008). While these terms are sometimes used interchangeably, Devine-Wright and Clayton (2010) contend that these terms are a matter of scope—that is, place identities refer to specific locales, whereas environmental identities refer more broadly to the natural environment in general. Still, the idea of an environmental (or ecological, or place) identity is relatively new and experiences a lack of consensus on definition or application.

Environmental identity has been defined in a number of ways. Clayton (2003) defines environmental identity as 'a sense of connection to some part of the non-human natural environment, based on history, emotional attachment, and/or similarity, that affects the ways in which we perceive and act toward the world' (pp. 45–46). This identity serves to inform people of who they are, as well as their understanding of nature and environmental issues. Environmental identity also has been defined as the combination of a person's perceptions and evaluations of the environment and nature, as well as their relevance to everyday life (Linneweber et al., 2003). Like other social identities, environmental identity is contextual, varies by situation, and must compete with other 'selves'. However, Linneweber et al. argue that this relationship is not unequivocal, and is mediated by the relative salience of environmental representations and the social positions people hold relevant to environmental actions. Weigert (1997) also adopts this more sociological approach, defining environmental identity as the 'experienced social understanding of who we are in relation to, and how we interact with, the natural environment as other'.

Weigert's approach to identity is grounded in symbolic interactionism, which holds that society and self are mutually determinative (Mead, 1934; Blumer, 1969). Blumer describes three fundamental premises of symbolic interactionism: (i) humans act toward things based on the meanings they have for them; (ii) the meanings of things are derived from social interactions with others; and (iii) these meanings are understood and modified through an interpretive process that is ongoing throughout various encounters. While the original proponents of symbolic interactionism held that it was a human-only phenomenon, Cerulo (2009) argues for the inclusion of non-humans in social interaction. Cerulo describes five main tenants or requirements of the human-only tradition, namely that social interaction requires: (i) human-only capabilities such as consciousness; (ii) intention; (iii) self-identity of those involved; (iv) other-orientation; and (v) active negotiation and definition of situations. In her rejection of these claims, she cites other recent studies and theories that challenge these notions of human-only social interaction, including studies with pets, deities, deceased loved ones, and technological creations such as avatars. Clayton (2003) also argues that from a psychological standpoint, non-social or non-human objects are often overlooked in their contributions to the development of identity, and that the natural environment may be a rich source of self-relevant beliefs.

Drawing on the Meadian tradition of symbolic interactionism, Weigert examines knowledge of the natural environment from a grounded, pragmatic, and social constructionist perspective, where the natural environment is a social construction that takes at least five forms. The first form is the cosmic environment, or the all-encompassing world. This is followed by the organismic environment, or those aspects of the cosmic environment that are physically or behaviorally experienced. Third, Weigert identifies the institutionalized environment, or that which is objectified and constructed by social organizations. Fourth, the encultured environment encompasses the 'taken-for-granted, naturally occurring environment'. Finally, the selfed environment is that which is constructed by

meaningful gestalts and identities of the self. From this framework, Weigert provides the following conceptual understanding of the natural environment:

> natural environment = what are taken as patterned facts within cosmic and organic processes; socio-cultural worldviews explaining and justifying such facts; institutional definitions of the factual aspects of experience; and personal experience of events interpreted as such facts. (Weigert, 1997, p. 150)

In their identity theory, Stryker and Serpe (1982) merge the symbolic interactionist approach with role theory, where identities are 'reflexively applied cognitions that answer the question "who am I?"' The answer to this question is phrased in terms of an individual's position in the organized structures of social relationships and social roles, and reflects the notion that just as society is organized, the self must be too. The number of identities a person may have is relatively unlimited as it corresponds with the number of distinct sets of structured relationships in which the individual is involved. Stryker and Serpe theorize that identities are organized within a salience hierarchy, which is based on the probability that an identity will be invoked in a given situation. Identity salience, in turn, rests on commitment to an identity, or 'the degree to which a person's relationship to specified sets of others depends on their being a particular type of person' (p. 207). Thus, commitment leads to salience, which leads to role performance. In terms of an environmental identity, then, the enactment of environmental behaviors would depend not only on the existence of an environmental identity, but also on commitment to and salience of that identity. In another interpretation, Cerulo (2009) tells us 'we can think of self-identity as a reflexiveness that results in the recognition of one's personhood—an understanding of who one is with reference to biography and situation'. With this understanding of identity in mind, it is easy to see how an environmental identity would be useful to both person and society.

Simply put, identity informs us of who we are. In the case of environmental identity, this sense of who we are is informed by the natural environment. In other words, the natural environment influences the ways we think about ourselves and about what it means to be human (Clayton, 2003). While much of the initial research into environmental identity and related concepts looked at those who had chosen pro-environmental careers and lifestyles (Thomashow, 1996; Chawla, 1999), it should be emphasized that having an environmental identity is not limited to these types of people. Clayton suggests that environmental identity arises from emotional connections to the natural environment through life experiences, special places, memories, and values. It is further suggested that our tendency to select natural environments over built environments underscores the human need for and valuation of nature. Clayton evokes the biophilia hypothesis as one potential explanation for this tendency.

While not always termed as such, the fundamental nature of an environmental identity or of human's relationship with nature is woven throughout the existing research and theorizing in the field. The biophilia hypothesis, for example, asserts an inherent human need to affiliate with and value nature, going so far as to propose a 'deprived and diminished existence' for those who are unable to fulfill this need (Kellert, 1993). Kellert proposes a typology of nine differing values of the natural environment. Numerous studies have demonstrated that while these nine values vary in their intensity and frequency across cultures, they all consistently appear, assisting Kellert and Wilson in their claim of human affiliation with the natural environment.

Each of these values corresponds with worldviews, or ways of knowing the natural environment. The moralistic value, for example, is thought to be closely aligned with indigenous perspectives (Kellert, 1993), and also tied to ideas of environmental identity: '…they [indigenous populations] emphasize a fundamental belief in the natural world as a living and vital being, a conviction of the continuous reciprocity between humans and nature, and the certainty of an inextricable link between human identity and the natural landscape' (p. 54). While the biophilia hypothesis is explored in greater detail earlier

in this chapter, it is important to note here both the interconnectedness of these ideas and the historical extent to which environmental identity can be traced in varying populations.

In all cases, it is clear that identity is implicated in the ways we perceive and act toward the natural world. When compared with environmental attitudes, environmental identity has been shown to be a better predictor of pro-environmental behavior (Linneweber *et al.* 2003; Stets and Biga, 2003; Devine-Wright and Clayton, 2010). Thus, the study of environmental identities may be beneficial in terms of both land preservation and also human health.

Psychologically 'deep' and extraordinary experiences

Whether a place they played in as children, a special location visited with a loved one, or a setting where everything seemed to be positive, many people experience a deep and significant connection with the natural environment. In addition, these connections are often linked to an activity (physical), a feeling (affect), or an aesthetic view (emotional/spiritual) and constitute an important component of 'being human'. Given the importance and impact of many of these experiences, a number of theories and theoretical constructs have been linked to the human health and natural environments paradigm. Moreover, these connections are often psychological rather than physical, implying that the visit to the local park is likely more intended for stress relief or relaxation rather than simply physical exercise. Roger Mannell began studying psychologically deep experiences in the 1970s and wrote that psychologically deep experiences are 'a transient psychological state, easily interrupted, and characterized by a decreased awareness of the passage of time, decreased awareness of the incidental features of physical and social surroundings and accompanied by positive affect' (Mannell, 1980, p. 76). He later defined psychologically deep experiences as 'psychological experiences [are those] that people experience and label as special, out-of-the-ordinary, or meaningful' (Mannell, 1996, p. 406).

Indeed, a strong and growing body of knowledge, whether empirically based, qualitatively generated, or anecdotal evidence, supports the contention that direct contact with natural environments can lead to increased mental health and psychological health (Davis, 2008). The importance of the link between health and natural environments cannot be overstated, with Frumkin and Louv (2007) suggesting that land conservation and management should now be viewed as a public health strategy. At the time, Frumkin was the Director of the National Center for Environmental Health at the US Centers for Disease Control and Prevention and Richard Louv is the author of *Last Child in the Woods: Saving Our Children from Nature-Deficit Disorder*. Park *et al.* (2011) recently found supporting evidence on the restorative effects of natural environments in their study of the psychological responses to natural and urban environments and the physical variables that characterize those environments. Variables such as tension, anxiety, depression, anger, vigor, confusion, fatigue, and mood disturbance all moved in a positive direction upon exposure to the natural setting. In another example, and from a physical activity perspective, Hug *et al.* (2008) compared the restorative effects of physical activities in forest and indoor settings. While physical exercise in either location was viewed as beneficial by the respondents, those exercising in a forest landscape reported a greater improvement in their mental balance and release from everyday hassles. In addition, those using the forest setting for exercise were more reluctant to leave the exercise area and had a heightened expectation of returning for the next session.

Following this line of reasoning articulating the relationship between humans and natural environments, Jefferies and Lepp (2012) describe the concept of 'extraordinary experiences' as experiences that are highly memorable, special, emotionally charged, and potentially life changing. In addition, Arnould and Price (1993) list three dimensions that characterize the extraordinary experience: (i) a deeply felt connection with a physical setting (they define this as harmony with nature); (ii) a deeply felt connection with other members in that physical setting; and (iii) a sense of personal growth and renewal. Louv (2011) in his book, *The Nature Principle*, alludes to the concept of extraordinary experiences by proposing a 'loose parts' theory in which exposure to the many parts of nature such as trees, topography, weather, biology, etc. can encourage a greater sensitivity to patterns and structure. Jefferies and Lepp (2012) identified four themes that study participants reported as characterizing extraordinary experiences: accomplishment, group identity, spontaneity, and walking outside. In addition, awe, relaxation, inspiration, introspection, and physical movement were associated with this factor, with nature being central to these components and a variable not easily duplicated in a gymnasium or indoor track. Thus, extraordinary experiences are often linked to exposure to nature, with resources such as protected natural areas featuring trails and places to experience nature often providing both extraordinary experiences and lasting benefits related to human health (Kaczynski *et al.*, 2008).

Flow

Related to extraordinary experiences and characteristic of many encounters with natural environments is the concept of flow. Originally proposed by Mihaly Csikszentmihalyi (1975), 'flow' can be defined as a mental state in which a person is performing an activity and engaged in single-minded immersion in that activity and experiencing feelings of joy, deep focus, and absorption. Considered a construct related to positive psychology, Nakamura and Csikszentmihalyi (2009) have identified six factors related to flow: (i) intense and focused concentration; (ii) a merging of action and awareness; (iii) a loss of reflective self-consciousness; (iv) a distortion or loss of time; (v) a sense of control over the situation or activity or at least an ability to influence the outcome; and (vi) engaging in the activity becomes an autotelic or intrinsically rewarding experience.

Flow occurs when someone is paying attention to an activity for its own sake such as when someone is doing a favorite activity or something that is deeply personally intriguing. This ability to engage in flow and its intrinsic rewards leads to happiness (Csikszentmihalyi, 2003). Csikszentmihalyi maintains that one can develop autotelic personality traits or the ability to be fully present in an activity without regard for a later reward or goal. Autolelic people seem to naturally find an optimal experience in situations that others would find boring or anxious through self-integrating enough complexity into the activity. For many people challenge helps to focus concentration and offers immediate feedback.

In this regard Csikszentmihalyi (1997) suggests flow can be linked to a balance between the challenge an individual faces and the skill level they possess relative to the challenge. One example of this within a natural environment setting would be a hiker evaluating the length and difficulty of a particular hiking trail and her/his personal level of hiking and physical skills. When the two (challenges and skills) are matched the hiker can go into 'flow'. Other than being in balance or 'in flow', Csikszentmihalyi lists a number of mental states associated with the flow experience when there is incongruity between challenges and skills, including: apathy, worry, anxiety, arousal, control, relaxation, and boredom. Engeser and Rheinberg (2008) found that when a person subjectively perceives an activity to be important flow can be achieved even if the demands are low, and that flow often predicts performance.

Csikszentmihalyi said that an autotelic person:

> experience(s) flow in work, in family life, when interacting with people, when eating, even when alone with nothing to do, they are less dependent on the external rewards that keep others motivated to go on with a life composed of dull and meaningless routines. They are more autonomous and independent because they cannot be as easily manipulated with threats or rewards from the outside. (Csikszentmihalyi, 1997, pp. 117–118)

As previously mentioned, encounters with nature often provide many of the prerequisite factors enjoined with flow such as experiences allowing intense focus, loss of the temporal experience, or a merging of action and personal awareness, and as a result can provide for a psychologically positive experience. Thus the natural environment can serve as a powerful catalyst in providing a setting for an individual to achieve a state of flow. In addition to providing a psychological positive experience autonomously, the flow experience can also be linked to another well-known positive psychological experience, the peak experience.

The peak experience

Developed and popularized by Maslow (1964) in his book, *Religions, Values, and Peak Experiences*, peak experiences can be described as joyous and exciting moments involving intense feelings of well-being, wonder, and/or awe. Maslow pointed out that peak experiences often come on suddenly and are inspired by meditation, reflection, or the overwhelming beauty of nature. They can become permanent features in an individual's memory and can be therapeutic by increasing a sense of self-determination, creativity, and empathy. History is replete with individuals who had key mystical experiences in natural environments, including Moses, Jesus, Buddha, and Mohammed (Scott, 1974). Moreover, it has been widely recognized in contemporary society that natural environments can be conducive to achieving a peak experience (Davis *et al.*, 1991). For example, McDonald *et al.* (2009), using a sample of wilderness users in Australia, found that wilderness settings can provide a combination of aesthetic qualities and a sense of renewal that leads to spiritual expression and growth.

Once again, nature and natural environments can serve as both a catalytic setting and an activity base from which individuals experience intrapersonal feelings of awe, mysticism, and happiness. And it is at these moments of happiness and fulfillment that individuals often develop some important meaning or insight that is both transcendent and often permanent. Many of these experiences that are embedded in the natural environment can be thought of transcendent. As has been alluded to in the previous paragraphs, whether a flow experience or a peak experience, both constructs often lead to the transcendent experience.

Transcendent experiences

Similar to the flow and the peak experience, transcendent experiences can be thought of as moments of extreme happiness with feelings of freedom and harmony with the whole world, and which are totally absorbing and feel of great importance (Williams and Harvey, 2001). Indeed, all three constructs, flow, peak experience, and transcendence, share some common characteristics including the following:

- a strong positive affect;
- a sense of connection with the whole (e.g. universe, world, community, etc.);
- absorption into the moment; and
- accompanied by a sense of timelessness.

As suggested by a growing list of authors and researchers (e.g. Laski, 1961; Chenoweth and Gobster, 1990; Suedfeld, 1992), there is a close and often profound connection between natural environments and transcendent experiences. This connection involves the aesthetic and restorative functions of nature and is usually attributed to the qualities of the physical environment in nature rather than just the activities. Among the many reasons for this transcendence through nature, three hypothetical concepts have emerged.

First, Schroeder (1996) suggests that the mystery that often surrounds nature serves to intensify the emotional connection that humans have with natural environments. From Schroeder's perspective, forests and other natural areas provoke this mystery because they 'hide what lies within' (p. 92). In addition, many of the attributes contained in natural environments such as trees, water, and other geologic structures are often richly endowed with symbolic meaning. In turn, these symbolic meanings can often serve as a catalyst to trigger a transcendent moment for the individual (Mandondo, 1997).

A second explanation for the link between natural environments and transcendental experiences is the sense of place that nature provides. While somewhat related to the mystery of nature presented previously, in this case, sense of place refers to the complex interaction between the person and the setting (Altman and Rogoff, 1992). Thus, natural environments are effective at producing transcendent experiences, which in turn can be positive, health-enhancing events because of the social and physical environments contained within the natural landscape. In a sense, natural environments can be transcendent because of the components of the environment and who people are with when experiencing that environment (Fredrickson and Anderson, 1999).

A third explanation draws much of its substance from Csikszentmihalyi's (1992) 'flow' theory by suggesting that specific activities performed in natural environments can create transcendent experiences. Thus, natural settings provide the physical context through which individuals experience transcendence because of the intense focus and absorption in the activity they are engaged in. Related to the activity focus, Mitchell (1983) distinguishes between active and passive experiences. His example is mountaineering where he describes an active experience as one where there is an 'active merging with mountains through the dynamics of climbing' (p. 147). Conversely, a passive experience in mountaineering is one where the individual is precluded from actively climbing because of weather or other circumstance and must otherwise simply be in the mountains. While usually not able to achieve the focus or intensity of climbing, the passive experience often results in feelings of awe or feeling small and insignificant compared with the grandeur of the landscape. Gallagher (1993) calls this the diminutive effect.

In sum, we have discussed the description and evolution of the transcendent experience as it relates to natural environments. Like flow and the peak experience, transcendent experiences represent a unique and complex interaction of people, environment, perceptions, and behaviors. In addition, these deep psychological and extraordinary experiences may be either manifestations or part of a broader picture that includes evolutionary-based or restorative-based theories of natural settings and human health. We start with several widely acknowledged restorative-based theories that connect human health issues with natural environments.

Restoration and Restorative Environments

From a clinical and epidemiologic perspective, several questions emerge in considering the effect of natural environments upon human health. First, what is meant by restoration and what constitutes a restorative environment? In this book, we are generally referring to restoration as a process through which people recover personal resources that they have diminished in their efforts to meet the demands of everyday life or other events (Hartig, 2004). Many of these demands result from issues related to stress. In this case, stress is operationalized as shown in Fig. 4.1. Thus, we consider stress as essentially a negative attribute that often serves to degrade or impair performance. Moreover, stress can be thought of as an individual's response in psychological, physiological, and/or behavioral terms to a situation that challenges or threatens an individual's sense of well-being. Psychological effects of stress can be manifested in outcomes such as cognitive appraisals of fear, anger, depression, or inefficiencies in coping. In a similar fashion, physiological manifestations of stress can occur in a variety of outcomes related to the muscular, cardiovascular, or neuroendocrine systems. Behavioral outcomes associated with stress include avoidance, substance abuse, and declines in cognitive and behavioral performances.

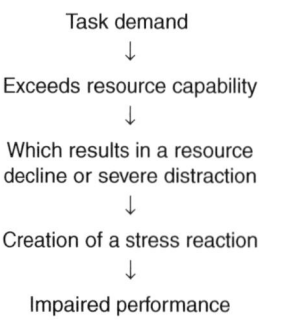

Task demand
↓
Exceeds resource capability
↓
Which results in a resource decline or severe distraction
↓
Creation of a stress reaction
↓
Impaired performance

Fig. 4.1. A generalized model of stress response.

Second, is there a relationship between exposure to natural environments and restoration? In other words, do people generally feel 'restored' or less stressed when exposed to natural environments? And third, how does this exposure actually work to reduce exposure? Are there certain dosage rates or specific characteristics of the exposure that are particularly effective in achieving a sense of restoration for the individual?

Psychological resources that serve to alleviate or lessen stress-related issues include social networks, escape (however temporary) from situations that deplete resources, and a reduction in the number and type of demands placed on an individual. Natural environments have often been characterized as having attributes that serve to provide restorative settings. Two primary theories have arisen that inform much of the discussion surrounding the restorative effects of natural environments and, through this connection, the promotion of positive health: psycho-evolutionary theory and attention restoration theory.

Psycho-evolutionary theory (PET)

As the title suggests, psycho-evolutionary theory (PET) implies a connection with the evolutionary process. Based on the proposition that affect precedes cognition (Ulrich, 1983), PET emphasizes emotions and the immediacy of affect or feeling. In turn, this connection between natural environments and affect or feelings serves to reduce stress reactions. That is, because of our ancestral direct connection to natural landscapes, humans can experience immediate reactions to natural environments well before they have had a chance to analyze these environments through cognitive processes. In the case of stress, recovery occurs in settings that evoke interest, pleasantness, and calm. And this is where biophilia and PET integrate; that is, humans are attracted to natural environments and this attraction helps reduce negative factors such as stress. Indeed, natural landscapes contain a number of dimensions that are associated with restoring scenes, including structure, depth, and content (Han, 2001).

Structural components associated with natural environments generally involve moderate complexity (unlike built environments which often contain high levels of complexity), order, and focality or places where individuals can focus their attention within a manageable level. Examples of depth would include a sense of spaciousness or openness, ground surface that has texture, unlike pavement or asphalt, and vistas which are inviting and aesthetic. Environmental content refers to areas that present perceived support or threat. For example, according to PET, humans are attracted to relatively calm water or relatively open vistas because they represent evolutionarily designated places of safety, shelter, or refreshment.

Likewise, the theory posits that, unlike situations commonly associated with stress and which involve increased negative emotions and heightened autonomic arousal responses (e.g. increased heart rate, blood pressure, or perceptions of stimulus overload), scenes involving those attributes listed above and which are often encountered in natural environments can evoke sensations of mild to moderate interest, pleasantness, and relaxation (see Ulrich, 1983).

Thus, natural environments can serve as stress recovery settings and point to a number of important positive health outcomes for humans under stress. These dimensions involve replacing negative affect with positive feelings, inhibiting negative thoughts, and decreasing autonomic and cortical responses such as fight, flight, or freeze. PET can be conceptualized as shown in Fig. 4.2.

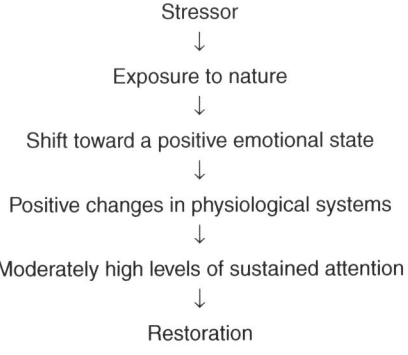

Stressor
↓
Exposure to nature
↓
Shift toward a positive emotional state
↓
Positive changes in physiological systems
↓
Moderately high levels of sustained attention
↓
Restoration

Fig. 4.2. A generalized model of psycho-evoluntionary theory (PET).

One of the most salient studies supporting the concept of PET investigated the impacts of viewing nature on recovery rates of surgery patients. After randomly assigning surgery patients to either a room with a view of nature or a hospital room without such a view, Ulrich (1984) found evidence that the patients with the view of nature had shorter recovery times, required less pain medication, and had lower incidence of medical issues associated with their surgery than the patients without the view of nature. Ulrich suggests that the results from this study point to a proposition that people benefit most from visual contact with nature, as opposed to urban environments lacking nature, i.e. built environments. Moreover, for individuals experiencing stress or excessive arousal, nature views appear to reduce arousal more effectively than urban scenes and hence are more beneficial in a psychophysiological sense. Thus, natural environments offer an evolutionary advantage over those of built, non-natural environments in terms of restoration and stress reduction.

This advantage results in a preference for certain environments and PET postulates that this preference is chronologically first, in front of cognitive awareness. That is, people prefer natural landscapes before they cognitively evaluate or compare that specific landscape (Parsons, 1991). Moreover, since restoration is linked to specific types of affective responses (e.g. pleasantness, sense of safety) that individuals express as a preference, the connection between restoration and natural environments through a mechanism such as PET seems logical.

Thus, central to PET as a means to restoration is the belief that positive changes in perceptions and subsequent emotions such as a sense of calm or joy as opposed to negative feelings can be facilitated through the presence of natural environments, primarily because of evolutionary learnings and background. In turn, these positive changes can be effective in enhancing adaptive behaviors and reducing health-related issues such as stress.

But while Ulrich's PET centers on restoration from negative emotional states, or physiological depletion of resources such as energy levels due to stress, the attention restoration theory emphasizes restoration as a way to improve everyday functioning through recovery from attention fatigue.

Attention restoration theory (ART)

Based on the earlier work of William James (1842–1910), Kaplan and Kaplan (1989) through their attention restoration theory (ART) presuppose that attention is either voluntary (directed) or involuntary. Directed attention requires effort, plays a central role in achieving and maintaining focus, is usually under a cognitive control process, controls distraction through the use of inhibition, is important in the problem-solving process, and is susceptible to fatigue. It is this susceptibility to fatigue that speaks to the underlying role of natural environments through ART. Essentially, the ART theoretical model assumes the components shown in Fig. 4.3.

According to Kaplan and Kaplan (1989), restorative settings, as provided through natural environments, can be effective in aiding in the recovery of directed attention fatigue through the following properties:

1. *Fascination*—engages attention effortlessly thus allowing directed attention to rest.
2. *Being away*—physically or conceptually different from one's usual environment.
3. *Extent*—setting is rich and coherent enough to engage the mind and promote exploration.
4. *Compatibility*—implies a good fit between one's inclinations/purposes and the activities supported by the setting.

Directed attention requires effort to maintain focus and avoid distractions
↓
Directed attention fatigues and results in difficulty in focusing, irritability, and distraction
↓
Situations that use involuntary attention result in a replenishment of directed attention mechanisms and ability

Fig. 4.3. A generalized model of attention restoration theory (ART).

Fascination relates to the quality of certain settings or events that can attract and hold a person's interest or attention (Kaplan *et al.*, 1998). In addition, fascination can take two forms: hard fascination and soft fascination. Hard fascination refers to experiences or activities that are intense, riveting, and with limited space or time for reflection or cognitive activity. Soft fascination involves activities or experiences that are moderate in intensity and can focus an individual's attention while allowing for reflection and cognition. Peaceful, natural settings are often thought to be good locations for experiencing soft fascination. Thus, the extent of fascination can be characterized by the function is serves, its overall intensity, and the amount of pleasure it provides. While often necessary, fascination by itself is generally not considered sufficient for restoration (Kaplan, 1995).

As the term suggests, being away implies achieving some psychological distance from the tasks, duties, or ongoing goals that an individual experiences in normal everyday living and that require directed attention (Hartig *et al.*, 2011, p. 151). Conceptually, being away involves three components: (i) escape from unwelcome stimuli such as noise, traffic, or crowding; (ii) leaving one's routine concerns or activities; and (iii) temporarily setting aside one's ongoing pursuit of goals or aspirations. Finally, being away can be a physical phenomenon, a psychological transformation, or a combination of both (Han, 2001).

Extent involves having a large enough setting in which an individual can experience in either physical or psychological terms without using directed attention. Thus, the individual remains fascinated and interested but without using energy to focus directed attention.

Finally, compatibility refers to the match between what a person wants or tries to do and what the environment allows them to do (Hartig *et al.*, 1997). High compatibility often serves to facilitate reflection which, in turn, contributes to recovery from mental fatigue and is an important step toward restoration.

It should be noted that while many environments can provide experiences of being away, extent, fascination, and compatibility, natural environments can be particularly effective in allowing for this array. Aesthetic scenery, open space, the presence of water and vegetation, and other variables all contribute to the development of 'soft' fascination and reflection which, in turn, facilitate restorative outcomes.

A substantial amount of research has demonstrated the efficacy of the attention restoration construct. For example, people exposed to natural settings generally report better performance on attention-demanding tasks (Taylor *et al.*, 2001, 2002). Natural views from home are positively related to effective functioning and feeling at peace and negatively related to distraction (Kaplan, 2001). People living near nature report improved interpersonal relations and effectiveness at handling major life issues (Kuo, 2001). Exposure to natural environments may be related to improved self-control, reduced symptoms of attention deficit disorder, and improved care in treating dementia (Bossen, 2010).

Combining the theories mentioned above with practice, the next section discusses a model to help people reliably gain benefits while being in nature.

Intentionally Designed Experiences (IDEs)

The literature and the authors' personal experiences generally point to a connection between human health and natural environments. Does the same connection carry over to experiences and activities associated with adventure programming? For example, aspects of quality of life that are often influenced by natural environments include: perceived safety, physical health and exercise, emotional well-being, a sense of community, social interaction, enjoyment, appropriate levels of stimulation, and a sense of autonomy and control. Beyond the obvious fact that many adventure activities occur in natural or outdoor environments, are there some underlying commonalities that link adventure programming, natural landscapes, and human health? Louv's (2005) book,

Last Child in the Woods, effectively promotes the idea that nature is desirable and even necessary for the proper development of children but is relatively silent on the connection between adventure education and health-related issues. However, as mentioned in Louv (2005, p. 225), Taylor and Kuo point out that some of the most exciting findings of a link between contact with greenspace and developmental outcomes come from studies examining the effects of outdoor challenge programs on children's self-esteem and sense of self.

Drawing back to Berlyne's (1960) arousal theory, to be restorative, levels of arousal must be sufficiently but not overly complex or intense, and have understandable consequences that are somewhat under the control of the participant. Likewise, participation in adventure programs has often been associated with developmental outcomes such as personal growth, enhanced interpersonal skills, and group development (Ewert and Garvey, 2007; Passarelli *et al.*, 2010). McKenzie (2000) ascribes these outcomes to four attributes common to many adventure program experiences: (i) the unfamiliar nature of the physical environment; (ii) the incremental and progressive sequencing of the challenges presented through the adventure program experience; (iii) the 'processing' of the experience in order to identify and organize meaning to the participant; and (iv) the use of small groups to facilitate aspects such as reciprocity, group cohesiveness, interpersonal relationships, and balance between group belonging and individual autonomy. Thus, many of the attributes associated with adventure program activities and experiences are similar to those attributed to healthy lifestyles or behaviors.

Nor is the field of adventure programming without an historical literature base supporting the efficacy of these areas in promoting human health. For example, Charles Eastman wrote a piece in 1921 extolling the importance of out-of-door activities in maintaining good health, especially the physical and nervous systems. Fast forward to the mid- to late 2000s, and one can find a plethora of studies and papers looking at the impacts of adventure-based activities upon a number of variables often associated with either actual health conditions (e.g. level of fitness) or health promotion (e.g. variables associated with health behavior decision-making such as smoking cessation). For example, Ewert and Yoshino (2011) found increases in levels of resilience of college students following an adventure education program.

In addition, Grocott and Hunter (2009) found that participation in an adventure-based sailing program enhanced both global and domain-specific attributes of self-esteem. Both self-esteem and resilience are often linked to both health behaviors and healthy lifestyles. Moreover, it is not just the adventure activities that serve as the vehicles to influence human health attitudes and behaviors; the location in which these adventures take place can also play a critical role. Cole and Hall (2010) investigated the connection between wilderness landscapes and restoration. Their findings pointed to substantial reductions in levels of stress and increases in mental rejuvenation following visits to wilderness areas. Miles (1987) provided the foundation of this work by examining the way that wilderness or wilderness-like areas contribute to physical, emotional, or spiritual health.

Not surprisingly, research has demonstrated a consistent link between physical exercise and human health (Pretty et al., 2005a,b). In addition, there are a number of physiological health-related benefits linked to participation in adventure programming. For example, Russell and Phillips-Miller (2002) found that hiking and physical exercise were critical components in a wilderness therapy program. Jelalian et al. (2006) found that an adventure therapy program patterned after Outward Bound served as an effective adjunct therapy for reducing weight in adolescents. In fact, the adventure therapy participants lost four times as much weight as a comparable group exercising indoors for the same 16-week period. In a similar fashion, Caulkins et al. (2006) reported that physical exercise plays an important role in wilderness therapy for adolescent women.

It seems clear that the physical activity inherent in most adventure programming activities contributes to human health in a variety of ways, such as enhancing fitness, aerobic training, and balance. However, the often short-term and episodic nature of these activities suggests that the true power of adventure programming in effecting health and wellness lies not just in the physicality of the activities, but also in the promotion of health-enhancing behaviors. For example, Bowler et al. (2010) found less consistency in the evidence supporting the idea of natural environments impacting variables such as blood pressure, hormone levels, or immune functioning, and much more consistency in positive changes in self-reported variables such as revitalization, reductions in anger and anxiety, enhanced levels of attention, and lower depression. Marcus and Forsyth (2008) suggest that the most important variables related to health behaviors such as physical activity include self-efficacy, a sense of achievement, and social support systems. Thus, the true value of adventure programming may lie in the affective outcomes it often produces in addition to the movement and activity inherent in the experiences.

With few exceptions, however, the research and subsequent literature strongly suggest that specific programming using natural environments can be linked to human health in a variety of ways, such as physicality, emotional health, spiritual development, and psychological well-being. One key component in this linkage between health and outdoor programming is the use of intentionally designed experiences (IDEs). While PET, biophilia, and ART provide a foundation of understanding relative to the relationship between human health and natural environments, they remain somewhat passive. That is, they provide a backdrop for understanding the impact of people simply being in a natural environment. IDEs, on the other hand, provide an active component; that is, people are engaged in programmed and structured activities in natural environments. Thus, the IDE builds on the power of the natural environment to enhance its benefits to human health. Figure 4.4 illustrates this model.

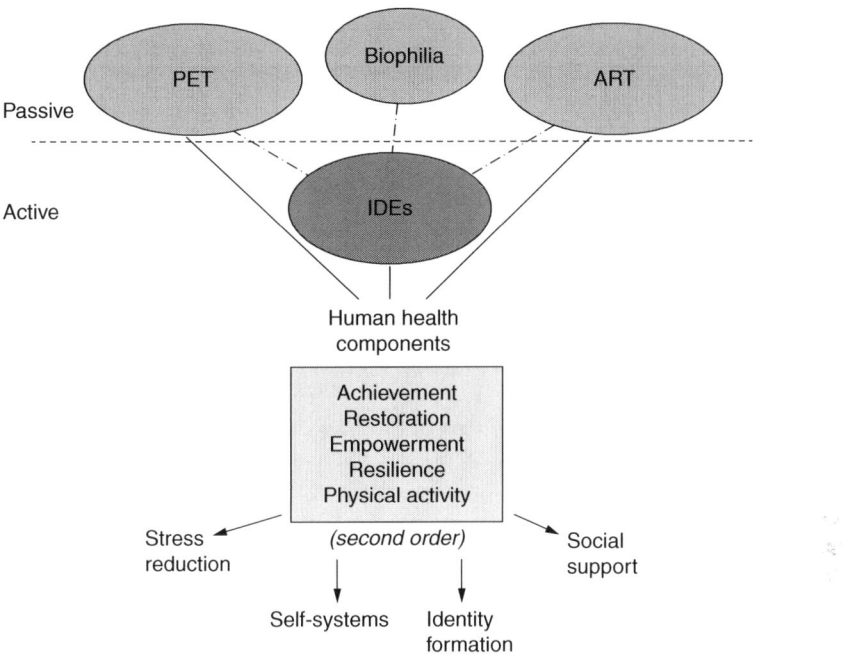

Fig. 4.4. A model of the relationship between intentionally designed experiences (IDEs) and natural environment theories (PET, psycho-evolutionary theory; ART, attention restoration theory). (From Ewert *et al.*, 2010).

As can be seen from Fig. 4.4, PET, biophilia, and ART serve as the foundational theories for understanding the connection between natural environments and human health; the very places where most adventure programming takes place. These inputs are passive, however, in that they provide a possible explanation of what happens when individuals are exposed to natural environments. While natural environments can be powerful inducements for positive benefits to the person, IDEs are often critical to combining the effect of the natural setting with the effectiveness of structured and purposely designed experiences. In turn, these designed experiences result in a set of first-order outcomes that provide a foundation for additional subsequent outcomes. These primary outcomes include achievement, restoration, empowerment, resilience, and physical activity. For example, return to the Cole and Hall (2010) study that found that wilderness environments can provide for restoration and mental rejuvenation. Second-order outcomes that emerge from the primary

outcomes and have connections to human health and well-being include self-systems, identity formation, social support, and stress reduction. For example, Holman and McAvoy (2005) report that beneficial outcomes from participation in integrated adventure programs using natural settings included achievement and self-awareness.

Chapter Summary

In sum, this chapter has explored a variety of concepts and models commonly associated with human health and natural environments. Salient theories that were discussed and serve to provide a theoretical underpinning for the health–environment relationship included PET, biophilia, and ART. Also included in this discussion were concepts such as naturalistic intelligence, indigenous consciousness, environmental identity, psychologically deep and extraordinary experiences such as flow, the peak experience, and transcendent experiences. All of these theories and concepts serve to inform our thinking about being able to conceptually explain how natural environments impact variables related to health and, as such, provide us with a 'roadmap' for how, why, and when a particular factor such as the natural environment can and does impact human health issues such as level of stress, sense of well-being, and numerous physiological parameters.

Finally, the chapter concluded with a discussion regarding ways to strengthen the connection between natural environments and health. More specifically, the concept of IDEs was introduced and represents a model in which the setting (natural environment) is combined with IDEs (e.g. programs) and together a number of outcomes are realized including resilience, empowerment, sense of achievement, and restoration. These first-order variables may then provide the psychological 'strength' for an individual to then achieve a number of other, often health-related goals such as weight reduction, stress reduction, belief that one can positively impact his or her health-related conditions, etc. The underlying assumption to the IDE concept is that, while often powerful determinants for human health on their own, natural settings can be augmented by the inclusion of programs and experiences specifically designed to beneficially impact human health.

References

Altman, I. and Rogoff, B. (1992) World views in psychology: trait, interactional, organismic, and transactional perspective. In: Stokols, D. and Altman, I. (eds) *Handbook of Environmental Psychology*. Krieger Publishing, Melbourne, Australia, pp. 7–40.

Appleton, J. (1975) *The Experience of Landscape*. Wiley, London.

Arnould, E.J. and Price, L.L. (1993) River magic: extraordinary experience and the extended service encounter. *Journal of Consumer Research* 20, 24–45.

Berlyne, D. (1960) *Conflict, Arousal, and Curiosity*. McGraw-Hill, New York.

Bixler, R.D. and Floyd, M.F. (1997) Nature is scary, disgusting, and uncomfortable. *Environment and Behavior* 29(4), 17–22.

Blumer, H. (1969) *Symbolic Interactionism: Perspective and Method*. Prentice-Hall, Englewood Cliffs, New Jersey.

Bossen, A. (2010) The importance of getting back to nature for people with dementia. *Journal of Gerontological Nursing* 36(2), 17–22.

Bowler, D.E., Buyung-Ali, L.M., Knight, T.M. and Pullin, A.S. (2010) A systematic review of evidence for the added benefits to health of exposure to natural environments. *BMC Public Health* 10, 456.

Carson, R. (1998) *The Sense of Wonder*. HarperCollins Publishers, New York.

Caulkins, M., White, D. and Russell, K. (2006) The role of physical exercise in wilderness therapy for troubled adolescent women. *Journal of Experiential Education* 29(1), 18–37.

Cerulo, K. (2009) Nonhumans in social interaction. *Annual Review of Sociology* 35, 531–552.

Chawla, L. (1999) Life paths into effective environmental action. *Journal of Environmental Education* 31(1), 15–26.

Chenoweth, R.E. and Gobster, P.H. (1990) The nature and ecology of aesthetic experiences in the landscape. *Landscape Journal* 9(1), 1–8.

Clayton, S. (2003) Environmental identity. In: Clayton, S. and Opotow, S. (eds) *Identity and the Natural Environment: The Psychological Significance of Nature*. MIT Press, Cambridge, Massachusetts, pp. 45–66.

Cole, D.N. and Hall, T.E. (2010) Experiencing the restorative components of wilderness environments: does congestion interfere and does length of exposure matter? *Environment and Behavior* 42(6), 806–823.

Csikszentmihalyi, M. (1975) *Beyond Boredom and Anxiety: The Experience of Play in Work and Games*. Jossey-Bass, San Francisco, California.

Csikszentmihalyi, M. (1992) *Flow: The Psychology of Happiness*. Rider, London.

Csikszentmihalyi, M. (1997) *Finding Flow: The Psychology of Engagement with Everyday Life*. Basic Books, New York.

Csikszentmihalyi, M. (2003) *Good Business: Leadership, Flow and the Making of Meaning*. Penguin Putnam Inc., New York.

Davidson, L. (2012) The calculable and the incalculable: narratives of safety and danger in the mountains. *Leisure Sciences* 34(4), 298–313.

Davis, J. (2008) Psychological benefits of nature experiences: an outline of research and theory with special reference to transpersonal psychology. Available at: http://www.soulcraft.co/essays/PSYCHOLOGICAL_BENEFITS_OF_NATURE_EXPERIENCES.pdf (accessed 6 April 2013).

Davis, J., Lockwood, L. and Wright, C. (1991) Reasons for not reporting peak experiences. *Journal of Humanistic Psychology* 31(1), 86–94.

DeLoache, J.S. and LoBue, V. (2009) The narrow fellow in the grass: human infants associate snakes and fear. *Developmental Science* 12(1), 201–207.

Devine-Wright, P. and Clayton, S. (2010) Introduction to the special issue: Place, identity and environmental behaviour. *Journal of Environmental Psychology* 30(3), 267–270.

Engeser, S. and Rheinberg, F. (2008) Flow, performance and moderators of challenge–skill balance. *Motivation and Emotion* 32(3), 158–172.

Ewert, A. and Garvey, D. (2007) Philosophy and theory of adventure education. In: Prouty, D., Panicucci, J. and Collinson, R. (eds) *Adventure Education: Theory and Applications*. Human Kinetics, Champaign, Illinois, pp. 19–32.

Ewert, A. and Yoshino, A. (2011) The influence of short-term adventure-based experiences on levels of resilience. *Journal of Adventure Education and Outdoor Learning* 11(1), 35–50.

Ewert, A., Overholt, J., Voight, A. and Wang, C.C. (2010) Understanding the transformative aspects of the wilderness and protected lands experience upon human health. In: Watson, A., Murrieta-Saldivar, J. and McBride, B. (comps) *Science and Stewardship to Protect and Sustain Wilderness Values. Ninth World Wilderness Congress Symposium*, Merida, Yucatan, Mexico, 6–13 November 2009. *Proceedings RMRS-P-64*. US Department of Agriculture, Forest Service, Rocky Mountain Research Station, Fort Collins, Colorado, pp. 140–146.

Fenton, L. (2008) Evolutionary tails of the dis-stressing response to nature and future research directions. Presented at *Canadian Parks for Tomorrow: 40th Anniversary Conference*, University of Calgary, Calgary, Alberta, Canada, 8–11 May 2008.

Fredrickson, L.M. and Anderson, D.H. (1999) A qualitative exploration of the wilderness experience as a source of spiritual inspiration. *Journal of Environmental Psychology* 19(1), 21–39.

Fromm, E. (1964) *The Heart of Man*. Harper and Row, New York.

Frumkin, H. and Louv, R. (2007) *The Powerful Link between Conserving Land and Preserving Health*. Land Trust Alliance, Washington, DC.

Gallagher, W. (1993) *The Power of Place*. Poseiden Press, New York.

Gardner, H.E. (2000) *Intelligence Reframed: Multiple Intelligences for the 21st Century*. Basic Books, New York.

Garst, B.A., Williams, D.R. and Roggenbuck, J.W. (2010) Exploring early twenty-first century developed forest camping experiences and meaning. *Leisure Sciences* 32(1), 90–107.

Grinde, B. and Patil, G.G. (2009) Biophilia: does visual contact with nature impact on well-being? *International Journal of Environmental Research and Public Health* 6(9), 2332–2343.

Grocott, A. and Hunter, J. (2009) Increases in global and domain specific self-esteem following a 10 day developmental voyage. *Social Psychology of Education* 12(4), 443–459.

Han, K. (2001) A review: theories of restorative environments. *Journal of Therapeutic Horticulture* 12, 30–43.

Hartig, T. (2004) Restorative environments. In: Spielberger, C. (ed.) *Encyclopedia of Applied Psychology*. Academic Press, San Diego, California, pp. 273–279.

Hartig, T., Korpela, K., Evans, G.W. and Gärling, T. (1997) A measure of restorative quality in environments. *Scandinavian Housing and Planning Research* 14(4), 175–194.

Hartig, T., van den Berg, A.E., Hagerhall, C.M., Tomalak, M., Bauer, N., *et al.* (2011) Health benefits and nature experiences: psychological, social and cultural processes. In: Nilsson, K., Sangster, M., Gallis, C., Hartig, T., de Vries, S., *et al.* (eds) *Forests, Trees and Human Health*. Springer, New York, pp. 127–168.

Holman, T. and McAvoy, L. (2005) Transferring benefits of participation in an integrated wilderness adventure program to daily life. *Journal of Experiential Education* 27(3), 322–325.

Hug, S., Hansmann, R., Monn, C., Krütli, P. and Seeland K. (2008) Restorative effects of physical activity in forests and indoor settings. *International Journal of Fitness* 4(2), 25–38.

Jefferies, K. and Lepp, A. (2012) An investigation of extraordinary experiences. *Journal of Park and Recreation Administration* 30(3), 37–51.

Jelalian, E., Mehlenbeck, R., Lloyd-Richardson, E.E., Birmaher, V. and Wing, R.R. (2006) 'Adventure therapy' combined with cognitive-behavioral treatment for overweight adolescents. *International Journal of Obesity* 30(1), 31–39.

Kaczynski, A.T., Potwarka, L.R. and Saelens, B.E. (2008) Association of park size, distance and features with physical activity in neighborhood park. *American Journal of Public Health* 98(8), 1451–1456.

Kaplan, R. (2001) The nature of the view from home: psychological benefits. *Environment and Behavior* 33(4), 507–542.

Kaplan, R. and Kaplan, S. (1989) *The Experience of Nature: A Psychological Perspective*. Cambridge University Press, Cambridge, UK.

Kaplan, R., Kaplan, S. and Ryan, R.L. (1998) *With People in Mind: Design and Management of Everyday Nature*. Island Press, Washington, DC.

Kaplan, S. (1995) The restorative benefits of nature: toward an integrative framework. *Journal of Environmental Psychology* 15(3), 169–182.

Kellert, S.R. (1993) The biological basis for human values of nature. In: Kellert, S.R. and Wilson, E.O. (eds) *The Biophilia Hypothesis*. Island Press, Washington, DC, pp. 42–69.

Kellert, S.R. (2003) *Kinship to Mastery: Biophilia in Human Evolution and Development*. Island Press, Washington, DC.

Korpela, K. (1989) Place-identity as a product of environmental self-regulation. *Journal of Environmental Psychology* 9(3), 241–256.

Kuo, F.E. (2001) Coping with poverty: impacts of environment and attention in the inner city. *Environment and Behavior* 33(1), 5–34.

Laski, M. (1961) *Ecstasy: A Study of Some Secular and Religious Experiences*. The Cressett Press, London.

Lear, L. (ed.) (1998) *Lost Woods: The Discovered Writing of Rachel Carson*. Beacon Press, Boston, Massachusetts.

Linneweber, V., Hartmuth, G. and Fritsche, I. (2003) Representations of the local environment as threatened by global climate change: toward a contextualized analysis of environmental identity in a coastal area. In: Clayton, S. and Opotow, S. (eds) *The Psychological Significance of Nature*. MIT Press, Cambridge, Massachusetts, pp. 227–246.

LoBue, V. and DeLoache, J.S. (2008) Detecting the snake in the grass attention to fear-relevant stimuli by adults and young children. *Psychological Science* 19(3), 284–289.

Louv, R. (2005) *Last Child in the Woods: Saving our Children from Nature-Deficit Disorder*. Algonquin Books of Chapel Hill, Chapel Hill, North Carolina.

Louv, R. (2011) *The Nature Principle: Human Restoration and the End of Nature-Deficit Disorder*. Algonquin Books of Chapel Hill, Chapel Hill, North Carolina.

Mandondo, A. (1997) Trees and spaces as emotion and norm laden components of local ecosystems in Nyamaropa communal land, Nyanga District, Zimbabwe. *Agriculture and Human Values* 14(4), 353–372.

Mannell, R. (1980) Social psychological techniques and strategies for studying leisure experience. In: Iso-Ahola, S.E. (ed.) *Social Psychological Perspectives on Leisure and Recreation*. Charles C. Thomas, Springfield, Illinois, pp. 62–88.

Mannell, R. (1996) Approaches in the social and behavioural sciences to the systematic study of hard-to-define human values and experience. In: Driver, B.L., Dustin, D., Baltic, T., Elsner, G. and Peterson, G. (eds) *Nature and the Human Spirit: Toward and Expanded Land Management Ethic*. Venture Publishing, Inc., State College, Pennsylvania, pp. 405–416.

Marcus, B. and Forsyth, L. (2008) *Motivating People to be Physically Active*. Human Kinetics, Champaign, Illinois.

Martinez, D.E. (2008) Indigenous consciousness and the production of knowingness. Presented at the *American Sociological Association Annual Meeting*, Boston, Massachusetts, 31 July 2008.

Maslow, A.H. (1964) *Religion, Values, and Peak-Experiences*. The Ohio State University Press, Columbus, Ohio.

McDonald, M.G., Wearing, S. and Ponting, J. (2009) The nature of the peak experience in wilderness. *The Humanistic Psychologist* 37(4), 370–385.

McKenzie, M. (2000) How are adventure education program outcomes achieved? A review of the literature. *Australian Journal of Outdoor Education* 5(1), 19–28.

Mead, G.H. (1934) *Mind, Self, and Society*. University of Chicago Press, Chicago, Illinois.

Miles, J. (1987) Wilderness as healing place. *Journal of Experiential Education* 10(3), 4–10.

Mitchell, R.G. (1983) *Mountain Experience: The Psychology and Sociology of Adventure*. University of Chicago Press, Chicago, Illinois.

Mitten, D. (2010) Friluftsliv and the healing power of nature: the need for nature for human health, development, and wellbeing. In: *Henrik Ibsen: The Birth of 'Friluftsliv'. A 150 Year International Dialogue*

Conference Jubilee Celebration. North Troendelag University College, Levanger, Norway, Mountains of Norwegian/Swedish Border, 14–19 September 2009. Available at: http://norwegianjournaloffriluftsliv.com/doc/122010.pdf (accessed 8 September 2013).

Nakamura, J. and Csikszentmihalyi, M. (2009) Flow: theory and research. In: Snyder, C.R. and Lopez, S.J. (eds) *Oxford Handbook of Positive Psychology*. Oxford University Press, New York, pp. 195–206.

Orians, G.H. (1980) Habitat selection: general theory and applications to human behaviour. In: Lockard, J.S. and Lowenthal, D. (eds) *Landscape Meanings and Values*. Allen and Unwin, London, pp. 86–94.

Park, B.J., Furuya, K., Kasetani, T., Takayama, N., Kagawa, T., *et al.* (2011) Relationship between psychological responses and physical environments in forest settings. *Landscape and Urban Planning* 102(1), 24–32.

Parsons, R. (1991) The potential influences of environmental perception on human health. *Journal of Environmental Psychology* 11(1), 1–23.

Passarelli, A., Hall, E. and Anderson, M. (2010) A strength-based approach to outdoor and adventure education: possibilities for personal growth. *Journal of Experiential Education* 33(2), 120–135.

Pretty, J., Griffin, M., Peacock, J., Hine, R., Sellens, M., *et al.* (2005a) *Countryside for Health and Wellbeing: The Physical and Mental Health Benefits of Green Exercise*. Countryside Recreation Network, Sheffield, UK.

Pretty, J., Peacock, J., Sellens, M. and Griffin, M. (2005b) The mental and physical health outcomes of green exercise. *International Journal of Environmental Health Research* 15(5), 319–337.

Proshansky, H., Fabian, A. and Kaminoff, R. (1983) Place-identity: physical world socialization of the self. *Journal of Environmental Psychology* 3(1), 57–83.

Russell, K. and Phillips-Miller, D. (2002) Perspectives on the wilderness therapy process and its relation to outcome. *Child and Youth Care Forum* 31(6), 415–437.

Schroeder, H.W. (1996) Psyche, nature, and mystery: some psychological perspectives on the values of natural environment. In: Driver, B.L., Dustin, D., Baltic, T., Elsner, G. and Peterson, G. (eds) *Nature and the Human Spirit: Toward and Expanded Land Management Ethic*. Venture Publishing, Inc., State College, Pennsylvania, pp. 81–96.

Scott, N.R. (1974) Toward a psychology of wilderness experience. *Natural Resources Journal* 14, 231–237.

Shorb, T.L. and Schnoeker-Shorb, Y.A. (2010) *What's Nature Got to Do with Me: The Kellert–Shorb Biophilic Values Indicator—a Continuing History of its Creative Development and Design*. Prescott College, Prescott, Arizona.

Stets, J. and Biga, C. (2003) Bringing identity theory into environmental sociology. *Sociological Theory* 21(4), 398–423.

Stryker, S. and Serpe, R.T. (1982) Commitment, identity salience, and role behavior: theory and research example. In: Ickes, W. and Knowles, E.S. (eds) *Personality, Roles, and Social Behavior*. Springer, New York, pp. 192–216.

Suedfeld, P. (1992) Extreme and unusual environments. In: Stokols, D. and Altman, I. (eds) *Handbook of Environmental Psychology*. Krieger Publishing, Melbourne, Australia, pp. 863–887.

Swiderski, M.J. (2011) Maria Montessori. In: Smith, T. and Knapp, C.E. (eds) *Sourcebook of Experiential Education: Key Thinkers and Their Contributions*. Routledge, New York, pp. 197–207.

Taylor, A.F., Kou, F.E. and Sullivan, W.C. (2001) Coping with ADD: the surprising connection to green play settings. *Environment and Behavior* 33(1), 54–77.

Taylor, A.F., Kou, F.E. and Sullivan, W.C. (2002) Views of nature and self-discipline: evidence from inner city children. *Journal of Environmental Psychology* 22(1–2), 49–63.

Thomashow, M. (1996) *Ecological Identity: Becoming a Reflective Environmentalist*. MIT Press, Cambridge, Massachusetts.

Ulrich, R.S. (1983) Aesthetic and affective response to natural environment. In: Altman, J. and Wohlwill, F. (eds) *Behavior and Natural Environment: Advances in Theory and Research*, vol. 6. Plenum, New York, pp. 85–125.

Ulrich, R.S. (1984) View through a window may influence recovery from surgery. *Science* 224(4647), 420–421.

Ulrich, R.S. (1999) Effects of gardens on health outcomes: theory and research. In: Marcus, C.C. and Barnes, M. (eds) *Healing Gardens: Therapeutic Benefits and Design Recommendations*. John Wiley and Sons, New York, pp. 27–86.

Ulrich, R.S., Simons, R.F., Losito, B.D., Fiorito, E., Miles, M.A., *et al.* (1991) Stress recovery during exposure to natural and urban environments. *Journal of Environmental Psychology* 11(3), 201–230.

Virden, R.J. and Walker, G.J. (1999) Ethnic/racial and gender variations among meanings given to, and preferences for, the natural environment. *Leisure Sciences* 21(3), 219–239.

Weigert, A. (1997) *Self, Interaction, and Natural Environment: Refocusing Our Eyesight*. State University of New York Press, Albany, New York.

Weigert, A. (2008) Pragmatic thinking about self, society, and natural environment: Mead, Carson, and beyond. *Symbolic Interaction* 31(3), 235–258.

Williams, K. and Harvey, D. (2001) Transcendent experience in forest environments. *Journal of Environmental Psychology* 21(3), 249–260.

Wilson, E.O. (1984) *Biophilia: The Human Bond with Other Species*. Harvard University Press, Cambridge, Massachusetts.

5

Human Development and Nature*

Adventure is the best souvenir.
Woodswomen, Inc., 1982

We know a great deal about child development (Erikson, 1998; Siegel, 1999; Schaefer and DiGeronimo, 2000; Shonkoff and Phillips, 2000; Kail and Cavanaugh, 2010; Bukatko and Daehler, 2011), as well as a great deal about the benefits of children being in nature (Wells, 2000; Fjørtoft, 2004; Kuo and Taylor, 2004; Chawla, 2006; James *et al.*, 2010; Matsuoka, 2010; Asah *et al.*, 2011; Cheng and Monroe, 2012). In fact, children and nature is a popular topic receiving both public notice and research dollars. In this chapter children's contact with nature is linked to healthy lifelong human development. Research from several different fields shows that nature is critical for healthy development and, more specifically, intertwined with human attachment from infancy.

The goal in this chapter is to integrate knowledge from attachment theory, developmental science, learning theory, and nature involvement research. Combining the need for primary attachment to a human caregiver and an attachment or connection to nature, we look at developmental theories and considerations for parenting. After setting the stage for the importance of childcare practices and the

need for physical nature contact for healthy development, attachment theory and pertinent developmental and learning theories are reviewed. Several developmental models are discussed showing how specific ties to time and activities in nature positively impact children's comprehensive development. Finally a sequential, research-based timeline of nature-related experiences within each developmental stage recommended for healthy development is offered. Implications from this conceptual understanding reach into early childhood development, youth development, and young adult development as well as later age considerations. It is common knowledge that it would be hard if not impossible to survive without the natural world; we depend on it for food, shelter, and energy. However, we do not often think about the necessity to spend time in nature in order develop as healthy, responsible human beings.

A simple explanation for introducing children to nature in age-appropriate ways is that humans tend to like that with which we are familiar. No different, young children tend to develop emotional attachments to what is familiar and comfortable for them, so the more personal, positive, and appreciative their experiences with nature are the more

---

* This chapter was written with input from Chiara D'Amore.

© CAB International 2014. *Natural Environments and Human Health*
(A.W. Ewert, D.S. Mitten and J.R. Overholt)

environmentally aware and active they will likely become (Chawla, 1999; Schultz *et al.*, 2004). This environmental appreciation and understanding helps the later adult make decisions that support environmental protection and stewardship and in turn increase the chances of human survival.

While children and nature is a current topic in a number of developed countries, an interesting historical perspective is that in the US there have been people speaking to this topic since the 1800s (Ellen Swallow). A fascinating example is Elwood Shafer's response to the 1960s' and 1970s' revival of the understanding of our need for nature, often labeled as the first modern-day green movement. Shafer, of the US Forest Service's Pinchot Institute of Environmental Forestry Research, worried about the impact of children and adolescents spending less time in nature, gathered professionals from different disciplines who worked with children. Representatives from local, state, and federal health, recreation, and park offices, and practitioners and academics in the fields of child development and environmental studies constituted most of the presenters. The resulting 1975 proceedings of the Children, Nature, and the Urban Environment Symposium concluded that children need opportunities to explore wild places and learn about nature for healthy maturation, and these opportunities were lacking. Along with ground-breaking environmental protection legislation many school systems adopted environmental education programs and, supported in part by federal agencies, several curricula were developed including Project WILD, Project WET, and Project Learning Tree.

Twenty-five years later the question about children needing and having opportunities to explore wild places and to learn about nature for healthy maturation was addressed again by Kahn and Kellert (2002) in their book *Children and Nature*. They had gathered academics who were ecologists, biologists, and psychologists, and the findings were consistent with the practitioners' of 1975. Children were being denied the opportunity to explore wild places and to learn about nature (Kahn, 2002). Kellert (2002) said that a child's direct and ongoing experience of accessible nature is an essential, critical, and irreplaceable dimension of healthy maturation and development. This chapter ties together the work of many professionals from many diverse fields who are working to generate a worldview that understands the need for children to have appropriate time in nature in order to foster healthy development.

What Do Children Need for Healthy Development?

The National Research Council and Institute of Medicine's Committee on Integrating the Science of Early Childhood Development is charged with reviewing the knowledge about early child development and early childhood experiences and integrating the science in these areas. The 2000 National Research Council report, *From Neurons to Neighborhoods: The Science of Early Childhood Development*, represents a milestone in understanding the research from a multidisciplinary standpoint and supports an integrated approach to understanding child development. This two-and-a-half-year study sought to: (i) update scientific knowledge about the nature of early development and the role of early experiences; (ii) disentangle such knowledge from erroneous popular beliefs or misunderstandings; and (iii) discuss the implications of this knowledge base for early childhood policy, practice, professional development, and research (Shonkoff and Phillips, 2000). Ten years later, the Committee on 'From Neurons to Neighborhoods' (Olson, 2012) held an update workshop to continue the documentation of important health issues for children, with important additions such as noting the significant increase in time that children spend with technology. Integration of the domains of development continues to be a focus of policy as well as a preventive approach, and these will be shown to be aided by time in nature. The original study focused on the growing volume of research in neurobiological, behavioral, and social sciences over the past several decades, which has led to major advances in understanding the conditions that influence whether children have a promising or a challenging start in life. As the knowledge generated by interdisciplinary developmental science has grown and evolved, a number of core concepts have come to frame understanding of the nature of early human development (Shonkoff and Phillips, 2000, p. 3):

- Human development is shaped by a dynamic and continuous interaction between biology and experience.
- Culture influences every aspect of human development and is reflected in child-rearing beliefs and practices designed to promote healthy adaptation.
- The growth of self-regulation is a cornerstone of early childhood development that cuts across all domains of behavior.
- Children are active participants in their own development, reflecting the intrinsic human drive to explore and master one's environment.
- Human relationships are the building blocks of healthy development.
- The broad range of differences among young children often makes it difficult to distinguish normal variations and maturational delays from transient disorders and persistent impairments.
- The development of children unfolds along individual pathways whose trajectories are characterized by continuities and discontinuities, as well as by a series of significant transitions.
- Human development is sharpened by the ongoing interplay among sources of vulnerability and sources of resiliencies.
- The timing of early childhood experiences can matter, but people remain vulnerable to risks and open to protective influences throughout the early years of life and into adulthood.
- The course of development can be altered in early childhood to achieve more adaptive outcomes by effective interventions that change the balance between risk and protection.

Based on these core concepts, the study committee came to the following fundamental conclusions. All children are born wired for feelings and ready to learn. From the time of conception to the first day of kindergarten, development proceeds at a pace exceeding that of any subsequent stage of life. What happens during the first months and years of life is significant because this period of development creates either a sturdy or a fragile foundation for what follows, which impacts lifelong well-being. Early environments matter and nurturing relationships are essential: virtually every aspect of early human development, from the brain's evolving circuitry to the child's capacity for empathy, is affected by cumulative exposure to environments and experiences from the prenatal period through the early childhood years. The science of early development is clear about the specific importance of parenting and of regular caregiving relationships in impacting healthy development (Shonkoff and Phillips, 2000). Children grow and thrive in the context of close and dependable relationships that provide love and nurturance, security, responsive interaction, and encouragement for exploration. Without at least one such relationship, development is disrupted and the consequences can be severe and long lasting (Shonkoff and Phillips, 2000).

These findings from the 'From Neurons to Neighborhoods' study focus on the way in which early relationships with primary caregivers play a major role in child development, including social relationships throughout life, support the tenets of attachment theory and constructivist theory of learning, and provide a link to how time in natural environments plays a major role in healthy development. Table 5.2 at the end of this chapter summarizes the complementary interaction of the critical needs in child development and time in nature.

After providing evidence for children's need for time in nature through providing philosophical and practical background information, the importance of caregiver attachment is discussed, preparing the reader to conceptualize how attachment to a primary caregiver and the environment may be related. A section about learning theory provides the foundation to interpret the ways that children integrate their experiences in nature into their development. Several constructs of discrete stages of human development advanced by psychological and psychoanalytic theorists as well as by educators and others working in the realm of child development complement the findings of this report and are summarized in preparation for the specific outlining of developmentally appropriate experiences in nature.

Why is Time with
Nature Critical for Children?

Since the mid-20th century in the US experts from a number of fields have argued that a child's experience of nature exerts a crucial and irreplaceable effect on physical, cognitive, emotional, and spiritual development. In the 1950s Edith Cobb from the field of social work, Rachel Carson who trained as a biologist, and Harold Searles, a practicing psychologist, all concurred that ample time in the outdoors was a key for healthy maturation. Research now has validated the wisdom and benefits of living in a way that fosters human connection to nature beginning in early childhood. While Carson (1962) may be most remembered for her germinal book *Silent Spring*, her life's work and passion are a precursor to the modern environmental movement and to what today we call place-based education (Warren and Wapotich, 2011). Carson understood the teaching and healing power of nature from both biological and developmental aspects. In her writing, such as her 1956 article, 'Help your child to wonder' (Lear, 1998), Carson attempted to show that nature was the teacher, creating a sense of wonder in children and adults alike, maintaining nature's necessity for children's spiritual development. She encouraged parents to take children outside and help them enjoy nature, understanding that children's connection to place facilitated healthy identity construction. Carson brought the downside of chemical use to light because she understood how connected humans are to natural processes and that having healthy outdoor environments was necessary for children's development.

Cobb's work is an example of early research supporting ecopsychology and demonstrating this connection between healthy human development and nature. Cobb trained in social work in the 1950s with an interest in the natural world, child development, and adult psychology, and undertook a massive research project wherein she collected and analyzed more than 250 autobiographies. In *The Ecology of Imagination in Childhood*, Cobb (1998) established the importance of children's deep experience of the natural world to their adult cognition and psychological well-being. She suggested that a sense of place (a tree, a stream, a knoll) is vital to a child's evolving personality and connected happy childhood experiences and time in nature with adult creativity (Cobb and Mead, 1977). Chawla (2002) expanded on Cobb's work and reinforced the importance of children spending time in the natural world in an unthreatening way that encourages a bond based on connection rather than on fear. An overwhelming majority of the environmentalists whom Chawla interviewed recalled 'positive experiences of natural environments in childhood and adolescence, and family role models who demonstrated an attentive respect for the natural world' (Chawla, 2002, p. 212).

Psychologist Harold Searles (1960) studied and wrote about transference. He said that we both long to merge with nature and fear being swallowed up by nature. He cites numerous examples where transference feelings, projections, and identifications with the natural world, especially with animals, occur. Setting the stage for future research he said:

> the nonhuman environment, far from being of little or no account to human personality development, constitutes one of the most basically important ingredients of human psychological existence. It is my conviction that there is within the human individual a sense, whether at a conscious or unconscious level, of relatedness to his nonhuman environment, that relatedness is one of the transcendentally important facts of human living, that – as with other very important circumstances in human existence – it is a source of ambivalent feelings to him, and that, finally, if he tries to ignore its importance to himself, he does so at peril of his psychological well-being. (Searles, 1960, p. 6)

Contributing to the argument that children need time in nature for healthy development are ecofeminism and ecopsychology; from grass-roots organizing these areas explore how our psychological health is related to the ecological health of planet Earth. Eco-psychology blended ecofeminism, environmental philosophy, ecology, and psychology, and evolved into a mental health profession. These fields also helped pave the way to add to our current understanding about emotional and spiritual connection with nature and how to improve our reciprocal relationship with nature, as well as the crucial importance of integrating nature time into children's lives in order to promote healthy development and maturation.

Supporting the work of ecofeminists and ecopsychologists, as well as the practice of traditional ecological knowledge, in terms of physical connection, and the theories of intra-indigenous consciousness (IIC) and psycho-evolutionary theory (PET) for physical, mental, and social connections, E.O. Wilson (1984) and Stephen Kellert (1993) described the biophilia hypothesis, the idea that humans are 'hard wired' to be connected to nature. Like those before him, he asserted that there is the existence of a biologically based, inherent human need to affiliate with life and lifelike processes. He went on to say that human identity and personal fulfillment depend on our relationship to nature as does humans' positive emotional, cognitive, aesthetic, and spiritual development, again suggesting the need for time in nature for healthy human maturation.

Nature appears to be necessary for healthy maturation and development in humans as well as for our sustained well-being. Nature's potential impact on human development is important because developmental gaps can lead to illness and dysfunction. Research and theories now guide us to conclude that healthy and successful accomplishment of different developmental milestones may be positively influenced by time in nature, while not enough time in nature can have negative effects. This chapter uses the current research to trace the specific need for nature during the different developmental periods, including the kind of activities and time needed.

While some 'modern' societies in such places as Scandinavia and Iceland have cultures in which embracing nature is a way of life for much of their population, it is more and more common in most Westernized societies for nature to be pushed to the margins if not explicitly feared and avoided (Mitten, 2009). In particular, in many Western cultures children have few opportunities for meaningful contact with the natural world. A culture of fear that being outdoors exposes children to danger in combination with highly structured schedules of education and activities have led the physical boundaries of children's play to be dramatically reduced (Malone and Tranter, 2003; Kimbro and Schachter, 2011). As a result, children's opportunity for direct and spontaneous contact with nature has become greatly diminished (Kellert, 2002; Chawla, 2006). Pyle (2002) calls this the 'extinction of experience', which breeds apathy toward the environment that we need for the survival of the human species. Kellert (2002) said society today has become 'so estranged from its natural origins, it has failed to recognize our species' basic dependence on nature as a condition of growth and development' (p. 118). The loss of children's contact with the natural world negatively impacts their development and sets the stage for a continuing loss of the natural environment. Kahn (2002) described the unsettling process of environmental amnesia where generations of children see their surrounding natural environment and their exposure or lack of exposure to it as normal. In each generation, as there is less time in nature and less nature in which to spend time, children grow up not cognitively missing what they never experienced. This becomes a problem when they are less healthy because of the lack of time in nature and yet, in a sense, they do not know what they are missing. However not knowing what one is missing does not inoculate one from the harmful effects of absence. For example, one might not know that many vitamins are necessary for health but not having them still leads to diminished health. Insufficient vitamin C causes scurvy. In that same vein not having contact with nature leads to developmental challenges and diminished health. For example, vitamin D, often called the sunshine vitamin, is crucial for healthy development and sunlight is the original source. Because vitamin D allows the body to make use of the calcium in the diet, among other functions, a lack of vitamin D can lead to soft bones and teeth and rickets. Other negative health conditions linked to a lack of vitamin D include inflammatory bowel disease, obesity, depression, chronic fatigue, and type I diabetes, as well as a diminished immune system allowing more colds and flu, and a higher incidence of asthma. At this point in time, a number of researchers have shown that positive exposure to nature is necessary for healthy human development in the physical, emotional, social, spiritual, and intellectual domains of health, and helpful in mitigating harmful results from urbanization. In essence, we need to attach to nature early in life. Human brains may be wired for attachment to nature, which increases our chances of survival on many levels. However, if we do not have experiences in nature early in our lives, as with a primary caregiver, we may lose or weaken some of our ability to attach with nature later in life.

Certainly mothers from the 1950s and 1960s were on to something positive when they would tell their children to play outdoors until dinner. Germinal thinkers such as Carson (1962; Lear, 1998), Cobb (1998), and Chawla (2006) agree that people across all ages are physically, mentally, spiritually, socially, and emotionally healthier when their childhoods include regular time outdoors in natural environments. Our connection to nature is interrelated with our need for human attachment, which is discussed next.

Attachment Theory

Human attachment in infancy has been shown to be necessary for healthy maturation. Attachment theory, originally proposed in the literature by John Bowlby, states that infants have an adaptive need to seek and sustain closeness to another person. Bowlby (1969) described a theory of attachment, calling it a 'lasting psychological connectedness between human beings'. Bowlby saw attachment as an adaptive behavior that contributes to human survival, much like other species

have adaptive behaviors such as migration or sleeping in trees to help them survive (St. Antoine, 2012). In the past several decades, research informed by Bowlby and Mary Ainsworth, who developed the means to assess the quality of the parent–child attachment, has confirmed that a secure parent–child attachment is a crucial foundation for a child's later competence and well-being (Karen, 1998). Through his research, Bowlby (1951) found that deprivation of a close maternal presence causes depression, acute conflict, and hostility in children, decreasing their ability to form healthy relationships in adult life. The importance of early bonds is supported by James Prescott's (1975) research, which found a clear link between disruption to the child–mother bonding processes and the emergence of violence and fear-based behavior in young primates. Unable to conduct the same research on human subjects, Prescott conducted cross-cultural studies of first contact observations of aboriginal societies. He found that he could predict with a high degree of accuracy the emergence of violence and hierarchical power in a society, based on the treatment of mothers and children (Prescott, 1975).

Recent studies suggest that approximately 65% of children in the general US population may be classified as having a secure pattern of attachment, with the remaining 35% being classified in one of three types of insecure attachments (Prior and Glaser, 2006). These insecurely attached children are at greater risk of experiencing anxiety, insecurity, behavioral problems, and relationship difficulties throughout their lifetimes than are securely attached children. In contrast, a secure attachment supports the child's development of trust, confidence, empathy, and self-regulation, which are important qualities for effectively connecting with the greater world (St. Antoine, 2012). Dr Karen Walant (1999) conducted research on disconnected parenting approaches in the US that emphasize the need for children to become totally self-reliant and autonomous at early ages, as compared to parenting approaches that nurture children's capacity to form close, loving, intimate relationships with others. She determined that a parenting style focusing on autonomy from a young age encourages and models to children not to connect deeply to other people, resulting in more intense attachment to material things (Walant, 1999). This replacement of human connection with a connection to material goods can be seen as a driver behind the rampant consumerism of many Western cultures and its environmental consequences, including the lack of an ability to feel connected to animals, plants, and nature in general. This lack of attachment to humans, to other sentient beings, and to the natural world in general highly impacts healthy human development and therefore the health and well-being of the human species, and has implications for long-term species survival. Nel Noddings (1984) using a lens of the ethic of care advised us to practice and encourage the development of care for others and nature in ourselves and children. This begins with secure attachment in infancy to a primary caregiver, and then extending that attachment to nature. Doing so has the potential to increase long-term positive connections as well as the ethical and moral nature of our relationships with humans and the Earth, which can lead to increased health and well-being for both people and the planet.

Learning Theories

Learning theories relate to the specific processes that lead to the accomplishment of a developmental milestone. The *From Neurons to Neighborhoods* report (Shonkoff and Phillips, 2000) indicated that humans are born wired to learn. This section describes the theories that illustrate how that learning takes place. The primary category of learning theories that we use in this text are constructivist theories including cognitive and social aspects. These theories effectively explain the act and process of learning and constructivist learning dovetails well with the developmental influences of the natural environment. Constructivist theories are based on the premise that through interaction with physical and social environments children construct knowledge and values. Constructing knowledge is how we make sense of the world; it means we have an understanding beyond simply the memorization and repetition of facts.

It could be argued that a child's ability to construct conceptual understandings or knowledge is related to either a genetic predisposition that prepares the child to learn or to the modeling by adults about how to construct knowledge. Whatever the cause, constructivist learning theories (Fosnot, 2005) often divide into cognitive constructivism and social constructivism. Experience is a central tenant in constructivism and learning through experience is probably the earliest form of learning for humans. People learned to gather food or construct shelters by living and working alongside their parents, siblings, and other group members. This experiential learning is demonstrated through the ages in apprentice systems used since ancient times. Socrates is classically thought of as a cognitive constructivist as he dialogued with his students encouraging them to think for themselves and come to their own conclusions, including understanding the weaknesses in their thinking.

Continuing to assert that learning is an active process and that knowledge is constructed rather than acquired, Marie Montessori furthered learning theory and child development theory through her understanding that children are curious and actively involved in learning through discovery. She asserted that if children were encouraged to develop all of their senses at an early age, then they would be better able to engage in self-learning or constructivism. Montessori believed that each child is different and that predicting exactly when and how a child should learn hampers learning (she would disagree with our industrial-influenced educational model of dictating what a child should learn in each grade). She said that when the senses are developed the child begins to explore areas and wants to learn, e.g. reading and math, at an age appropriate to that child. Pre-dating Howard Gardner's theory of multiple intelligences (Gardner, 2000), Montessori understood the importance of nurturing musical, bodily-kinesthetic, spatial, interpersonal, intrapersonal, intuitive, as well as the more commonly nurtured linguistic and logical-mathematical intelligences, and put her theories into praxis in 1900 and beyond (Swiderski, 2011). Her work influenced Piaget, Dewey, Gardner, Goleman, Kolb, and many others who focus on helping educators understand the value of experience in constructing knowledge and learning (Dewey, 1938/1997; Kolb, 1983; Goleman, 1995/2006). Her work continues today in the form of Montessori Schools. This variance in the timing of children's development observed and valued by Montessori may have offered an evolutionary advantage when humans lived physically closer to nature, in that not all children would be vulnerable at the same time, or that as children gained competencies in different areas at different rates they could help each other and learn the value of social and collaborative interactions.

Vygotsky reinforced Montessori's understanding that learning is a social, collaborative activity and that children develop and mature well in social or group settings (Newman and Holzman, 2013). The group settings to which he is referring are often multi-aged, but also sometimes include single-aged groups. He explained that the culture provides learning tools such as cultural history, social context, and language, all of which can be used to construct knowledge; culture and social context are critical/necessary for healthy learning and human development. Discussion, negotiation, and sharing are effective learning venues, supporting the value of group work in learning. The child seeks to understand the actions or instructions provided by the caregiver or teacher; in this quest for understanding, often involving experimentation and experience, knowledge is constructed. Additionally, the child internalizes behavioral information as well as environmental manipulation knowledge, using it to guide or regulate her or his own performance in the future.

Psychologist Albert Bandura's (1994) social learning theory builds on Vygotsky's work, asserting that children develop new skills and acquire new information by observing the actions of others, including parents, caregivers, and peers, as well as through intrinsic reinforcements such as a sense of pride, competency, satisfaction, and

accomplishment. Current research about mirror neurons (Keysers, 2011) in the brain indicates that humans may be wired to mimic or mirror others, supporting this concept of tutoring. Earliest societies did not separate children and adults, therefore children were gifted with the modeling and tutelage of successful adults. There is concern in current Western society that children have been left in children's groups too much and without intergenerational experiences, resulting in a loss of wisdom or positive behavioral learning.

Some constructivists such as Piaget failed to see the importance of language and the collaborative or social nature of learning; Piaget advocated concrete stages that are currently used in classroom teaching in most Western schools, failing to integrate Montessori's advice that children differ in their timing for development, which causes both overestimation and underestimation of children's abilities at any given age. For example, Piaget would say not to teach reading and math until age 7, at which point these subjects must be taught, because, in his understanding, children reach their concrete operational stage at this age and are prepared to learn reading and math. This leads to an overall caution to readers about using models as static absolutes. Models may work for a majority of people; however, in mimicking nature we understand that encouraging diversity, including developmental diversity, can increase the overall strength of the human species. There is individual variability in the timing of developmental milestones that we encourage parents, caregivers, and educators to value. In this text we see the social constructivism and cognitive constructivism as interrelated and refer to the construction of knowledge as constructivism; constructivism is the active mental constructions of children resulting in the ways in which children organize and act on their knowledge and values.

Developmental Theories

A set of ideas offered in order to explain behavior and development of individuals as they age and mature makes up a developmental theory (Kail and Cavanaugh, 2010). Academic interest in the field of child development emerged early in the 20th century, leading to the creation of several major theories that attempt to describe human development, often using a stage approach. From birth to early adulthood children go through developmental processes that include cognitive, emotional, spiritual, physical, and social growth, which sets the foundation for their lifelong relationships with themselves, other people, and the natural world.

To think that there may be one key to health or healthy human development is misleading. For healthy development a combination of time in nature, positive attachment, appropriate diet, constructive learning opportunities, and more are necessary. Therefore, we need to use a systems approach when thinking about needs for healthy human development. Before delving into the practical application of being in nature to reinforce positive human development the nature versus nurture debate is examined from a post-duality perspective and several developmental theories are discussed.

Child development specialists began dialoging about the causes or prompters of human development decades ago and most people have heard about or even taken part in the nurture versus nature debate about the influences on child development. Child development theories based on nature propose that the mechanism for human development lies within each individual, meaning that it is our nature to develop in a certain manner, and may be classified as *endogenous theories*. These mechanisms include genetics, wired conditions in the brain, and other causes of innate behavior. This means that humans automatically behave in ways that promote survival and that such behavior has been genetically programmed through an evolutionary process of selection based on the ability to adapt. Even within endogenous theories descriptions of this internal mechanism humans supposedly have differs among experts, resulting in conflicting beliefs. Some people believe that the survival instinct is a competitive model causing humans to want to dominate other humans physically, socially, financially, and intellectually. Other people believe that humans are genetically predisposed or wired

for relationships and that survival depends on developing humans' ability to experience empathy and be spiritually connected. This spirituality focus is based on aspects such as wisdom, altruism, and compassion and presupposes that, with healthy attachment to humans and nature, this survival instinct and positive behavior regarding how we interact with others is realized. Theories such as ecopsychology, naturalistic intelligence, and biophilia may be related to genetics and be realized and further developed through nurture.

Theories based on nurture, called *exogenous theories*, propose that development occurs through external mechanisms in our environments such as how we are parented and interactions with other people, as well as exposure to the environment, including positive and negative time in nature as well as the presence of toxins and other harmful scenarios. Therefore behavior is acquired through attachment, conditioning, and construction of learning, and is also a reaction to rewards, punishments, stimuli, and reinforcement (Pavlov, 1960/2003; Skinner, 1974/1976; Watson, 1923/2012) combined with indoor and outdoor environmental exposure. Theories such as IIC, friluftsliv, and PET might be related to our interactions with our environments and genetics, indicating the flexibility of genes and the challenge of drawing lines between nature and nurture.

Most developmental theories evolved using a combination of these schools of thought, with some depending more on exogenous theories and others more on endogenous theories, but generally assuming an integration of biological- and genetic-based aspects with one's environmental experiences to determine human development. Additionally, new research showing the possibility of genetic changes during one's lifetime, known as epigenetics, continues to blur the lines between nature and nurture. Fosar and Bludorf (2011) report research demonstrating that the alkalines in human DNA follow a regular grammar, including syntax (the way in which words are put together to form phrases and sentences) and semantics (the study of meaning in language forms), possibly indicating that language is a reflection of our DNA and that genetic information can be altered using sound and frequency. This research helps to understand why certain bird songs and nature sounds, music, and other types of sound meditation have healing effects for some people (Aung and Lee, 2004; Halstead and Roscoe, 2002; Docksai, 2011).

Most of the human development theories applied today are influenced by learning theories, taking an interactionist position of causes. For example, cognitive psychologists assume that behavior is the result of information processing, believing that people use the information in their environment; therefore behavior is influenced by learning and experience as well as direct interactions with the environment. Later in this book, the benefits of the learning environment created by being in the natural world are discussed. Cognitive psychologists believe that behavior is influenced by ways people have learned to manipulate or process information as well as people's brains' innate capacities as information processors, e.g. the inherent differences in the types of brains people have, including our genetics. The following developmental theorists use a constructivist approach and they inform the practical applications described later.

- *Erik Erikson*: Building on Freud's psychodynamic theory 'that personality emerges from conflicts that children experience between what they want to do and what society wants them to do' (Kail and Cavanaugh, 2010, p. 11), Erickson proposed a popular comprehensive lifespan view. Erikson's eight-stage theory claims that the order of the stages is biologically fixed, that at some point in each stage the person meets the challenge, and when these challenges are met successfully the individual progresses to the next stage. 'In essence there is interplay between an internal maturation plan and external societal demands' (Kail and Cavanaugh, 2010, p. 11). For example, the primary conflict during preschool years is the assertion of power and control over the world through directing play and other social interactions. Erikson suggested that success or failure in dealing with the seeming dualities at each stage can impact long-term development and functioning. According to

Erikson, the environment in which a child lives is crucial to providing growth, adjustment, a source of self-awareness, and identity (Schickendanz, 2001).

- *Maria Montessori's planes of development*: Montessori, physician and educator, saw human behavior as guided by universal, innate characteristics in human psychology. She observed four distinct cycles, or 'planes', in human development, extending from birth to 6 years, from 6 to 12 years, from 12 to 18 years, and from 18 to 24 years. Some of these planes were also delineated into more discrete stages. Montessori saw different characteristics, learning modes, and developmental imperatives active in each of these planes and stages, and called for holistic educational approaches specific to each period that integrated learning in the physical, emotional, social, and spiritual realms.

- *Eric Berne's and Pam Levin's transactional analysis development cycles*: Transactional analysis is a theory of interpersonal communication, development, growth, and change that originated in the fields of psychotherapy and counseling. Developed by Eric Berne, transactional analysis is based on mutual respect, acceptance, and the belief that everyone has the ability to learn and the potential to change. Challenging the modern, technological perspective that presents development as linear, Berne refined Erickson's theory by noting that developmental processes begun in childhood remain active and important throughout our lives and are probably triggered cyclically by parenting or other major life changes. Pam Levin built on this theory with the concept of cycles of development, and it was further expanded and adapted by several key thinkers such as Jean Illsley Clark (Clarke and Dawson, 1989/1998). The transactional analysis viewpoint is more positive and less based on duality of struggles than is Erickson's theory. This concept presents a lifelong development process in which people respond to an internal developmental clock or organizational pattern that prescribes the tasks and skills that need to be learned. The overarching task in each of the eight stages is to find or create an appropriate answer to each of four questions: Who am I? Who are you? Who am I in relationship to others? How do I get what I need? (Clarke and Dawson, 1989/1998). Over the course of adulthood people return to issues of the earlier stages with new opportunities to grow and complete the developmental tasks and issues each stage represents—this is thought of as spirals within spirals of development. The development of attachment is understood to be a transactional process (Bowlby, 1969) and is basic to the foundation of this developmental theory.

As shown in Table 5.1, there is variability particularly with the details of infancy and toddlerhood in the stages identified by these three theories, with increasing consistency as childhood progresses.

With the information presented thus far we are building a concept that healthy human development begins with secure attachment to a primary caregiver, positive time in nature, and the opportunity to construct learning in social and physical environments, which all combine to help humans navigate the developmental tasks. Kahn (1997) found that through an individual's interaction with the social and natural world one's perspective and understanding of one's relationship with nature widens, which can also transform negative affiliations with nature into life-affirming orientations. Therefore, if social constructivists are correct that knowledge transforms and builds on

itself over time and that during the human development process children discover the relative adequacy of knowledge, then having positive affiliation with nature during all stages of development is beneficial for survival and health and well-being. That being the case, we should study the development of children's relationships with nature. The following section integrates research about benefits of time in nature with human development stages, primarily using Berne and Levin's stage designations. The section provides human development considerations and concepts intertwined with developmental stages that exemplify our relationship with nature and the stages we go through to become a mature person who is resilient, contributes to the common good, takes individual responsibility, and understands her or his connection with the natural world in positive and practical ways. This work of outlining time and activities in a way that dovetails with developmental stages is evolving. As more research is completed we will have more data and perhaps know how to better interpret some of the findings thus far. For example, young children show a strong positive response to water (Zube et al., 1983). While this adds to the evidence that we are prone to be near nature and specifically water, and it correlates with Ulrich's (1981) findings about humans' preference for water scenes in pictures, we do not know why this preference is there for young children or adults. Perhaps it is because humans are mostly water, because water is a necessity for human survival, or perhaps because water has a calming effect, as noted by some research.

Table 5.1. A basic chart of the stages and ages for three popular developmental theories from Erickson, Montessori, and Berne and Levin.

Stage	Erickson	Montessori	Berne & Levin
Infancy	0–1 years	0–3 years	0–6 months & 6–18 months
Early childhood/toddlerhood	1–3 years	0–3 years	18 months–3 years
Preschool/early childhood	3–6 years	3–6 years	3–6 years
School age/elementary	6–11 years	6–12 years	6–12 years
Adolescence/early adolescence	12–20 years	12–18 years	12–19 years
Early adulthood	20–24 years	18–24 years	Adulthood
Second stage adult	25–64 years	–	Adulthood
Old age	≥65 years	–	Towards death

A goal in this book is to help readers understand that with healthy attachment to the natural world, originating with secure primary caregiver attachment, one tends to the natural world relationship with the understanding that human health and well-being are intimately connected with the health of the Earth. This foundation of secure attachment to the primary caregiver and then to nature lays the foundation for humans feeling and practicing the ethic of care and understanding our interrelationship with everything in the world. As we come to know nature within ourselves and experience the interconnectedness and interdependence between nature and ourselves, we will be more likely to develop a worldview that supports choosing to live in ways that sustain all life forms, a behavior necessary to sustainably support human life. So exactly how ought children's connections with nature be encouraged?

The Practical Application of Theory: Time in Nature and Developmental Stages

In this section the ways in which time in natural environments reinforces, directs, and/or satisfies human developmental needs is foremost in our minds. This section divides the discussion about relationship practices with nature into stages that are parallel with the human development stages described by Berne and Levin, with some reference to Montessori and Erikson. Pairing current research findings about health benefits from time in nature with developmental needs ascribed to eight stages of human development, the associated activities or kinds of nature involvement that will trigger or reinforce positive development of the stage milestones are integrated in the stage descriptions. Given what we know about child development and our need throughout the lifespan for time in natural environments, it seems likely that the theories of biophilia, naturalistic intelligence, IIC, and PET explain some of the wiring with which humans are born. Biophilia, naturalistic intelligence, and IIC are theories that may explain that we *are* connected to nature. PET describes a process

by which we may have developed biophilia, naturalistic intelligence, and IIC. However, research shows that we need time in nature in order to reap the benefits ascribed to nature and probably the earlier in life, the better for development. Perhaps humans are born with neurons in the brain that will develop into naturalistic intelligence or IIC, but like language neurons, if they are not used early in life then they are pruned and these skills and attitudes that create positive relations with natural environments are much harder to develop later in life.

Children are born wired to learn and they are born with an innate affiliation with nature described by IIC, biophilia, and naturalistic intelligence. A question for caregivers is how can we initiate an awakening of indigenous consciousness and naturalistic intelligence that will encourage and enhance positive affiliation with nature? As we introduce infants to nature and continue this relationship throughout childhood, we recommend using a scaffolding approach. The infant is introduced to being in natural environments in a safe secure manner, mimicking her or his developmental needs. As the child grows, more and varied activities, including more independent activities, are added, in line with development. While there does not seem to be an upward limit of time spent in nature (except in areas where outdoor air, water, or land pollution is a concern), minimum times are suggested for exposure or immersion. Remember, historically humans spent essentially all of their time outdoors. Today some cultures such as the Nharo Bushmen in Africa; continue to live all or most of their time outdoors in Scandinavia there continue to be Friluftsliv Schools from preschool through college where children are outdoors the entire school day.

Our goal as caregivers is to spark these already latent nature-seeking aspects in children, starting in infancy. The amount of time in nature varies with climate and air quality (a consideration recent in necessity, perhaps first understood in the 1800s in Britain with coal pollution and now a common reality in large cities such as Tokyo, Los Angeles, and others). At different locations factors such as season, time of day, and specific weather patterns can influence when a child is taken outdoors. The basic concept is that fresh air, nature sounds and smells, and visual stimuli trigger the innate brain activity that has evolved through the millennia, according to PET, to help us survive and thrive in concert with nature. Keep in mind that if someone has grown up with little or no contact with nature that these developmental stages ought to be experienced in the order described. For example, if a nature center wants to offer nature programming for inner-city youth then providing activities that allow the youth to experience nature as they would have in each of the previous stages from infancy to the youths' current developmental stage is paramount. This exposure can be brief and adapted for the population, but the experience is crucial. While some people who have not had previous opportunities to be in nature thrive during their first experience, usually too much immersion in nature too soon can result in fear and a nature phobic attitude. This need for graduated immersion can be especially true for populations where nature has been a place of danger in modern times, including the last 500 years.

For each of these sections, starting with pregnancy and going through adulthood, please refer to Box 5.1 with affirmations from Clarke (Clarke and Dawson, 1989/1998). These affirmations are messages that caregivers impart through actions and words to their unborn children through adulthood. Additionally, the kind of time or activities in nature suggested in this chapter reinforce these developmentally appropriate messages from the caregiver to the child, in early stages, and reinforce earlier developmental messages for the child in later stages. These messages, reinforced through specific activities or quiet time in nature, are useful for older children or adults who have not been exposed to nature following the order discussed below.

Pregnancy to birth: becoming

The health of the mother impacts the health of the growing fetus. Appropriate time in nature reinforces the mother's health, including her physical, mental, emotional, spiritual, and social health. The mother is embarking on a journey encompassing her whole being

and needs to experience the stress reduction (Hartig et al., 1997; Wells and Evans, 2003) and restorative effect of being in nature (Kaplan and Kaplan, 1989; Van den Berg et al., 2007). Quiet time in nature is recommended. While following doctors' recommendations, most pregnant women can continue their same activity level as before pregnancy but usually not begin higher levels of exercise (Mitten, 2008). Therefore if a pregnant woman has been hiking, kayaking, canoeing, and camping, these activities can most likely continue, though during the first trimester some health practitioners recommend limiting vigorous activity. During the first trimester as hormones are readjusting and during early fetus development, including limb development, trips requiring altitude acclimatization may not be recommended (unless the woman lives at altitude). In early times women probably did not gain or lose altitude in great quantities as they might today with the travel options available.

Through spending time in nature the mother is transmitting important developmental messages to the unborn child, especially 'your needs and safety are important to me' and 'I celebrate that you are alive'. These messages are given through the boost in health through vitamin D, the physical health that typically increases with time in nature (Kimbro and Schachter, 2011) partly because of opportunities for exercise (Marti et al., 2002; Gasser and Kaufmann-Hayoz, 2004; Pretty et al., 2005a,b; Dyment and Bell, 2008; Cooper et al., 2010; Wheeler et al., 2010) and the mood states that typically increase with time in nature (Hartig et al., 1999; Nisbet and Zelenski, 2011), including a reduction in depression and anxiety (Yankou, 2002), as well as overall improvement in mental health (Howell et al., 2011; Nisbet and Zelenski, 2011; Nisbet et al., 2011; Cervinka et al., 2012) and specifically an increase in ability to express joy and satisfaction (Hartig et al., 1999; Kaplan, 2001; Korpela et al., 2002).

In Berne and Levin's developmental theory, once a child moves into a subsequent stage the messages or affirmations from earlier stages need to continue into adulthood. In that same vein, once specific benefits are named and described, such as the opportunity for

Box 5.1. Personal affirmations related to human developmental stages from Jean I. Clarke (1998)

1. These messages are for parents to say to their children, from pregnancy through to adulthood.
2. They can be used by adults who may need to re-cycle their own developmental stages.
3. They can be useful messages in preparing to go into a new situation or relationship.
4. People need all of the affirmations from birth up to their current age.

Age	Developmental stage	Examples of messages
Pregnancy to birth	Becoming	• I celebrate that you are alive • Your needs and safety are important to me • We are connected and you are whole • You can make healthy decisions about your experiences • You can be born when you are ready • Your life is your own
0 to 6 months	Being	• I love you just the way you are • You belong here • I'm glad you are you • I'm glad you are alive • You can grow at your own pace • I love and care for you willingly • You can feel all of your feelings • What you need is important to me
6 to 18 months	Doing	• You can know what you know • You can explore and experiment and I will support and protect you • You can do things as many times as you need to • I love you when you are active and when you are quiet
18 months to 3 years	Thinking	• I'm glad you are starting to think for yourself • You can learn to think for yourself and I will think for myself • You can think and feel at the same time • It's okay for you to be angry and I won't let you hurt yourself or others • You can know what you need and ask for help • You can become separate from me and I will continue to love you • You can say no and push and test limits as much as you need to
3 to 6 years	Identity, Power	• I love who you are • You can find out the results of your behavior • All of your feelings are okay with me • You can learn what is pretend and what is real • You can explore who you are and find out who other people are • You can be powerful and ask for help at the same time • You can try out different roles and ways of being powerful
6 to 12 years	Structure	• You can learn when and how to disagree • You can think before you say yes or no and learn from your mistakes • You can trust your intuition to help you decide what to do • I love you even when we differ; I love growing with you • You can find a way of doing things that works for you • You can learn the rules that help you live with others • You can think for yourself and get help instead of staying in distress

Box 5.1. Continued.

Age	Developmental stage	Examples of messages
13 to 19 years	Identity, Sexuality, Separation	• You can learn to use old skills in new ways • You can develop your own interests, relationships and causes • I look forward to knowing you as an adult • My love is always with you. I trust you to ask for my support • You can know who you are and learn and practice skills for independence • You can learn the difference between sex and nurturing and be responsible for your needs and behavior • You can grow in your maleness or femaleness and still be dependant at times
Adulthood	Interdependence	• Your needs are important • You can be uniquely yourself and honor the uniqueness of others • You can trust your inner wisdom • You can be creative, competent, productive, and joyful • You can be independent and interdependent • You are lovable at every age • You can say your hellos and goodbyes to people, roles, dreams, and decisions • You can be responsible for your contributions to each of your commitments • Your love matures and expands • You can finish each part of your journey and look forward to the next • Through the years you can expand your commitments to your own growth, to your family, your friends, your community, and to all humankind • You can build and examine your commitments to your values and causes, your roles and your tasks

exercise, mood enhancement, and stress reduction, these benefits will be there at subsequent stages, too, though they may not be specifically named again or described in detail.

Infancy: 0 to 6 months old

During this stage children need to attach positively to their caregiver and be in an environment where they know that their needs will be met. In this stage infants begin to learn about place from outside the womb and generally become aware and familiar with surroundings. Virtually the infant learns to be in the environment. It is usually healthy

for infants to spend time outdoors dressed according to weather. Given a healthy baby, cooperative weather, and appropriately low pollution levels, parents can walk with the baby snuggled to their chest for at least an hour and up to unlimited time. It is a win/win for a parent to walk several hours per day while the baby sleeps or is awake because the parent gets exercise and the baby gets outdoor time. The physical, mental, emotional, and spiritual benefits referenced for the mother above continue for the caregiver and are enjoyed by the infant. A sling or 'snuggly' is recommend if the caregiver is strong or healthy enough because having the baby physically close while walking outside addresses the need of the baby in this *being* stage to know that the caregiver cares and protects her or him, and allows the baby to hear the human heart. The infant has spent nine months *in utero* traveling with the mother everywhere and sensing what is happening in the environment. To deprive the child of this sensatory time with the mother or other caregiver would be an abrupt change for the child. The movement of walking with the caregiver while in a snuggly is closer to the movements to which the child is accustomed, rather than lying in a crib or being in a stroller. This time outdoors reinforces attachment, aligns with Berne and Levin's stage 1 of 'being' in the world and needing to feel secure, wanted, and loved, and helps encourage the trust deemed necessary by Erickson's first stage. In infancy the development of a trusting attachment with primary care providers lays the foundation for future healthy relationships. From the moment of birth people look to those around them for physical and emotional support. The family unit is usually the first support group for a newborn. The quality of this support establishes the bonds of attachment that provide the groundwork for all later relationships (Schaefer and DiGeronimo, 2000), including those with the more than the human world. According to attachment theory, the deep mutual bond between a child and her or his primary caregivers is adaptive in that it is critical for the survival of both the individual child and the species (Bowlby, 1969). The more this attachment period includes time outside, in the appropriate conditions described above, often the stronger and healthier the child and caregiver can be. Being outside as a family strengthens family bonds (Freeman and Zabriskie, 2002); therefore including all family members is encouraged.

An infant relies on experience for learning, including the experience of care and having needs attended to, the experience of language, the experience of love and other emotions from caregivers, and an ever-increasing experience of exploring the environment. These needs can be well met by time outdoors with a caregiver, and research has shown that being outside in nature encourages nurturing characteristics. Because the infant cannot see well, this concentration on the sensual being validates that the child is valued just as she or he is, and she or he learns other valuable forms of communication that come from touch and feeling the breeze, the sunshine, rain, and other weather characteristics. These experiences tap into the child's IIC, initiating the development of naturalistic intelligence and intuition, and encouraging biophilia. As sight develops, the child seeing the mother's eyes strengthens attachment and creates a sense of safety for the child in the outdoors. In essence, the time in nature strengthens attachments and sets the stage for a positive physical and spiritual engagement with nature.

Caregivers can help infants experience and explore specific aspects of nature by handing them safe objects as they start to grasp, usually at about 4 weeks old, sitting on the ground with the infant letting her or him explore, and allowing the child to taste safe objects. The caregiver sets the stage by taking the child outside to a safe area, meaning an area safe from poisonous plants, cliffs, too much wind, and the like, where it is safe for the baby to touch and taste and see things within reach. In proper moderation, letting the baby experience snow, rain, sun, and wind helps develop senses. Babies are learning quickly at this age, their brain is triggered in many areas. They will begin to understand the outdoor environment as safe and they begin to attach to nature. Having safe yet varied experiences positively impacts future development. During the early childhood years, it is important to help children discover what has been termed their ecopsychological self, the child's natural sense of self in relation to the natural world (Phenice and Griffore, 2003); this begins in infancy, evolving from time spent outdoors as appropriate for the age and child, and continuing outdoor time through the developmental phases.

Infancy: 6 to 18 months old

At 6 months the infant begins to move into the doing stage, as defined by Berne and Levin. The child likely will crawl at around 5 or 6 months and by 1 year may be standing and walking, and able to explore on her or his own. The caregiver continues taking the child outside to safe areas, with the added factor of movement on the child's part. If a child is fussy, take her or him outdoors, if the weather is appropriate. If the weather is inclement, look through the window and watch for a while. Catching the light in prisms is another activity to perform with children that connects them to the awe of nature. Caregivers provide security and reassurance, and plenty of safe spaces for free exploration with defined natural boundaries. Children play peek-a-boo, which evolutionarily might have helped with hiding from predators later. It also helps them discover that they are a person surrounded by other beings. Children start to explore even more with their mouth. Having safe natural objects to handle, including organic food sources, helps in development. Historically, children may have been with parents and given a variety of food to put in their mouths to suck, which helped develop taste buds.

This stage begins the reciprocal learning that continues throughout the child's development. The caregiver stays attentive to the child so that the child continues to feel secure and builds trust. As the child explores, the caregiver acts as an audience. If possible, vary places such as a grassy area (native plants with no chemicals being preferable), a sand box, or a rocky area (no cliffs!). This play time helps the child know she or he can be creative and active. Crawling gives the child access to more territory and children can crawl quite quickly. Crawling allows them to explore while helping their physical body and brain develop adequately for standing and walking. As important as exploring is the message that it is okay to be quiet; it is okay to not be energetic and not move around for a while. Reinforced by allowing the child to relax outdoors, nap in a snuggly while the caregiver is walking, and participate in other quiet activities, this quiet time acceptance helps healthy development of self-regulation. While outdoors, the caregiver offers to be quiet with the child just as the caregiver offers to be active with the child. An important component of being quiet outdoors is allowing spiritual development and giving children a sense of peace and oneness with the world (Cobb and Mead, 1977; Crain, 2011) that

will help establish a sense of place for children and a sense of being restored in nature. Children learn to stop, look, and listen. In this way they have opportunities to experience the awe of the natural world. Additionally, this sets the stage for self-regulation later when for reasons of safety the child literally has to stop, look, and listen. Montessori says that the child cannot be overly directed at this age and in outdoor environments caregivers follow the child's lead as safety allows; the child who has plenty to explore and areas in which to rest will likely have enough fresh air and be tired enough, but not exhausted, to sleep through the night. The concept of having children go outside to work off energy so that they will be able to sit in school and be quiet indoors undermines an accurate understanding of the deep and varied benefits from outdoor play and activity. Nondirected outdoor exploration helps create a secure identity and develops creativity (Cobb and Mead, 1977). This action of being with the child as she or he explores continues to build trust as deemed necessary by Erickson's stages. According to Bowlby, during this time attachment becomes clear. One behavior to notice, but not necessarily call attention to yet, is the child venturing off a few feet and then returning to the caregiver. This behavior, resulting in short forays into the larger world, reflects secure attachment and was beneficial when humans lived outdoors. Because the child was securely attached to the caregiver, it was safer in that the child would not wander too far and become lost. Children will continue to explore and as developmentally appropriate go further from the caregiver. It is important not to encourage premature independent exploration because of safety as well as attachment considerations. Having a child return to you like a homing pigeon is safer than prematurely encouraging a child to be independent and who then might not search regularly for her or his caregiver. Fear first surfaces as an emotion at about 6 months old, but usually does not manifest in concrete ways until about 18 months when the child may start to experience stranger anxiety, another developmental milestone likely with evolutionary roots. Being attached and preferentially responsive to the primary caregiver makes it harder for the child to wander off in nature or to toddle over to an unfriendly adult. In our US culture today some of these same safety concerns exist. Observations of young children show that they move back and forth from their caregiver to the attractions of the world around them, pivoting around the caregiver as a secure base that the child keeps in sight and often returns to touch (Ainsworth *et al.*, 1978).

Early childhood/toddlerhood: 18 months to 3 years old

At 18 months to 3 years, according to Berne and Levin, children start to reason things out, think for themselves, and solve some of their own problems. They continue to use direct experience as their prime way of learning. Children need areas in which to explore where they might get a few bumps and bruises, but are safe from harm. During this sort of exploration outside, children are starting to understand how they interact with their environment on many levels and they begin to gain competencies in manipulating themselves in relationship to their environment as well as other people. Historically this behavior may have helped develop skills for food gathering and hunting and shelter construction later in life. Today it sets the stage for children to do the same, as well as to be able to contribute to problem solving in our complex world. Instead of the terrible twos, these are the terrific twos and kids want to explore, have opinions, and continue to taste things they find. This may be the start of learning to find their own food to feed themselves. Continue with as much outside time as practical, from the child exploring on her or his own two feet to riding in a bicycle trailer (with a helmet) to accompanying caregivers on camping, backpacking, and canoeing trips. According to Bowlby, children continue to use their parent as a safe base from which to explore. Exploration is greater when the caregiver is present because the child feels safe knowing the caregiver is in sight and therefore is relaxed and feels free to explore (Colin, 1996). However, if healthy attachment has not been secured, anxiety, fear, illness, and fatigue will cause a child to increase attachment behaviors (Karen, 1998). By allowing the child plenty of closely supervised outdoor time to explore,

accompanied with walks in a snuggly, a backpack, or a stroller as the child gets larger, usually healthy attachment follows. If a child displays anxiety, continue to be attentive and enjoy outside time with the child, because children go through phases that are not necessarily linear and textbook. The theory of object relations explains how a positive emotional foundation enables children to explore the world with confidence. When children learn that they are noticed responsively and their needs are accurately read, they gain confidence to look outward and respond openly to their surroundings. In contrast, experiencing fear and anxiety lead children to be self-preoccupied.

During this stage children act like many other mammals; they tumble and rough house, learning about each other and getting stronger—and maybe learning to avoid or escape from other animals. The Montessori school of thought encourages caregivers to continue to allow non-directed exploration or play. Children between 18 months and 3 years begin to demonstrate even more emotions. As they experiment they can become agitated and strongly object when others impose limits on them in an area they feel they want to control or they can become intensely frustrated when they cannot accomplish a task. Ambivalence may also be expressed as toddlers learn whether to trust or distrust their knowing or thinking. Outdoor environments are excellent places for toddlers to work out their understanding of limits, control, and competency needed at this time in life. In an outdoor environment a caregiver may need to set boundaries of exploration and activities. This is healthy for the child. In general if the child has a large area in which to explore but is heeded back by a parent when wandering too far afield, the child learns self-restraint and continues to build security. The challenge for the caregiver is to know that this behavior is normal and to consistently herd the child inbound with a neutral or smiling attitude. With a continued positive attitude the child likely will not develop unnecessary fears or feel constrained against taking bold actions. Erickson says that children at this developmental stage need to start to have some personal control over physical skills, which leads to some independence. Spending time in natural environments, where children can explore and experiment without having to feel as constricted as one might in a play pen or house and being worried about breaking things, can enhance this developmental stage. Having play areas with natural boundaries can help the child feel secure without having to be directed by a caregiver. Too much constriction or not enough ability to experiment can lead to other side of Erikson's duality of feelings of failure, leading to shame and doubt rather than security and confidence. In a rich natural outdoor environment the child development tasks take place in a more compact space because the child does not have as much need to keep moving outside the local environment to find creative ways to explore and develop competency. Children live positive messages needed for healthy development such as 'you can think and feel at the same time', 'I'm glad you are starting to think for yourself', 'it's okay for you to be angry and I won't let you hurt yourself or others', 'you can know what you need and ask for help', 'you can venture out from me and I will continue to love you', and 'you can say no and push and test limits as much as you need to'. The children are building a strong foundation; research shows that children who play outside have stronger immune systems (Wells, 2000) and outdoor time increases life expectancy through lower mortality (Takano *et al.*, 2002; Hu *et al.*, 2008; Li *et al.*; 2008, Mitchell and Popham, 2008). Another study revealed that outdoor time in nature may prevent myopia (Rose *et al.*, 2008).

Preschool/early childhood: 3 to 6 years old

During the ages of 3 to 6 years children continue to develop competency and a sense of self. In healthy maturation children decide that it is okay to have their own view of the world, and to be who they are. To achieve this development, children test the consequences of their behavior and their capacity to influence others. Children may set up disagreements, start or repeat false rumors, and say things they know to be untrue; this behavior helps them begin to discern between

fantasy and reality. Montessori tells us that children become more social, leading to more receptivity to adult influence as well as to the desire to spend more time with other children. Reciprocal learning, important in social learning theory, becomes more evident. In fact, being outside stimulates social interactions among children (Moore, 1986; Bixler *et al.*, 2002). While some children develop fears earlier, commonly children at about 3 years old develop fears of the dark or of large animals, which can surprise some caregivers since these fears may not have been expressed earlier. When humans lived closer to nature this may have helped the exploring toddler not to wander off at night or not to run up to large animals that may step on or eat the toddler.

Personality can become more evident, coupled with the development of a personal worldview. This worldview, according to Bowlby, includes a system of thoughts, memories, beliefs, expectations, emotions, and behaviors about the self and others that regulate, interpret, and predict attachment-related behavior in the child and the attachment figure. This system, called the 'internal working model of social relationships', continues to develop with time and experience and enables the child to handle new types of social interactions; knowing, for example, that an infant should be treated differently from an older child, or that interactions with teachers and parents share certain characteristics. In part, the varied environments in the outdoors support this development.

Children under 5 years old seemingly have endless energy, enthusiasm, and creativity, engage in exploration, and often stay in a state of wonder. One cause of this ability, which we can now measure, may be that children's brains stay more in an alpha state, which is related to higher levels of joy and creativity. As humans mature the beta state becomes the norm. Being in natural environments stimulates alpha states, which helps children of all ages as well as adults (Douillard, 2004, p. 309) feel energetic and joyful. In addition to inducing alpha states, research demonstrates that outdoor play sparks creativity and imagination (Cobb and Mead, 1977; Crain 2001; Taylor *et al.*, 2001).

Children continue to venture further away from caregivers and for longer periods of time. Having complex natural environments that offer the child seemingly secluded areas, such as a place on the back side of a tree, or exciting areas, such as a tunnel, can help children develop independence and competency, while remaining in a safe area. Sobel (2008) refers to the ages 4 to 7 years as *early childhood*. He confirms that during this time children's homes fill the center of their world and they spend much of their time within sight or earshot of home; this is the time to allow free time in the child's own yard or on a trip to the woods with an adult.

While in outdoor environments children need to continue to be closely monitored because they do not understand the difference between pretend and real. Though during this time children begin to separate fantasy and reality, they do not necessarily achieve this understanding until the age of 6 to 7 years. Playing with other children in outdoor environments can enhance development of social skills and competency. Children are able to incorporate reciprocal learning and make great strides as they watch other children and adults manipulate their environment and are able to comprehend verbal instructions better, too. Collaborative learning occurs as children at different levels and with different gifts and skills help each other. Because children are forming a sense of self through competency—or, as some developmentalists say, forming identity and power (Clarke and Dawson, 1989/1998)—they often go through a period of hitting trees with sticks, killing insects, and by 6 years old maybe even killing and dissecting small animals. The dissecting of animals can signal a curiosity for life processes. When humans lived closer to the Earth these sorts of behaviors were useful in helping the child learn to gather and hunt food. Through playing outside, agility and coordination increase as well as motor abilities in general (Fjørtoft, 2004). Caregivers can help children learn about light and shadows, which helps children to discern changes in their environment, including when to come in at night. They may learn about seasons and cycles, day lengthening and shortening, and migratory animals. Teaching children about their outside environment and nature at a young age leads to an awareness of place and encourages local knowledge, leading to a kind of security of knowing where one is from.

Additionally, children are learning to be powerful and ask for help at the same time, as noted in Box 5.1. This skill helps in community development later on, so when these children are adults they can cooperate. The affirmation that 'you can try out different roles and ways of being powerful' also is achieved during supervised play. These behaviors help children understand their environment better as well as their individual competencies and their interrelationship with others.

While learning to be competent, these young people need to continue to develop and practice empathy. According to Sobel (1998), empathy between the child and the natural world should be a main objective during early childhood because this stage is characterized by a lack of differentiation between the self and the other. For example, children feel implicitly drawn to baby animals and a child feels pain when someone else scrapes her or his knee. Rather than encourage separateness, a sense of connectedness should be cultivated such that it can become the emotional foundation for the more abstract ecological concept that everything is connected to everything else. Encouraging relationships with animals, both real and imagined, fosters empathy. Stories, songs, moving like animals, celebrating seasons, and fostering Carson's (1998) 'sense of wonder' are primary activities during this stage.

School age/elementary: 6 to 12 years old

The developmental stage starting at about age 6 years lasts for six years until about age 12 years during which children need and learn about structure and acquire the skills, tools, and values they need to thrive in the world. School and play environments continue to be important because the child's internal understanding of the world, as well as the values that will guide her or his behavior, is becoming more fixed. This includes the rules the child will or will not follow. By school age reciprocal learning has many venues including between teacher and student, and between students collaborating in environmental education activities and field studies. Students at different levels and with different gifts and skills help each other in the physical, emotional, and intellectual learning.

Much of the development will be children trying on behaviors, copying behaviors from adults and other children that seem to fit in their value structure; this gives the appearance of children wanting to do things their own way. While spending more time with peers is normal, regular caregiver presence is needed to reinforce healthy attachment. With secure caregiver attachment, children will imitate and try on peer behavior but not attach to peers in a manner that is unhealthy.

By age 6 years children are less likely to develop new fears and most childhood fears of big animals and imaginary beings disappear by puberty, perhaps an age in earlier times when the youth would begin foraging without close adult supervision. Fear is a positive protective emotion: when youth learn to recognize and work with fear they can be safer.

Contact with animals continues to be important because caring for animals provides a (perhaps essential) link by which children develop caring relationships with the natural world (Myers and Saunders, 2002). Additional literature suggests that this formative importance of animals may be particularly pronounced during early and middle childhood (Searles, 1960; Shepard, 1996). If it works for a family, fully integrating children of this age in the responsibility of taking care of animals and gardening helps with the development of positive empathy as well as self-reliance, interdependency, and responsibility.

In middle childhood free play in the natural world becomes highly important for personal development because children need to experiment and make mistakes and learn how to have their own morals and methods. The outdoors provides an incubator for childhood experimenting. While all along children who play outdoors have been exposed to natural consequences such as a hot stove top, natural consequences can have a different feel and meaning for the child when outside. It is exciting and adventurous to see a frog or a turtle or to try to cross a stream or climb an embankment. The caregiver is further removed from the process and may not be present for all of the adventures. The child by now probably understands that there is a power greater than humans or that humans do not control nature,

which puts the caregiver(s) in perspective and sets the stage for the differentiation that happens in the next developmental stage.

This time can be especially important in developing connections with nature and learning to feel at ease and a part of the out-of-doors. Many children who grow up with natural environments nearby may already have a special place, such as a tree or rock that they use as a touchstone. Following what Cobb found, children bond and remain bonded to this place or piece of nature and will return throughout their childhood or lifetime if they do not move away, or the tree or rock does not get cut down or moved. As they explore further into the environment they remember and use their special place as symbolic of safe nature, a specific place to which they are attached.

Encouraging children to explore nearby areas and get to know their place continues to strengthen the bond with the Earth that began in infancy. Research substantiates that an affinity to and love of nature, along with a positive environmental ethic, grow out of children's regular contact with and play in the natural world (Chawla, 1999; Kellert, 2002; Sobel, 2004). It is likely that the kind of time and amount of time in natural environments influences children's development of an ethic of stewardship because this is the time when children's experiences give form to the values, attitudes, and basic orientation toward the world that they will carry with them throughout their lives (Knapp, 1999). It is also during this time that children's geographical ranges expand rapidly and their focus shifts from the home to the landscapes around them. If this landscape includes cities and malls the youth will be there, if it includes nature then they will explore natural landscapes. Sobel (1998) labels this stage in his nature-related development as the *elementary years*; children from the ages of 8 to 11 years, characterized by rapidly expanding their radius of comfort in the outdoors. These children wander out of reach of parents and play 'grown-up'. However, children are able to engage in a larger radius because the foundation is laid and the bond already there by these elementary or middle childhood years.

Bonds with family continue to play a key role in children's social and emotional development. Parents and other caregivers who encourage a child to share her or his thoughts and feelings, listen attentively, and respond appropriately to the child's needs will help that child feel more confident and secure into adolescence and beyond (St. Antoine, 2012). Activities that caregivers can engage in with their children or that children will engage in on their own include making forts, creating small imaginary worlds, hunting and gathering, searching for treasures, following streams and pathways, exploring the landscape, and shaping the earth—all can be primary activities during this stage. These activities continue to build spatial and problem-solving skills, and teach children about the processes of the natural world. These activities help secure developmental outcomes such as multiple components of self-regulatory and executive capacities, and the ability to make friends and engage with others as contributing members of a group.

Adolescence: 13 to 19 years old

During this stage youth are forming their adult identity, understanding and accepting their sexual identity, and differentiating from their caregivers and others. The word differentiation is purposefully chosen rather than separation. Western countries, especially the US, tend to place separateness and individualism as a developmental goal. However, the objective is to be an individual within a connected and interdependent system. With exposure to nature from infancy through adolescence, the child is likely to have learned that humans are interdependent with natural processes as well as among humans. This connection and attachment is needed in order to have the care and responsibility that leads to being a member of a sustainable community. At this time in life having a comfortable relationship with nature can be a huge comfort to a growing adolescent.

Many earlier societies had rituals to honor the transformation from childhood to adolescence. The rituals often involved nature in some manner and honored the child's passage. Given a dearth of rituals in Western societies, a connection with nature can help buffer some of the change and turmoil often felt by adolescents.

This period is turbulent for teens, partly because of the dramatic changes in hormones and partly because of the different expectations from caregivers and other adults. In this adolescent period children are going through major hormonal changes possibly accompanied by large mood swings and changes in energy. They mature in a more healthy manner if they understand that it is okay to be who they are, they have opportunities to be interested in and try many things, and they also receive affirmations when they are quiet. Successful negotiation of these two stages prepares adolescents for forming a strong identity with a healthy sense of personal power, followed by the structure stage where the adolescent will decide the rules to follow and how to be successful.

During this time of change and reformation the adolescent can find nature receptive, accepting, calming, and nurturing. Nature is okay with their changes, the mood swings, and energy shifts. Teens worry about their identity, sexuality, appearance, and their future, but when in nature they can get a reprieve from demands and feel like nature accepts them as they are, without judgment and without special treatment.

Kaplan and Kaplan (1989) found that during this developmental stage teens were not necessarily feeling positively attached to nature or calmed by nature. All relationships with nature, just like with other humans, are under scrutiny and feel up for grabs. Youth will go to their old stomping grounds and see the area differently, such as not realizing how small that rock really is or that the stream now seems narrower than before. They go with friends camping, hiking, or mountain biking, going much further from home than ever before. The role of the parental figures is to be available when needed while the adolescent makes these more substantial excursions into the outside world.

Success in this stage leads to an ability to stay true to one's self, and previous time in nature will have provided a strong foundation for spiritual support on difficult matters. Emotional support from parents, teachers, and other important adults is crucial for these children and so is the support of nature.

Nature helps children find the balance between not too large, e.g. having a large ego, and not too small, e.g. knowing that all are needed and valued. This happens though the general acceptance that people tend to feel in nature's presence. It happens during impressive downpours, brilliant sunrises or sunsets, watching a butterfly emerge, seeing a tornado in the distance, or any other number of magnificent and powerful natural events. While this has happened throughout the child's life, the child now sees the natural events through adolescent eyes. Natural consequences from nature play have happened all along, but now as the adolescent ventures further with no parental backup, the consequences can take on a more serious tone. Getting wet and going inside because you are in the backyard is different from being 3 miles out on a trail and unable to go directly inside. With preparation since infancy the adolescent is ready for this next step and because the adolescent has periods of poor judgment, nature will continue to teach humbleness. Some youth go to summer camp. In these environments under the stars or around a campfire, discussions on topics of interest such as drugs, sex, and music can be less loaded; philosophical questions and social issues that may become more important can be talked about in a wide open space reinforcing open minds. The strength derived from having that strong connection with the environment and knowing that one belongs, helps youth know that they have a place among grown-ups and can succeed. Some adolescents attach to peers, but many do not feel like their peers understand them, so having nature or pets to spend time with is helpful during this searching and settling period.

As children start to discover the 'self' of adolescence and feel their connectedness to society, they naturally incline toward wanting to save the world. Managing school recycling programs, passing town ordinances, testifying at hearings, and planning and going on school expeditions are all appropriate activities at this point (Sobel, 1998). Developmental outcomes mentioned in the last stage continue in this stage, including multiple components of self-regulatory and executive capacities, and the ability to make friends and engage with others as contributing members of a group.

Adulthood

With continued time in nature since infancy, in adulthood we understand and live with/ incorporate the rhythms of nature—the rising and setting of the sun, the shifting of the seasons, the cold and the heat. An adult who is connected to nature understands the interrelationship of human behavior and biological processes. The adult understands, as Ellen Swallow did, that the Earth is our home. In 1892, as a chemist interested in pollution and public health hazards for humans, she chose the word oekology (oikos is the Greek word for home), later spelling it ecology, and encouraged homemakers and other scientists to study interrelationships with the environment so that we could understand how to keep our homes and the Earth healthy.

Adults are interdependent; the developmental milestone is to live in a manner in which we can be uniquely ourselves and honor the uniqueness of others. Adults learn to trust their inner wisdom and to be independent and interdependent. The work of an adult is to be creative, competent, productive, and joyful while being responsible for contributions and to commitments.

The list of adults who have written about their relationship with nature and how that connection adds to a healthy life for adults is long. Experiences involving nature are often intensely personal and include self-discovery (Chawla, 1988), and while this is true for every age, adults in times of transition often seek out additional time in nature.

Chapter 7 highlights the research about the health benefits for adults. As most adults know, the trials and tribulations of life continue into adulthood. As adults negotiate life, having the respite and solace found in nature helps in the seemingly ever-increasingly busy world.

Toward death

In the final stage of life the task is the integration of life experiences, culminating with making preparations for leaving or dying. Ideally humans continue to grow their whole life through celebrating the gifts received and the gifts given and sharing wisdom. Like in adulthood, nature is used as a place of peace; think, for example, of rocking on the outside porch in the evening. Even people who used to aggressively hunt game are content to sit in nature.

Some hospitals and nursing homes are beginning to have rooms that open to the outdoors.

Throughout life, being in the outdoors and feeling connected to nature is important for health, well-being, and development.

What about Pollution in the Outdoors?

An important issue regarding child development is the impact of toxic chemicals on healthy maturation. Some might argue that the natural environment where many children live is polluted to the point of danger. This is true in some areas. There is a constant challenge of deciding if being outside is better than being inside in terms of air pollution and soil and water contaminants. Unfortunately, the toxins are both inside homes and in the natural environment. In 2000, Greater Boston Physicians for Social Responsibility released a peer-reviewed report entitled *In Harm's Way: Toxic Threats to Child Development*. The project, launched in 1998, shed light on the complex interactions among genetic, environmental, and social factors that impact children during vulnerable periods of development. The report indicated that 17% of children were diagnosed with learning, behavioral, and developmental disabilities including attention-deficit hyperactive disorder (ADHD) and autism, preventing them from reaching their full potential. These developmental disabilities and their increase have negative societal implications from healthcare and education costs to the repercussions of criminal behavior. The report authors say that:

> Research demonstrates that pervasive toxic substances, such as mercury, lead, PCBs, dioxins, pesticides, solvents, and others, can contribute to neurobehavioral and cognitive disorders. Human exposure to neurotoxic substances is widespread. A review of the top twenty chemicals reported released under the 2000 Toxics Release Inventory reveals that nearly half are known or suspected neurotoxicants. Over 2 billion pounds of these neurotoxic chemicals were released on-site by facilities into the air, land or water. As our knowledge about these neurotoxic chemicals has increased, the 'safe' threshold of exposure has been continuously revised downward. Toxic exposures deserve special scrutiny because they are preventable causes of harm. (Schettler *et al.*, 2000, paragraph 1)

Another report published in 2006 revealed 18% of children under 18 years old were diagnosed with learning, behavioral, and developmental disabilities (Blanchard *et al.*, 2006). Even with this grim information, we also know that experiencing nature may help to quell ADHD (Taylor *et al.*, 2001). When indoor, built outdoor, and green outdoor environments were compared, only in green outdoor settings did activities reduce symptoms of ADHD regardless of social context. Research has also demonstrated an increase in attention span and cooperative behavior with a decrease in hostile and aggressive behavior for children diagnosed with autism, developmental disorder, ADHD, conduct disorder, and oppositional-defiant disorder, when children are encouraged to interact with animals (Breitenbach *et al.*, 2009; Prothmann *et al.*, 2009; Memishevikj and Hodzhikj, 2010; Wuang *et al.*, 2010). Children need non-toxic green areas in which to play and construct knowledge as they navigate their developmental milestones. This green time can supplement medication and help future generations avoid overwhelming social and financial debt related to developmental disabilities.

A Call to Action

Attachment to a caregiver along with time in nature lead to healthy development. Development as a healthy human being is about relationships and compassion. While a child is constructing learning through social and cognitive arenas, there needs to be guidance from adults in stimulating the neurons and creating the pathways for empathy and compassion. When we care for nature we care for ourselves and when we care for ourselves we care for nature.

Dr Elliot Barker (1987) has found a correlation between a consumer-based society where values such as immediate gratification, materialism, and wealth are raised above those such as family, connection, and altruism, and clinical psychopaths who have a consistent inability to trust, empathize, and form affectionate relationships. Barker (1987) argues that those who are most wired towards consumption are often those who are the least capable of achieving satisfaction from mutually caring, trusting interpersonal relationships and posits that these individuals may also be missing qualities of trust, empathy, and affection that arise from an inadequate nurturing experience in infancy and toddlerhood. This means that just being outside is not enough. Children need to be healthily and securely attached to a caregiver.

Carson (1962), Cobb (1998), and Chawla (2002) all agreed that children developed into healthier adults when their childhoods included regular time outdoors in natural environments and they also agreed that adults need to model healthy attitudes and behaviors toward the environment. Research supports that an affinity to and love of nature, along with a positive environmental ethic, grow out of children's regular contact with and play in the natural world (Bunting and Cousins, 1985; Chawla, 1988; Pyle, 1993; Wilson, 1993; Chipeniuk, 1994; Sobel, 1996, 2002, 2004; Hart, 1997; Moore and Wong, 1997; Kals *et al.*, 1999; Moore and Cosco, 2000; Fisman, 2001; Bixler *et al.*, 2002; Kellert, 2002; Kals and Ittner, 2003; Phenice and Griffore, 2003; Schultz *et al.*, 2004) and this positive affinity to the environment needs the support of early childhood positive attachment.

Entwined with regular experiences in nature for children is the need for adults to model enjoyment of, comfort with, and respect for nature (Phenice and Griffore, 2003). Chawla (1988) reviewed numerous studies that focused on understanding what happened in the childhoods of environmentalists to make them grow up with strong ecological values. She found a pattern in which most environmentalists attributed their commitment to a combination of two sources: 'many hours spent outdoors in a keenly remembered wild or semi-wild place in childhood or adolescence, and an adult who taught respect for nature' (Chawla, 1999).

In Chawla's (1999) research, environmental activists talked about childhoods in which a family member took them into woods or gardens and modeled appreciative attention. Chawla found that significant adults gave this attention to their surroundings in

four ways: (i) care for the land as a limited resource essential for family identity and well-being; (ii) a disapproval of destructive practices; (iii) simple pleasure at being out in nature; and (iv) a fascination with the details of other living things and elements of the earth and sky. The activists' stories suggested that the quality of the relationship that they shared with this adult as a child was as important as the quality of the experience with nature that child and adult shared together. It seems that when children have close connections with important adults and supported access to the natural world, they have a strong basis for environmentally connected and protective behavior (Chawla, 1999).

However, children are spending less and less time in nature and in play. Clements (2004) surveyed mothers in the US finding that 70% played outdoors every day when they were children, compared with only 31% of their children, and that when the mothers played outdoors, 56% remained outside for three or more hours compared with only 22% of their children. As important, if a child's affinity for nature is not developed during early years, it is possible that biophobia, or aversion to nature, can develop. Biophobia manifests with discomfort in natural places and disdain for anything not created and managed by humans. Biophobia can foster an attitude that nature is a disposable commodity to be used and discarded at whim (Kellert and Wilson, 1993).

With physical space for outdoor play shrinking (Francis and Devereaux, 1991; Kyttä, 2004) and more parents keeping children inside because of fear (Clements, 2004; Kimbro and Schachter, 2011), many children have little opportunity for direct and spontaneous contact with nature (Rivkin, 1990; Chawla, 1994, 2006; Kellert, 2002; Pyle, 2002; Kuo, 2003; Malone and Tranter, 2003). These same parents are working more, therefore have less time to spend outdoors with their children, and time with children is often in structured activities. Visits to national parks have been trending downward since 1981 (Pergams and Zaradic, 2008). It appears bleak in terms of combining the necessary ingredients for healthy maturation.

The loss of children's contact with the natural world negatively impacts their development and sets the stage for a continued loss of natural environments. Because humans need nature to survive and thrive in their physical emotional, intellectual, spiritual, and social domains, creating a generation that does not understand the value of nature could result in a worldwide calamity.

Regenerativity: Reconnecting with Nature

Growing research over the past decade has shown that positive exposure to nature is necessary for healthy human development and helpful in mitigating harmful results from urbanization. Research has identified many connections between nature play and children's cognitive, physical, and emotional development. Time outdoors, especially unstructured time in more natural settings, can reduce children's stress, increase their curiosity and creativity, improve their physical coordination, and reduce symptoms associated with attention-deficit disorder and other conditions (St. Antoine, 2012). This research and the experience of many people explain why we need to create numerous opportunities for children to feel nature and not just think about it.

There are reasons to be hopeful. Noddings (1984) is confident that the possibility of affection for both our children and the natural world is in everyone. We may move from not even knowing that all of our parts are connected, to being preoccupied with self, to knowing that there are other people and things, and finally to seeing ourselves as relational with ourselves and all our surrounding, including the natural world. With healthy development, people will tend to their relationship with the natural world with the understanding that humans' health and well-being is intimately connected with the health of the Earth (as we would tend to our relationships with others in the world if we were taught an ethic of care). As we come to know nature both within ourselves and experience the interconnectedness and interdependence between nature and ourselves, we will probably choose to live in ways that support all life forms.

The Children & Nature Network's (C&NN) Grassroots Survey results showed an increase in US children getting outside, perhaps indicating positive results from their work in creating a social movement that inspires families to spend time in nature together. Results revealed a cumulative total of 8 million children and 75 million youth of 17 years of age or younger in the US getting outside in nature in the period from 2009 to 2012, a growth from 1 million annually in 2009 to 3 million annually in 2011. If this growth can be maintained then a 20% threshold—the amount thought to be needed to sustain a cultural behavior—will be reached in 2015 (Charles, 2013).

In their international and national work, the C&NN, established in 2000, provides a clearinghouse of information, including research and application about the benefits of children spending time in natural environments. C&NN staff want to give teachers, families, and others the tools and resources to connect children with nature. Their site includes current research as well as networking assistance to aid in community building.

As we move forward we can also learn from the past. Many indigenous people continue to live in harmony with nature's rhythms, natural processes, and natural landforms. A number of aboriginal populations continue to use initiation or rite-of-passage ceremonies to aid in healthy development and maturation that have been used for thousands of years. Living close to the land, as many Aboriginal peoples do, can help people develop and value naturalistic intelligence. More research and inquiry is needed about the specific practices that might help Western countries regain some of the practices and attitudes of the past, which might influence a positive connection with nature. Because in most Westernized countries many people live estranged from nature, we consciously have to make time to be in nature. Developmentally, time in nature needs to be integrated at all stages. According to Charles (2013), 'We can build a movement that succeeds in reconnecting children and nature—and in that process inspires new generations to believe in a better future. We together can leave a legacy of leadership and an ecology of hope'.

Table 5.2. The relationship between child development key findings as described by the *From Neurons to Neighborhoods* report (Olson, 2012) correlated with the benefits of nature engagement.

From Neurons to Neighborhoods key findings on human development (HD)	How nature engagement can enhance and support HD (for selected resources listed, full citations are provided in the Bibliography)
HD is shaped by a dynamic and continuous interaction between biology and experience	Disciplines such as ecology, biology, and psychology find that the dynamic and continuous interaction between biology and experience is enhanced by time spent in natural environments. Humans are dependent on nature not only for biological and/or material needs (food, water, shelter, etc.) but also for psychological, emotional, and spiritual needs, with access to nature playing a vital role in human health and well-being (Wilson, 1984; Katcher and Beck, 1987; Friedmann and Thomas, 1995; Roszak *et al.*, 1995; Frumkin, 2001; Wilson, 1993). This is possibly because most of human evolutionary history was spent in close proximity with nature. Additionally, a substantive and growing body of research shows that spending time in nature is supportive of healthy physical, cognitive, psychological, emotional, and social development (Kahn, 1999; Ulrich, 1999; Kuo, 2004; Bell *et al.*, 2008; Muñoz, 2009). The inborn tendency to affiliate with nature often needs nurturing to fully materialize, thus biology and experience are bridged (Mitten, 2010). Chapter 4 describes a number of evolutionary theories (e.g. naturalistic intelligence, biophilia, intra-indigenous consciousness, ecopsychology) and constructivist theories (e.g. psycho-evolutionary and attention restoration) that indicate humans have a biologically adaptive propensity to seek affiliation with other forms of life and natural environments. Chapter 6 discusses adaptations and applications that consider and nurture the relationship between biology and experience (e.g. folk biology, traditional ecological knowledge, friluftsliv, and eco-feminism as adaptations; and spiritual connections, ecopsychology, conservation and development of a land ethic, medicine from nature and conservation medicine, and citizen science as applications). *Selected resources:* Wilson (1984); Katcher and Beck (1987); Wilson (1993); Friedmann and Thomas (1995); Roszak *et al.* (1995); Kahn (1999); Ulrich (1999); Frumkin (2001); Kuo (2004); Bell *et al.* (2008); DeLoache and LoBue (2009); Mayer *et al.* (2009); Muñoz (2009); Mitten (2010); Howell *et al.* (2011).
Culture influences every aspect of HD and is reflected in childrearing beliefs and practices designed to promote healthy adaptation	Worldviews, heavily influenced by culture, affect child-rearing practices (see Chapter 2). While childrearing beliefs and practices differ substantively over time and place, the health benefits of active, unstructured play outdoors for children's well-being are increasingly recognized (see Bratton *et al.*, 2005 for a discussion of special or unstructured play) as is the need to embed this belief and practice into the culture (Cobb and Mead, 1977; Chawla, 2006; Sobel, 2008). Because increased child obesity and other health issues stem from children's increasingly sedentary lifestyles, physical health benefits will accrue as cultural changes encourage outdoor play (Wells, 2000). One example of a cultural adjustment is the Centers for Disease Control and Prevention's (CDC) recognition of the need for children to engage in physical activity outside. The CDC now has set guidelines for physical activity by age group, resulting in funds for research and programs through the National Institutes for Health, Science, and Engineering. Children in these programs will acculturate to more active time outdoors (CDC, 2008).

Continued

Table 5.2. Continued.

From Neurons to Neighborhoods key findings on human development (HD)	How nature engagement can enhance and support HD (for selected resources listed, full citations are provided in the Bibliography)
	Campaigns to get children moving outdoors are now targeted to all parental demographics (Charles, 2013). A number of parenting practices recognize and emphasize the full range of adaptive benefits of a connection to nature (physical, emotional, social, cognitive, and psychological). These practices include 'natural parenting', which places a strong emphasis on connection with nature and environmentally aware lifestyles. Friluftsliv (see Chapter 6), a nature engagement philosophy and practice embedded in Scandinavian culture, is an example of how a society's broader culture impacts childrearing practices, as families in Scandinavia spend substantial amounts of time in nature and receive the associated benefits (Faarlund, 2010). *Selected resources:* Cobb and Mead (1977); Blue (1979); Doyne *et al.* (1983); Dustman *et al.* (1984); Dunn *et al.* (2001); Khatri *et al.* (2001); Kubesch *et al.* (2003); Phillips *et al.* (2003); Atlantis *et al.* (2004); Crews *et al.* (2004); Dunn *et al.* (2005); Bratton *et al.* (2005); Manger and Motta (2005); Stein (2005); Stella *et al.* (2005); Berlin *et al.* (2006); Chawla (2006); Blumenthal *et al.* (2007); CDC (2008); Sobel (2008); Faarlund (2010); Charles (2013).
The growth of self-regulation is a cornerstone of early childhood development that cuts across all domains of behavior	Self-regulation references people's ability to set limits for themselves, control their emotions and behavior, interact positively with others, and engage in independent learning. Engagement with the natural environment can optimize children's opportunities to learn self-regulation (Korpela 1995; Korpela, Kyttä *et al.*, 2002; Taylor *et al.*, 2002; Pryor *et al.*, 2006). Time in nature, and an associated sense of nature relatedness, is significantly correlated with psychological well-being and its six dimensions—autonomy, environmental mastery, positive relations with others, self-acceptance, purpose in life, and personal growth—aspects of well-being closely linked with self-regulation. Emotional benefits to time spent in nature include stress reduction, reduced aggression and increased happiness; social benefits include cooperation, flexibility, and self-awareness; and cognitive benefits include creativity, problem-solving, focus and self-discipline, which are also closely linked with self-regulation (Wells, 2000; Taylor *et al.* 2002; Chawla, 2006; Heintzman, 2009; Ryan *et al.*, 2010). *Selected resources:* Korpela (1995); Bronson (2000); Wells (2000); Korpela, Kyttä *et al.* (2002); Taylor *et al.* (2002); Burdette and Whitaker (2005); Chawla (2006); Pryor *et al.* (2006); Heintzman (2009); Nisbet *et al.* (2010); Ryan *et al.* (2010).
Children are active participants in their own development, reflecting the intrinsic human drive to explore and master one's environment	Maria Montessori noted that children play an active role in gaining knowledge of the world, an idea that has been adopted and adapted by many subsequent child development theorists. With this understanding comes the realization that children actively construct knowledge through interaction with the psychological, social, and physical world. Free play in nature, especially during the critical period of middle childhood, appears to be especially important for developing the capacities for creativity, problem-solving, and emotional and intellectual development (Cobb and Mead, 1977; Fjørtoft, 2001; Kellert, 2005; Maller and Townsend, 2006). Unstructured time in the natural environment is particularly useful for providing children with the opportunities to explore and develop a sense of agency (Kahn, 1999).

Table 5.2. Continued.

From Neurons to Neighborhoods key findings on human development (HD)	How nature engagement can enhance and support HD (for selected resources listed, full citations are provided in the Bibliography)
	Chapter 9 outlines how adventure-based activities positively affect children's drive to explore and master their environment, including their internal and social environment through intentionally designed experiences (IDEs). Achieving competence in activities that allow them to travel and thrive in the outdoors, such as hiking, canoeing, kayaking, and rock climbing, gives children more access to the outdoors and the benefits of spending time in nature. *Selected resources:* Cobb and Mead (1977); Kahn (1999); Kellert (2005); Maller and Townsend (2006); Kellert (2012).
Human relationships are the building blocks of healthy HD	Human relationships often are enhanced through time in nature (Fredrickson, 1996; Fredrickson and Anderson, 1999; Ewert and McAvoy, 2000; Paxton and McAvoy, 2000; Maller, 2005), which leads to increases in the capacity to care and form nurturing relationships for families or other groups spending time together in nature. A review of research found that time spent in nature in the form of school gardening enhanced student bonding (Blair, 2009). Nature exposure has also been found to increase intrinsic aspirations (personal growth, intimacy, and community) and decrease extrinsic aspirations (money, image, fame) (Weinstein *et al.*, 2009). Chapter 4 describes theories about the restorative properties of nature and Chapter 6 describes various applications of these theories. Specific findings indicate that time spent in nature diminishes mental fatigue and restores attention, and also increases family ties and reduces family violence (Kaplan and Kaplan 1989; Hartig *et al.*, 1991). *Selected resources:* Kaplan and Kaplan (1989); Hartig *et al.* (1991); Fredrickson (1996); Fredrickson and Anderson, 1999; Ewert and McAvoy (2000); Paxton and McAvoy (2000); Maller (2005); Blair (2009); Weinstein *et al.* (2009).
The broad range of differences among young children often makes it difficult to distinguish normal variations and maturational delays from transient disorders and persistent impairments	Children reach developmental goals in multiple ways and by following diverse pathways. Experiences outdoors in nature allow children to engage in developmental tasks, as they are ready and able. The outdoor environment provides important diversity in landscapes and opportunities that in turn allows for great variation in the ways children relate to that environment, thus allowing different developmental timing. The outdoor environment allows students across the spectrum of Gardner's (2000) multiple intelligences to excel, which increases self-acceptance and highlights the normalcy of variation. Exposure to nature can also help to mitigate some developmental challenges. For example, Taylor *et al.* (2001) and Taylor and Kuo (2009) surveyed parents of children with attention-deficit disorders (ADD) regarding their child's attentional functioning after activities in several settings. The results showed that ADD symptoms were milder for those children with greener play settings. For example, children who played in windowless indoor settings had significantly more severe symptoms than did children who played in grassy outdoor spaces with or without trees (Taylor *et al.*, 2002). As these developmental challenges are decreased children more often are able to participate positively in social situations.

Continued

Table 5.2. Continued.

From Neurons to Neighborhoods key findings on human development (HD)	How nature engagement can enhance and support HD (for selected resources listed, full citations are provided in the Bibliography)
	Time outdoors can help with the diagnosis of maturational delays from transient disorders and persistent impairments (Rose and Massey, 1993; McAvoy *et al.*, 2006). Intentionally Designed Experiences (IDEs), as discussed in Chapter 9, augment natural settings through the inclusion of programs and experiences specifically designed to beneficially impact human health. IDEs result in first-order outcomes such as achievement, restoration, empowerment, resilience, and physical activity, which then provide the psychological strength needed to achieve other health-related goals such as weight reduction, stress reduction, belief that one can positively impact her or his health-related conditions, etc. (Holman and McAvoy, 2005; Cole and Hall, 2010). Nature therapy and adventure therapists often use IDEs to focus on the interconnectedness of the psyche and the natural world to reconnect the bonds between humans and the environment. Individual psychological healing embeds psyches back into the natural world and eventually the illness embedded in Western culture will be healed (Mitten, 2009) (see Chapters 4, 6, and 9). *Selected resources:* Erikson (1950); Sameroff and Emde (1989); Cicchetti and Beeghly (1990); Rose and Massey (1993); Pumariega and Cross (1997); Gardner (2000); Taylor *et al.* (2001); Taylor *et al.* (2002); Holman and McAvoy (2005); McAvoy *et al.* (2006); Mitten (2009); Taylor and Kuo (2009); Cole and Hall (2010).
The development of children unfolds along individual pathways whose trajectories are characterized by continuities and discontinuities, as well as by a series of significant transitions	The process of development is essentially a process of change, or developmental transitions. The transitional phase often reflects both the stage that is ending and the stage that is beginning. Developmental transitions occur frequently in childhood and are times when the emotional communication between children and caregivers is particularly significant; as communication among family members is enhanced through time in nature, family time in the outdoors can aid children in their transitions. As children grow they relate to the outdoor environment differently, including going further from home in their explorations and connecting even more deeply to place (Cobb and Mead, 1977; Chawla, 2006). Safe outdoor space allows children to gradually experiment with increasing distance from their caretaker and the adult's willingness to trust the child's competence is essential for that differentiation to happen, perhaps especially important for children who live in small and crowded homes (White and Stoecklin, 1998). Children's connection to place aids in healthy transitions, especially as children internalize who they are and develop a sense of self and grow in differentiation (Wells, 2000). *Selected resources:* Cobb and Mead (1977); Wells (2000); White and Stoecklin (1998); Chawla (2006).
HD is sharpened by the ongoing interplay among sources of vulnerability and sources of resiliencies	Positive and negative risk factors influence development, whether inherent or environmental. Biological and environmental risk produces vulnerability and impacts the extent to which change can occur. A child's own expectations and those of the significant people in her or his life often play an important role in maintaining or changing the child's direction. Experiences in nature and physical activity positively influence self-efficacy, an individual's belief in her/his ability to take control of life situations and take appropriate action, because of the interaction between success and challenge that they provide the participant (Bandura, 1986; Mitten, 1992; Kelley *et al.*, 1997; Propst and Koesler, 1998; McAuley *et al.*, 2001; Maddux and Gosselin, 2003; Sibthorp, 2003; Sheard and Golby, 2006; Havitz *et al.*, 2013).

Table 5.2. Continued.

From Neurons to Neighborhoods key findings on human development (HD)	How nature engagement can enhance and support HD (for selected resources listed, full citations are provided in the Bibliography)
	The relationship between humans and places allows us to better understand how humans connect to the natural environment. Sense of place invokes the concept of human relationships with non-human nature, and has been shown to be important to child and adolescent development, self-regulation, attention restoration, self-esteem, self-concept, and identity (Korpela *et al.* 2001; Korpela, Kyttä *et al.*, 2002; Taylor & Kuo, 2009). Allowing children to develop a relationship with the natural world and a sense of place within their environment ensures that they are ready to think critically about solutions to ecological problems before asking them to tackle the myriad of environmental crises (Sobel, 1996). Focusing on one's place in nature can avoid psychological shutdown in response to major environmental destruction (Crompton and Kasser, 2009). Adventure education and IDEs, in addition to school gardens and other informal outdoor education in nature, can positively influence these developmental factors of vulnerability and resilience. Research demonstrates that adventure experience can be effective in enhancing resilience scores. For example, the inclusion of adventure programming in an anti-bullying initiative was found to positively impact resilience in youth (Beightol *et al.*, 2012) (see Chapters 8 and 9). *Selected resources:* Cobb and Mead (1977); Bandura (1986); Mitten (1992); Sobel (1996); Kelley *et al.* (1997); Propst and Koesler (1998); Kahn (1999); Korpela *et al.* (2001); McAuley *et al.* (2001); Korpela, Kyttä *et al.* (2002); Maddux and Gosselin (2003); Sibthorp (2003); Sheard and Golby (2006); Crompton and Kasser (2009); Taylor and Kuo (2009); Beightol *et al.* (2012); Kellert (2012); Havitz *et al.* (2013).
The timing of early childhood experiences can matter, but people remain vulnerable to risks and open to protective influences throughout the early years of life and into adulthood	While it is optimal for people to have opportunities to connect with the natural world starting in infancy and to have continued nature experiences throughout each stage of human development (toddlerhood, early childhood, childhood, adolescents, etc.), it is never too late for people to receive the protective influence of time in nature. The benefits of nature can be particularly supportive for children, who are, by their very nature, in a state of rapid development of capacities, such as the ability to adjust to changing circumstances and be resilient when experiencing challenges and risks. For example, childhood stress caused by rapidly increasing school and extracurricular demands in the US can have negative impacts on a child's development. Research has shown that time outdoors can help to relieve stress for children. A particular study (Wells and Evans, 2003) determined that contact with nature not only decreased the stress of rural children, but higher amounts of exposure to nature were linked to lower levels of stress. Looking at the protective influences of nature later in life, a well-known study (Ulrich, 1984) compared a matched set of adult hospital patients and found that those in rooms with a view of nature had notably faster improvements in their path to recovery, including shorter hospital stays and less pain medication requirements. Chapter 5 outlines the importance of and methods for remaining connected to nature throughout the life span. Of particular importance is the concept of a socio-ecological approach to human health, which creates ways for nature to be accessible to populations on a daily bases (Maller *et al.*, 2009) (see Chapters 5 and 6).

Continued

Table 5.2. Continued.

From Neurons to Neighborhoods key findings on human development (HD)	How nature engagement can enhance and support HD (for selected resources listed, full citations are provided in the Bibliography)
	Selected resources: Ulrich (1984); Kaplan and Kaplan (1989); Hartig *et al.* (1991); Sobel (1998); Chawla (1999); Knapp (1999); Kellert (2002); Wells and Evans (2003); Sobel (2004); Maller *et al.* (2006); Maller *et al.* (2009).
The course of HD can be altered in early childhood to achieve more adaptive outcomes through the use of effective interventions that change the balance between risk and protection	Environments that facilitate a child's sense of competence and sense of personal efficacy are more likely to motivate positive action and to foster children who do well cognitively and emotionally (see Chapter 9). Time in nature can be an effective intervention to tip the scales away from risk and towards protective factors (see Chapter 7 and Cobb and Mead, 1977; Korpela, 1995; Taylor *et al.*, 2001; Kuo and Taylor, 2004; Maller, 2005; Chawla, 2006). Specifically, a socio-ecological approach to human health (Pryor *et al.*, 2005; Maller *et al.*, 2006) should be considered because it incorporates protective factors including community parks, trees, and close access to nature, resulting in less violence and family abuse, improved cognitive outcomes, and improved adjustment for children and for people who have immigrated to the US (Herzog *et al.*, 1997; Wong, 1997; Taylor and Kuo, 2009). *Selected resources:* Cobb and Mead (1977); Korpela (1995); Herzog *et al.* (1997); Wong (1997); Taylor *et al.* (2001); Kuo and Taylor (2004); Pryor *et al.* (2005); Chawla (2006); Maller *et al.* (2006); Taylor and Kuo (2009).

Resources for Table 5.2

Atlantis, E., Chow, C.M., Kirby, A. and Fiatarone Singh, M. (2004) An effective exercise-based intervention for improving mental health and quality of life measures: a randomized controlled trial. *Preventive Medicine* 39(2), 424–434.

Berlin, A.A., Kop, W.J. and Deuster, P.A. (2006) Depressive mood symptoms and fatigue after exercise withdrawal: the potential role of decreased fitness. *Psychosomatic Medicine* 68(2), 224–230.

Blair, D. (2009) The child in the garden: an evaluative review of the benefits of school gardening. *Journal of Environmental Education* 40(2), 15–38.

Blue, F. (1979) Aerobic running as a treatment for moderate depression. *Perceptual and Motor Skills* 48(1), 228.

Blumenthal, J.A., Babyak, M.A., Doraiswamy, P.M., Watkins, L., Hoffman, B.M., *et al.* (2007) Exercise and pharmacotherapy in the treatment of major depressive disorder. *Psychosomatic Medicine* 69(7), 587–596.

Bratton, S.C., Ray, D., Rhine, T. and Jones, L. (2005) The efficacy of play therapy with children: a meta-analytic review of treatment outcomes. *Professional Psychology, Research and Practice* 36(4), 376–390.

Bronson, M.B. (2000) *Self-regulation in Early Childhood: Nature and Nurture.* Guilford Press, New York.

Centers for Disease Control and Prevention (2008) 2008 Physical Activity Guidelines for Americans. Available at: http://www.cdc.gov/physicalactivity/everyone/guidelines/index.html (accessed 15 October 2013).

Cicchetti, D. and Beeghly, M. (1990) Developmental perspectives on the self: a historical review. In: Cicchetti, D. and Beeghly, M. (eds) *The Self in Transition: Infancy to Childhood.* University of Chicago Press, Chicago, Illinois, pp. 1–15.

Cobb, E. and Mead, M. (1977) *The Ecology of Imagination in Childhood.* Columbia University Press, New York.

Crews, D.J., Lochbaum, M.R. and Landers, D.M. (2004) Aerobic physical activity effects on psychological well-being in low-income Hispanic children. *Perceptual and Motor Skills* 98(1), 319–324.

Doyne, E.J., Chambless, D.L. and Beutler, L.E. (1983) Aerobic exercise as a treatment for depression in women. *Behavior Therapy* 14(3), 434–440.

Dunn, A.L., Trivedi, M.H. and O'Neal, H.A. (2001) Physical activity dose–response effects on outcomes of depression and anxiety. *Medicine and Science in Sports and Exercise* 33(6 Suppl.), S587–S597.

Dunn, A.L., Trivedi, M.H., Kampert, J.B., Clark, C.G. and Chambliss, H.O. (2005) Exercise treatment for depression: efficacy and dose response. *American Journal of Preventive Medicine* 28(1), 1–8.

Dustman, R.E., Ruhling, R.O., Russell, E.M., Shearer, D.E., Bonekat, H.W., *et al.* (1984) Aerobic exercise training and improved neuropsychological function of older individuals. *Neurobiology of Aging* 5(1), 35–42.

Erikson, E.H. (1950) *Childhood and Society*. Norton, New York.

Ewert, A. and McAvoy, L. (2000) The effects of wilderness settings on organized groups: a state-of-knowledge paper. In: McCool, S.F., Cole, D.N., Borrie, W.T. and O'Loughlin, J. (comps) *Wilderness Science in a Time of Change Conference—Volume 3: Wilderness as a Place for Scientific Inquiry*, Missoula, Montana, 23–27 May 1999. *USDA Forest Service Proceedings RMRS-P-15-VOL-3*. US Department of Agriculture, Forest Service, Rocky Mountain Research Station, Ogden, Utah, pp. 13–26.

Fjørtoft, I. (2001) The natural environment as a playground for children: The impact of outdoor play activities in pre-primary school children. *Early Childhood Education Journal* 29(2), 111–117.

Fjørtoft, I. (2004) Landscape as playscape: the effects of natural environments on children's play and motor development. *Children, Youth and Environments* 14(2), 21–44.

Fredrickson, L.M. (1996) Exploring spiritual benefits of person–nature interactions through an ecosystem management approach. Doctoral dissertation, University of Minnesota, Minneapolis, Minnesota.

Fredrickson, L.M. and Anderson, D.H. (1999) A qualitative exploration of the wilderness experience as a source of spiritual inspiration. *Journal of Environmental Psychology* 19(1), 21–39.

Friedmann, E. and Thomas, S.A. (1995) Pet ownership, social support, and one-year survival after acute myocardial infarction in the Cardiac Arrhythmia Suppression Trial (CAST). *American Journal of Cardiology* 76(17), 1213–1217.

Heintzman, P. (2009) Nature-based recreation and spirituality: a complex relationship. *Leisure Sciences* 32(1), 72–89.

Herzog, T.R., Black, A.M., Fountaine, K.A. and Knotts, D.J. (1997) Reflection and attentional recovery as distinctive benefits of restorative environments. *Journal of Environmental Psychology* 17(2), 165–170.

Katcher, A. and Beck, A. (1987) Health and caring for living things. *Anthrozoös* 1(3), 175–183.

Kellert, S.R. (2005) *Building for Life: Designing and understanding the Human–Nature Connection*. Island Press, Washington, DC.

Kellert, S.R. (2012) *Birthright: People and Nature in the Modern World*. Yale University Press, New Haven, Connecticut.

Kelley, M.P., Coursey, R.D. and Selby, P.M. (1997) Therapeutic adventures outdoors: a demonstration of benefits for people with mental illness. *Psychiatric Rehabilitation Journal* 20(4), 61–73.

Khatri, P., Blumenthal, J.A., Babyak, M.A., Craighead, W., Herman, S., *et al.* (2001) Effects of exercise training on cognitive functioning among depressed older men and women. *Journal of Aging and Physical Activity* 9(1), 43–57.

Korpela, K. (1995) *Developing the Environmental Self-Regulation Hypothesis*. University of Tampere, Tampere, Finland.

Korpela, K., Kyttä, M. and Hartig, T. (2002) Restorative experience, self-regulation, and children's place preferences. *Journal of Environmental Psychology* 22(4), 387–398.

Kubesch, S., Bretschneider, V., Freudenmann, R., Weidenhammer, N., Lehmann, M., *et al.* (2003) Aerobic endurance exercise improves executive functions in depressed patients. *Journal of Clinical Psychiatry* 64(9), 1005–1012.

Kuo, F.E. (2004) Horticulture, well-being, and mental health: from intuitions to evidence. In: Relf, D. (ed.) Proceedings of the XXVI International Horticulture Congress: Expanding Roles for Horticulture in Improving Human Well-being and Life Quality. *Acta Horticulturae* 639, 27–36.

Kuo, F.E. and Taylor, A.F. (2004) A potential natural treatment for attention-deficit/hyperactivity disorder: evidence from a national study. *American Journal of Public Health* 94(9), 1580–1586.

Maller, C. (2005) Hands-on contact with nature in primary schools as a catalyst for developing a sense of community and cultivating mental health and wellbeing. *Eingana* 28(3), 16–21.

Maller, C. and Townsend, M. (2006) Children's mental health and wellbeing and hands-on contact with nature. *International Journal of Learning* 12(4), 359–372.

Maller, C., Townsend, M., Pryor, A., Brown, P. and St. Leger, L. (2006) Healthy nature healthy people: 'contact with nature' as an upstream health promotion intervention for populations. *Health Promotion International* 21(1), 45–54.

Maller, C., Henderson-Wilson, C. and Townsend, M. (2009) Rediscovering nature in everyday settings: or how to create healthy environments and healthy people. *EcoHealth* 6(4), 553–556.

Manger, T.A. and Motta, R.W. (2005) The impact of an exercise program on posttraumatic stress disorder, anxiety, and depression. *International Journal of Emergency Mental Health* 7(1), 49–57.

Mayer, F.S., McPherson Frantz, C., Bruehlman-Senecal, E. and Dolliver, K. (2009) Why is nature beneficial? The role of connectedness to nature. *Environment and Behavior* 41(5), 607–643.

McAvoy, L., Holman, T., Goldenberg, M. and Klenosky, D. (2006) Wilderness and persons with disabilities. *International Journal of Wilderness* 12(2), 23–31.

Mitten, D. (1992) Empowering girls and women in the outdoors. *The Journal of Physical Education, Recreation, & Dance* 63(2), 56–60.

Muñoz, S.A. (2009) Children in the outdoors: a literature review. Available at: http://www.educationscotland. gov.uk/images/Children%20in%20the%20outdoors%20literature%20review_tcm4-597028.pdf (accessed 15 October 2013).

Nisbet, E.K., Zelenski, J.M. and Murphy, S.A. (2010) Happiness is in our nature: exploring nature relatedness as a contributor to subjective well-being. *Journal of Happiness Studies* 12, 303–322.

Paxton, T. and McAvoy, L. (2000) Social psychological benefits of a wilderness adventure program. In: McCool, S.F., Cole, D.N., Borrie, W.T. and O'Loughlin, J. (comps) *Wilderness Science in a Time of Change Conference— Volume 3: Wilderness as a Place for Scientific Inquiry*, Missoula, Montana, 23–27 May 1999. *USDA Forest Service Proceedings RMRS-P-15-VOL-3*. US Department of Agriculture, Forest Service, Rocky Mountain Research Station, Ogden, Utah, pp. 23–27.

Phillips, W.T., Kiernan, M. and King, A.C. (2003) Physical activity as a nonpharmacological treatment for depression: a review. *Complementary Health Practice Review* 8(2), 139–152.

Propst, D.B. and Koesler, R.A. (1998) Bandura goes outdoors: role of self-efficacy in the outdoor leadership development process. *Leisure Sciences* 20(4), 319–344.

Pryor, A., Carpenter, C. and Townsend, M. (2005) Outdoor education and bush adventure therapy: a social-ecological approach to health and wellbeing. *Australian Journal of Outdoor Education* 9(1), 3–13.

Pryor, A., Townsend, M., Maller, C. and Field, K. (2006) Health and well-being naturally: 'contact with nature' in health promotion for targeted individuals, communities and populations. *Health Promotion Journal of Australia* 17(2), 114–123.

Pumariega, A.J. and Cross, T.L. (1997) Cultural competence in child psychiatry. In: Noshpitz, J. and Alessi, N. (eds) *Basic Handbook of Child and Adolescent Psychiatry*, vol. 4. John Wiley & Sons, New York, pp. 473–484.

Rose, S. and Massey, P. (1993) Adventurous outdoor activities: an investigation into the benefits of adventure for seven people with severe learning difficulties. *Mental Handicap Research* 6(4), 287–302.

Ryan, R.M., Weinstein, N., Bernstein, J., Brown, K.W., Mistretta, L. and Gagne, M. (2010) Vitalizing effects of being outdoors and in nature. *Journal of Environmental Psychology* 30(2), 159–168.

Sameroff, A.J. and Emde, R.N. (eds) (1989) *Relationship Disturbances in Early Childhood: A Developmental Approach*. Basic Books, New York.

Stein, M.B. (2005) Sweating away the blues: can exercise treat depression? *American Journal of Preventive Medicine* 28(1), 140–141.

Stella, S.G., Vilar, A.P., Lacroix, C.C., Fisberg, M.M., Santos, R.F., *et al.* (2005). Effects of type of physical exercise and leisure activities on the depression scores of obese Brazilian adolescent girls. *Brazilian Journal of Medical and Biological Research* 38(11), 1683–1689.

Taylor, A.F. and Kuo, F.E. (2009) Children with attention deficits concentrate better after walk in the park. *Journal of Attention Disorders* 12(5), 402–409.

Taylor, A.F., Kuo, F.E. and Sullivan, W.C. (2002) Views of nature and self-discipline: evidence from inner city children. *Journal of Environmental Psychology* 22(1), 49–63.

Ulrich, R. (1984) View through a window may influence recovery from surgery. *Science* 224(4647), 420–421.

Weinstein, N., Przybylski, A.K. and Ryan, R.M. (2009) Can nature make us more caring? Effects of immersion in nature on intrinsic aspirations and generosity. *Personality and Social Psychology Bulletin* 35(10), 1315–1329.

White, R. and Stoecklin, V. (1998) Children's outdoor play & learning environments: returning to nature. *Early Childhood News*, March/April 1998 issue. Available at: http://www.whitehutchinson.com/children/articles/ outdoor.shtml (accessed 15 October 2013).

Wong, J.L. (1997) The cultural and social values of plants and landscapes. In Stoneham, J. and Kendle, D. (eds) *Plants and Human Well-Being*. The Federation for Disabled People, Gillingham.

References

Ainsworth, M.D.S., Blehar, M.C., Waters, E. and Wall, S. (1978) *Patterns of Attachment: A Psychological Study of the Strange Situation*. Erlbaum, Hillsdale, New Jersey.

Asah, S.T., Bengston, D.N. and Westphal, L.M. (2011) The influence of childhood: operational pathways to adulthood participation in nature-based activities. *Environment and Behavior* 44(4), 545–569.

Aung, S.K.H. and Lee, M.H.M. (2004) Music, sounds, medicine and meditation. *Alternative & Complementary Therapies* 10(5), 266–270.

Bandura, A. (1994) *Self-efficacy*. Academic Press, New York.

Barker, E.T. (1987) The critical importance of mothering. *Mothering Magazine* vol. 47. Available at: http://www.naturalchild.org/elliott_barker/mothering.html (accessed 14 October 2013).

Bixler, R.D., Floyd, M.F. and Hammitt, W.E. (2002) Environmental socialization: quantitative tests of the childhood play hypothesis. *Environment and Behavior* 34(6), 795–818.

Blanchard, L.T., Gurka, M. and Blackman, J.A. (2006) Emotional, developmental, and behavioral health of American children and their families: a report from the 2003 National Survey of Children's Health. *Pediatrics* 117(6), 1202–1212.

Bowlby, J. (1951) *Maternal Care and Mental Health*. World Health Organization, Geneva, Switzerland.

Bowlby, J. (1969) *Attachment and Loss*, 2nd edn. Basic Books, New York.

Breitenbach, E., Stumpf, E., Ferson, L.V. and Harald, E. (2009) Dolphin-assisted therapy: changes in interaction and communication between children with severe disabilities and their caregivers. *Anthrozoös* 22(3), 277–289.

Bukatko, D. and Daehler, M. (2011) *Child Development: A Thematic Approach*. Cengage Learning, Belmont, California.

Bunting, T.E. and Cousins, L.R. (1985) Environmental dispositions among school-age children. *Environment and Behavior* 17(6), 725–768.

Carson, R. (1962) *Silent Spring*. Houghton-Mifflin, Boston, Massachusetts.

Carson, R. (1998) *The Sense of Wonder*. HarperCollins Publishers, New York.

Cervinka, R., Roderer, K. and Hefler, E. (2012) Are nature lovers happy? On various indicators of well-being and connectedness to nature. *Journal of Health Psychology* 17(3), 379–388.

Charles, C. (2013) Cheryl Charles welcomes C&NN's new Executive Director. Web log comment, 27 June 2013. Available at: http://blog.childrenandnature.org/2013/06/27/cheryl-charles-welcomes-cnns-new-executive-director/ (accessed 9 September 2013).

Chawla, L. (1988) Children's concern for the natural environment. *Children's Environments Quarterly* 5(3), 13–20.

Chawla, L. (1994) Editors' note. *Children's Environments* 11(3), 175–176.

Chawla, L. (1999) Life paths into effective environmental action. *Journal of Environmental Education* 31(1), 15–26.

Chawla, L. (2002) Spots of time: manifold ways of being in nature in childhood. In: Kahn, P.H. Jr and Kellert, S.R. (eds) *Children and Nature: Psychological, Sociocultural, and Evolutionary Investigations*. MIT Press, Cambridge, Massachusetts, pp. 199–225.

Chawla, L. (2006) Learning to love the natural world enough to protect it. *Barn* 2, 57–78.

Cheng, J. and Monroe, M.C. (2012) Connection to nature: children's affective attitude toward nature. *Environment and Behavior* 44(1), 31–49.

Chipeniuk, R.C. (1994) Naturalness in landscape: an inquiry from a planning perspective. PhD thesis, University of Waterloo, Waterloo, Canada.

Clarke, J.I. (1998) *Self-Esteem: A Family Affair*. Hazelden Foundation, Center City, Minnesota.

Clarke, J.I. and Dawson, J. (1989/1998) *Growing Up Again: Parenting Ourselves, Parenting Our Children*, 2nd edn. Hazelden Foundation, Center City, Minnesota.

Clements, R. (2004) An investigation of the status of outdoor play. *Contemporary Issues in Early Childhood* 5(1), 68–80.

Cobb, E. (1998) *The Ecology of Imagination in Childhood*. Spring, Dallas, Texas.

Cobb, E. and Mead, M. (1977) *The Ecology of Imagination in Childhood*. Columbia University Press, New York.

Colin, V.L. (1996) *Human Attachment*. Temple University Press, Philadelphia, Pennsylvania.

Cooper, A.R., Page, A.S., Wheeler, B.W., Hillsdon, M., Griew, P., *et al.* (2010) Patterns of GPS measured time outdoors after school and objective physical activity in English children: the PEACH project. *International Journal of Behavioral Nutrition and Physical Activity* 7, 31.

Crain, W. (2001) How nature helps children develop. *Montessori Life* 9(2), 41–43.

Crain, W. (2011) *Theories of Development: Concepts and Applications*, 6th edn. Pearson Education, Inc., Upper Saddle River, New Jersey.

Dewey, J. (1938/1997) *Experience and Education*. Touchstone, New York.

Docksai, R. (2011) The sounds of wellness. *Futurist* 45(5), 13–16.

Douillard, J. (2004) *Perfect Health for Kids: Ten Ayurvedic Health Secrets Every Parent Must Know*. North Atlantic Books, Berkeley, California.

Dyment, J.E. and Bell, A.C. (2008) Grounds for movement: green school grounds as sites for promoting physical activity. *Health Education Research* 23(6), 952–962.

Erikson, E. (1998) *The Life Cycle Completed*. W.W. Norton & Company, New York.

Fisman, L. (2001) *Child's Play: An Empirical Study of the Relationship between the Physical Form of Schoolyards and Children's Behavior*. Yale School of Forestry & Environmental Studies, Hixon Center for Urban Ecology, New Haven, Connecticut.

Fjørtoft, I. (2004) Landscape as playscape: the effects of natural environments on children's play and motor development. *Children, Youth and Environments* 14(2), 21–44.

Fosar, G. and Bludorf, F. (2011) DNA can be influenced and reprogrammed by words and frequencies (taken from the book "Vernetzte Intelligenz" von Grazyna Fosar und Franz Bludorf and summarised by Baerbel), 13 June 2011. Available at: http://www.alchemyoflight.org/2011/06/dna-can-be-influenced-and-reprogrammed-by-words-and-frequencies-taken-from-the-book-"vernetzte-intelligenz"-von-grazyna-fosar-und-franz-bludorf-and-summarised-by-baerbel/ (accessed 25 September 2013).

Fosnot, C.T. (2005) *Constructivism: Theory, Perspectives and Practice*, 2nd edn. Teacher's College, New York.

Francis, M. and Devereaux, K. (1991) Children of nature. *UC Davis Magazine* 9(2). University of California–Davis, Davis, California.

Freeman, P.A. and Zabriskie, R.B. (2002) The role of outdoor recreation in family enrichment. *Journal of Adventure Education & Outdoor Learning* 2(2), 131–146.

Gardner, H.E. (2000) *Intelligence Reframed: Multiple Intelligences for the 21st Century*. Basic Books, New York.

Gasser, K. and Kaufmann-Hayoz, R. (2004) *Woods, Trees and Human Health & Well-being*. Interfakultäre Koordinationsstelle für Allgemeine Ökologie (IKAO), Bern.

Goleman, D. (1995/2006) *Emotional Intelligence: Why It Can Matter More Than IQ*, 10th anniversary edn. Bantam, New York.

Halstead, M. and Roscoe, S. (2002) Music as an intervention for oncology nurses. *Clinical Journal of Oncology Nursing* 6(6), 332–336.

Hart, R. (1997) *Children's Participation: The Theory and Practice of Involving Young Citizens in Community Development and Environmental Care*. Earthscan Publications Limited, Oxford, UK.

Hartig, T., Korpela, K., Evans, G.W. and Gärling, T. (1997) A measure of restorative quality in environments. *Scandinavian Housing and Planning Research* 14(4), 175–194.

Hartig, T., Nyberg, L., Nilsson, L.-G. and Gärling, T. (1999) Testing for mood congruent recall with environmentally induced mood. *Journal of Environmental Psychology* 19(4), 353–367.

Howell, A.J., Dopko, R.L., Passmore, H. and Buro, K. (2011) Nature connectedness: associations with well-being and mindfulness. *Personality and Individual Differences* 51(2), 166–171.

Hu, Z., Liebens, J. and Rao, K.R. (2008) Linking stroke mortality to air pollution, income, and greenness in northwest Florida: an ecological geographical study. *International Journal of Health Geographics* 7, 20.

James, J.J., Bixler, R.D. and Vadala, C.E. (2010) From play in nature, to recreation then vocation: a developmental model for natural history-oriented environmental professionals. *Children, Youth and Environments* 20(1), 231–256.

Kahn, P.H. Jr (1997) Developmental psychology and the biophilia hypothesis. *Developmental Review* 17(1), 1–61.

Kahn, P.H. Jr (2002) Children's affiliations with nature: structure, development, and the problem of environmental generational amnesia. In: Kahn, P.H. Jr and Kellert, S.R. (eds) *Children and Nature: Psychological, Sociocultural, and Evolutionary Investigations*. MIT Press, Cambridge, Massachusetts, pp. 93–116.

Kahn, P.H. Jr and Kellert, S.R. (2002) *Children and Nature: Psychological, Sociocultural, and Evolutionary Investigations*. MIT Press, Cambridge, Massachusetts.

Kail, R.V. and Cavanaugh, J.C. (2010) *Human Development: A Life-span View*, 5th edn. Wadsworth, Belmont, California.

Kals, E. and Ittner, H. (2003) Children's environmental identity, indicators and behavioral impacts. In: Clayton, S. and Opotow, S. (eds) *Identity and the Natural Environment, The Psychological Significance of Nature*. MIT Press, Cambridge, Massachusetts, pp. 135–157.

Kals, E., Schumacher, D. and Montada, L. (1999) Emotional affinity towards nature as a motivational basis to protect nature. *Environment and Behavior* 31(2), 178–202.

Kaplan, R. (2001) The nature of the view from home: psychological benefits. *Environment and Behavior* 33(4), 507–542.

Kaplan, R. and Kaplan, S. (1989) *The Experience of Nature: A Psychological Perspective.* Cambridge University Press, Cambridge, UK.

Karen, R. (1998) *Becoming Attached: First Relationships and How They Shape Our Capacity to Love.* Oxford University Press, New York.

Kellert, S.R. (1993) The biological basis for human values of nature. In: Kellert, S.R. and Wilson, E.O. (eds) *The Biophilia Hypothesis.* Island Press, Washington, DC, pp. 42–69.

Kellert, S.R. (2002) Experiencing nature: affective, cognitive, and evaluative development in children. In: Kahn, P.H. Jr and Kellert, S.R. (eds) *Children and Nature: Psychological, Sociocultural, and Evolutionary Investigations.* MIT Press, Cambridge, Massachusetts, pp. 117–151.

Kellert, S.R. and Wilson, E.O. (eds) (1993) *The Biophilia Hypothesis.* Island Press, Washington, DC.

Keysers, C. (2011) *The Empathic Brain.* CreateSpace Independent Publishing Platform.

Kimbro, R.T. and Schachter, A. (2011) Neighborhood poverty and maternal fears of children's outdoor play. *Family Relations* 60(4), 461–475.

Knapp, C.E. (1999) *In Accord with Nature: Helping Students Form an Environmental Ethic Using Outdoor Experience and Reflection.* ERIC/CRESS, Appalachia Educational Laboratory, Charleston, West Virginia.

Kolb, D.A. (1983) *Experiential Learning: Experience as the Source of Learning and Development.* Prentice-Hall, Upper Saddle River, New Jersey.

Korpela, K.M., Klemettilä, T. and Hietanen, J.K. (2002) Evidence for rapid affective evaluation of environmental scenes. *Environment and Behavior* 34(5), 634–650.

Kuo, F.E. (2003) Book review: Children and nature: psychological, sociocultural, and evolutionary investigations. *Children, Youth and Environments* 13(1).

Kuo, F.E. and Taylor, A.F. (2004) A potential natural treatment for attention-deficit/hyperactivity disorder: evidence from a national study. *American Journal of Public Health* 94(9), 1580–1586.

Kyttä, M. (2004) The extent of children's independent mobility and the number of actualized affordances as criteria for child-friendly environments. *Journal of Environmental Psychology* 24(2), 179–198.

Lear, L. (ed.) (1998) *Lost Woods: The Discovered Writing of Rachel Carson.* Beacon Press, Boston, Massachusetts.

Li, Q., Kobayashi, M. and Kawada, T. (2008) Relationships between percentage of forest coverage and standardized mortality ratios (SMR) of cancers in all prefectures in Japan. *The Open Public Health Journal* 1, 1–7.

Malone, K. and Tranter, P. (2003) Children's environmental learning and the use, design and management of school grounds. *Children, Youth and Environments* 13(2), 1–30.

Marti, B., *et al.* (2002) Bekanntheit, Nutzung und Bewertung des Vitaparcours: Vergleich zwischen 1997 und 2001 (Attitude towards use of the fitness trail Vitaparcours: comparison between 1997 and 2001). *Schweizerische Zeitschrift für Sportmedizin und Sporttraumatologie* 50(4), 161–163.

Matsuoka, R.H. (2010) Student performance and high school landscapes: examining the links. *Landscape and Urban Planning* 97(4), 273–282.

Memishevikj, H. and Hodzhikj, S. (2010) The effects of equine-assisted therapy in improving the psychosocial functioning of children with autism. *Journal of Special Education & Rehabilitation* 11(3–4), 67–67.

Mitchell, R. and Popham, F. (2008) Effect of exposure to natural environment on health inequalities: an observational population study. *Lancet* 372(9650), 1655–1660.

Mitten, D. (2008) Getting fit for hiking and backpacking. In: Goldenberg, M. and Martin, B. (eds) *Hiking and Backpacking.* Wilderness Education Association and Human Kinetics, Champaign, Illinois, pp. 21–50.

Mitten, D. (2009) Under our noses: the healing power of nature. *Taproot Journal* 19(1), 20–26.

Moore, R.C. (1986) The power of nature orientations of girls and boys toward biotic and abiotic play settings on a reconstructed schoolyard. *Children's Environments Quarterly* 3(3), 52–69.

Moore, R.C. and Cosco, N.G. (2000) Developing an earth-bound culture through design of childhood habitats. Presented at the *International Conference on People, Land and Sustainability,* University of Nottingham, Nottingham, UK, 13–16 September 2000.

Moore, R.C. and Wong, H.H. (1997) *Natural Learning: The Life History of an Environmental Schoolyard. Creating Environments for Rediscovering Nature's Way of Teaching.* MIG Communications, Berkeley, California.

Myers, O.E. Jr and Saunders, C.D. (2002) Animals as links toward developing caring relationships with the natural world. In: Kahn, P.H. Jr and Kellert, S.R. (eds) *Children and Nature: Psychological, Sociocultural, and Evolutionary Investigations*. MIT Press, Cambridge, Massachusetts, pp. 153–178.

Newman, F. and Holzman, L. (2013) *Lev Vygotsky: A Revolutionary Scientist*. Psychology Press, London.

Nisbet, E.K. and Zelenski, J.M. (2011) Underestimating nearby nature: affective forecasting errors obscure the happy path to sustainability. *Psychological Science* 22(9), 1101–1106.

Nisbet, E.K., Zelenski, J.M. and Murphy, S.A. (2011) Happiness is in our nature: exploring nature relatedness as a contributor to subjective well-being. *Journal of Happiness Studies* 12(2), 303–322.

Noddings, N. (1984) *Caring: A Feminine Approach to Ethics and Moral Education*. University of California Press, Berkeley, California.

Olson, S. (2012) *From Neurons to Neighborhoods: An Update. Workshop Summary*. Institute of Medicine and National Research Council of the National Academies, Washington, DC.

Pavlov, I.P. (1960/2003) *Conditioned Reflexes*. Dover, New York.

Pergams, O.R.W. and Zaradic, P.A. (2008) Evidence for a fundamental and pervasive shift away from nature-based recreation. *Proceedings of the National Academy of Sciences USA* 105(7), 2295–2300.

Phenice, L. and Griffore, R. (2003) Young children and the natural world. *Contemporary Issues in Early Childhood* 4(2), 167–178.

Prescott, J. (1975) Body pleasure and the origins of violence. *The Bulletin of the Atomic Scientists* (November), 10–20.

Pretty, J., Griffin, M., Peacock, J., Hine, R., Sellens, M., *et al.* (2005a) *Countryside for Health and Wellbeing: The Physical and Mental Health Benefits of Green Exercise*. Countryside Recreation Network, Sheffield, UK.

Pretty, J., Peacock, J., Sellens, M. and Griffin, M. (2005b) The mental and physical health outcomes of green exercise. *International Journal of Environmental Health Research* 15(5), 319–337.

Prior, V. and Glaser, D. (2006) *Understanding Attachment and Attachment Disorders: Theory, Evidence and Practice*. Jessica Kingsley Publishers, Philadelphia, Pennsylvania.

Prothmann, A., Ettrich, C. and Prottman, S. (2009) Preference for, and responsiveness to people, dogs, and objects in children with autism. *Anthrozoös* 22(2), 161–171.

Pyle, R. (1993) *The Thunder Trees: Lessons from an Urban Wildland*. Houghton Mifflin, Boston, Massachusetts.

Pyle, R. (2002) Eden in the vacant lot: special places, species and kids in the neighborhood of life. In: Kahn, P.H. Jr and Kellert, S.R. (eds) *Children and Nature: Psychological, Sociocultural, and Evolutionary Investigations*. MIT Press, Cambridge, Massachusetts, pp. 305–327.

Rivkin, M. (1990) *The Great Outdoors: Restoring Children's Rights to Play Outside*. National Association for the Education of Young Children, Washington, DC.

Rose, K.A., Morgan, I.G., Ip, J., Kifley, A., Huynh, S., *et al.* (2008) Outdoor activity reduces the prevalence of myopia in children. *Ophthamology* 115(8), 1279–1285.

Schaefer, C. and DiGeronimo, T. (2000) *Ages and Stages: Tips and Techniques for Building Your Child's Social, Emotional, Interpersonal and Cognitive Skills*. John Wiley & Sons, Inc., New York.

Schettler, T., Stein, J., Reich, F., Valenti, M. and Wallinga, D. (2000) *In Harm's Way: Toxic Threats to Child Development*. Physicians for Social Responsibility, Greater Boston Chapter, Somerville, Massachusetts.

Schickendanz, J. (2001) Theories of child development and methods of studying children. In: Schickendanz, J., Schickendanz, D.I., Forsyth, P.D. and Forsyth, G.A. (eds) *Understanding Children and Adolescents*, 4th edn. Allyn and Bacon, Boston, Massachusetts, pp. 12–13.

Schultz, P.W., Shriver, C., Tabanico, J.J. and Khazian, A.M. (2004) Implicit connections with nature. *Journal of Environmental Psychology* 24(1), 31–42.

Searles, H. (1960) *The Nonhuman Environment in Normal Development and in Schizophrenia*. International Universities Press, Madison, Connecticut.

Shepard, P. (1996) *The Others: How Animals Made Us Human*. Island Press, Washington, DC.

Shonkoff, J. and Phillips, D. (2000) *From Neurons to Neighborhoods: The Science of Early Childhood Development*. National Academy Press, Washington, DC.

Siegel, D. (1999) *The Developing Mind: How Relationships and the Brain Interact to Shape Who We Are*. The Guilford Press, New York.

Skinner, J.B. (1974/1976) *About Behaviorism*. Random House Vintage Editions, New York.

Sobel, D. (1996) *Beyond Ecophobia: Reclaiming the Heart in Nature Education*. The Orion Society and the Myrin Institute, Great Barrington, Massachusetts.

Sobel, D. (1998) Beyond ecophobia. *Yes! Magazine*, 2 November 1998. Available at: http://www.yesmagazine.org/issues/education-for-life/803 (accessed 12 April 2013).

Sobel, D. (2002) *Children's Special Places: Exploring the Role of Forts, Dens, and Bush Houses in Middle Childhood*. Wayne State University Press, Detroit, Michigan.

Sobel, D. (2004) *Place-based Education: Connecting Classrooms and Communities*. The Orion Society and the Myrin Institute, Great Barrington, Massachusetts.

Sobel, D. (2008) *Childhood and Nature: Design Principles for Educators*. Stenhouse Publishers, Portland, Maine.

St. Antoine, S. (2012) *Together in Nature: Pathways to a Stronger, Closer Family*. Children and Nature Network, Santa Fe, New Mexico.

Swiderski, M.J. (2011) Maria Montessori. In: Smith, T. and Knapp, C.E. (eds) *Sourcebook of Experiential Education: Key Thinkers and Their Contributions*. Routledge, New York, pp. 197–207.

Takano, T., Fu, J., Nakamura, K., Uji, K., Fukuda, Y., *et al.* (2002) Age-adjusted mortality and its association to variations in urban conditions in Shanghai. *Health Policy* 61(3), 239–253.

Taylor, A.F., Kou, F.E. and Sullivan, W.C. (2001) Coping with ADD: the surprising connection to green play settings. *Environment and Behavior* 33(1), 54–77.

Ulrich, R.S. (1981) Natural versus urban scenes: some psychophysiological effects. *Environment and Behavior* 13(5), 523–556.

Van den Berg, A.E., Hartig, T. and Staats, H. (2007) Preference for nature in urbanized societies: stress, restoration, and the pursuit of sustainability. *Journal of Social Issues* 63(1), 79–96.

Walant, K. (1999) *Creating the Capacity for Attachment: Treating Addictions and the Alienated Self*. Jason Aronson, Inc., Northvale, New Jersey.

Warren, K. and Wapotich, L. (2011) Rachel Carson. In: Smith, T. and Knapp, C.E. (eds) *Sourcebook of Experiential Education: Key Thinkers and Their Contributions*. Routledge, New York, pp. 197–207.

Watson, J.B. (1923/2012) *Behavior: An Introduction to Comparative Psychology*. Henry Holt and Company, New York.

Wells, N.M. (2000) At home with nature: effects of 'greenness' on children's cognitive functioning. *Environment and Behavior* 32(6), 775–795.

Wells, N.M. and Evans, G.W. (2003) Nearby nature: a buffer of life stress among rural children. *Environment and Behavior* 35(3), 311–330.

Wheeler, B.W., Cooper, A.R., Page, A.S. and Jago, R. (2010) Greenspace and children's physical activity: a GPS/GIS analysis of the PEACH project. *Preventive Medicine* 51(2), 148–152.

Wilson, E.O. (1984) *Biophilia: The Human Bond with Other Species*. Harvard University Press, Cambridge, Massachusetts.

Wilson, R. (1993) *Fostering a Sense of Wonder during the Early Childhood Years*. Greyden, Columbus, Ohio.

Wuang, Y.P., Wang, C.C., Huang, M.H. and Su, C.Y. (2010) The effectiveness of simulated developmental horse-riding program in children with autism. *Adapted Physical Activity Quarterly* 27(2), 113–126.

Yankou, D. (2002) MorningStar: a place of restoration. *MorningStar Adventures* 18(2), 3.

Zube, E.H., Pitt, D.G. and Evans, G.W. (1983) A lifespan developmental study of landscape assessment. *Journal of Environmental Psychology* 3(2), 115–128.

6

Adaptations and Applications

*The more clearly we can focus our attention
on the wonders and realities of the universe
about us, the less taste we shall have for
destruction.*

Rachel Carson, 'The real world
around us' (1954)

This chapter discusses a variety of adaptations and applications relative to human health and natural environments. Adaptation refers to the definition and negotiation of a wider set of beliefs within cultural practice. The presence and existence of cultural adaptations provide a philosophical foundation for connectedness to the natural environment. Humans evolved with nature and as humans physically distanced themselves from nature from the agriculture stage and forward, adaptations such as traditional ecological knowledge and friluftsliv may have become embedded into the culture in order to fill the emptiness of no longer having the time in nature; these practices help humans adapt to their new setting. It is our hope that these adaptations may be used as models for negotiating the demands and challenges of today's world. In this vein, a number of applications are also reviewed, demonstrating ongoing initiatives toward bridging the perceived gap between humans and the environment in the Western world. These applications represent progress toward a new understanding of the natural world, and arise from within a

variety of disciplines. Belief systems inform the applications, which demonstrate ways people have put ethics, values, and knowledge into practice. Citizen science, for example, provides healthful experiences for participants, engenders a sense of place, and provides natural history and monitoring data.

While the developed, Western world can be primarily characterized by a perceived disconnect from nature, there are pockets of geographical and cultural practices that position the relationship between humans and the natural environment differently. The cultural adaptations or ways of knowing and interacting with the natural environment discussed here illustrate belief systems. These beliefs regard nature as positive and contributing to health. Underpinning both adaptations and applications are the theories in Chapter 4, including indigenous and intra-indigenous consciousness, naturalistic intelligence, and the biophila hypothesis.

Cultural Adaptations

Cultural adaptations regarding nature and health and well-being refer to the ways that individual cultures define and negotiate a wider set of values and beliefs that influence how they think about the human and nature relationship and health. In the case of this book,

© CAB International 2014. *Natural Environments and Human Health*
(A.W. Ewert, D.S. Mitten and J.R. Overholt)

we refer to variations in the dominant discourse and worldview relative to the natural environment, including folk biology, traditional ecological knowledge, friluftsliv, and ecofeminism. Granted, each of these areas has an applied side; in that way the division between adaptation and application is somewhat arbitrary. Folk biology is a cultural adaptation shared by many past and current cultures. Traditional ecological knowledge and local and indigenous knowledge have a long history and practices are recognized as useful for current environmental problems. Friluftsliv, shared primarily by Scandinavian cultures, was a brilliant adaptation useful in securing Norway's independence as well as helping with ongoing health considerations. Ecofeminism, a concept that makes a connection between the way women are treated and the way the environment is treated, helped inform applications such as ecopsychology, ecotherapy, and nature therapy and related practices that integrate nature at the core of the therapeutic action. Finally, the socio-ecological approach to human health is an adaptation that encompasses the application of many practices.

Folk biology

For centuries humans have practiced what in retrospect has been called folk biology or folk medicine, relying on folk biology. Folk biology is practical applied knowledge gained through observation or intuition and has a cultural context, so practices from earlier times may not make sense to people today; it may be thought of as a way to classify and think about the natural world (Atran, 1998). This categorization and knowledge seems to have similarities across cultures and the propensity to engage in folk biology may be related to naturalistic intelligence described in Chapter 4. Folk biology also can be defined as the informal and intuitive ways we explain biological cause and effect (Coley *et al.*, 2002), usually seeing the natural world in a systems context. There is argument as to whether the practice of folk biology is biologically determined or culturally determined, or both (Atran, 1999); though often folk biology takes into account the contribution of cultures and worldviews in shaping beliefs

regarding causal relationships to explain biological phenomena. For example, bundling up with a hat and scarf may not protect against the germs that cause a common cold, but it does have a symbolic effect in terms of catching a cold and it may help one cope better with germs if they are present. A study with the Hua people in Papua New Guinea discovered culturally held beliefs that young men should eat fast-growing green vegetables to help them grow fast. Most likely these fast-growing green vegetables are packed with nutrition, but the growth rate of the plant does not necessarily accelerate the growth rate of young men (Nemeroff and Rozin, 2000). The conclusion of this and similar studies is that folk thinking is part of human nature and serves as a useful and culturally accessible guide for promoting hygiene, enhancing nutrition, avoiding contagions, and more. Throughout pre-industrial history, human cognition was shaped by close contact with plants and animals (Medin and Atran, 1999). This first-hand knowledge of the natural world helped shape cultural beliefs, which were also based on observation. Folk biology or folk medicine has a realistic base and may also be related to traditional or local ecological knowledge.

Traditional ecological knowledge (TEK)

Traditional ecological knowledge (TEK), sometimes known as traditional environmental knowledge or folk ecology and wisdom, is defined by the Convention on Biological Diversity, Article 8 (j):

> Traditional knowledge refers to the knowledge, innovations and practices of indigenous and local communities around the world. Developed from experience gained over the centuries and adapted to the local culture and environment, traditional knowledge is transmitted orally from generation to generation. It tends to be collectively owned and takes the form of stories, songs, folklore, proverbs, cultural values, beliefs, rituals, community laws, local language, and agricultural practices, including the development of plant species and animal breeds. Traditional knowledge is mainly of a practical nature, particularly in such fields as agriculture, fisheries, health, horticulture, and forestry.

However that definition is only part of the story. Deborah McGregor (2004) spoke of the dichotomy between the Aboriginal view of TEK and the Eurocentric view, which reflects colonial attitudes towards aboriginal peoples. When McGregor teaches a TEK class she starts with creation stories of the Anishinaabe and Haudenosaunee peoples. She says that she begins her class this way in order to convey the indigenous understanding of the relationship to creation as well as the philosophies, ethics, and values that flourished for thousands of years.

Leanne Simpson (2004) said that TEK has been of recent interest to Western scientists to supplement their ecological knowledge, but that the scientists have a long history of denial of the existence and validity of indigenous knowledge. She says that Western scientists too often want to use TEK to increase their control over the environment and do not value or include the spiritual foundations and the indigenous values and worldviews, leading to further marginalization of indigenous people. This is similar to the mining of medicine systems of indigenous people, such as Ayurvedic Medicine where parts of the system are appropriated for use, not understanding that the strength of the modality lies in the whole system approach.

Throughout this book systems theory has been mentioned (see Box 6.1). The concerns that Simpson and McGregor bring up relate to systems thinking. When conceptualizing human health and the natural environment we have shared ways that physical health as well as emotional, spiritual, and social health improves from contact with nature. However it is easy to see these as separate areas of health rather than a holistic system. For example, without the understanding of the creation aspect of indigenous values, the reciprocal nature of the relationship with the land is not felt. Without the reciprocal nature of the relationship with the land, TEK practices do not contribute to spiritual, social, and psychological health of people. Without considering the spiritual relationship with the land at the same time as the bounty received from the land, the full health benefits are not received by humans and by the earth.

In this same vein, using TEK in teaching environmental education can offer benefits to students (Feinstein, 2004); however respect and regard must be given to the originating cultures. The ecological principles need to be combined with the cultural aspects, including the conceptual frameworks of their ontology.

TEK connects land and culture because it is based on a spiritual connection to the land as well as using the natural resources for survival. In addition to recognizing, as modern ecology does, that all things are connected, TEK also incorporates a belief that all things are related, and as relatives are honored and treated with respect. An important feature of TEK is that it recognizes that adaptation to our natural environment is key to human survival and uses observation and practices spanning generations to understand the natural environment's response to human activities. Indigenous groups practicing TEK in their area offer perspectives, insights, and knowledge that can be different or complementary to what present-day ecologists and land managers understand. Using TEK to enhance ecosystem resilience is an area especially useful to conservation professionals.

Natural resource managers and many of those in the commercial sector now understand biodiversity's importance, not just for sustainable development, but also for human survival. Therefore, academics have begun to appreciate and study TEK. The largest academic-based ecology conference in the US, that of the Ecological Society of America (ESA), has had a section and presentation track dedicated to TEK since 2001. Using research methods drawn from the fields of anthropology and general biology, ethnobiology, ethnobotany, and human ecology are some of the disciplines that study and inform research about TEK. Additionally, TEK itself is multidisciplinary and based on reciprocal practices. Aspects of the origin of TEK may be genetic, learned as a cultural practice, or both.

As an example of practice, in Canada the Government of Nunavut makes selective use of Inuit TEK in its northern conservation and management policy (Wenzel, 2004). However, Wenzel (1999) has raised the ethical concern of appropriating the TEK practices without giving TEK data the same respect given by law to Western science data.

Box 6.1. Systems theory embraces principles of organization, patterns of collaboration, relationships, and interdependence

Systems theory (from Mitten, 2012) is a coherent scientific framework that understands that everything in the universe (and perhaps beyond) is connected to and affects everything else, and that everything known to humans is in effect one living system of which humans are a part. People use systems theory in systems thinking, which is the process of understanding how entities, regarded as systems, influence one another within a whole, and then use this understanding as a foundation for judgment and decision making. Finally, employing systems approaches, based on systems theory, means taking actions that affirm processes, interrelationships, and feedback loops and that avoid reductionist and mechanistic thinking. The belief is that the components of a system can best be understood in the context of relationships with the other elements of the system, as well as with other systems. Rather than looking at the parts of an entity in relationship to the whole, a systems approach looks at the linkages and interactions among the elements that compose the system as well as the cyclical nature of the system.

Systems that are within the universe that humans know about include: (i) ecosystems, in which elements of air, water, plants, soil, animals (including humans), and more interact in connection and coordination with each other; and (ii) human systems in which people may be part of a society, a movement, or an organization, and their interrelationships define the health and well-being of that community. Cartesian science might look at a car and conceptually divide it into pieces including the engine, tires, battery, and exhaust system, while a systems approach would see the exchange of labor in making the car, the exchange between the environment and the car exhaust, and others. Instead of concentrating on basic building blocks, the systems view concentrates on principles of organization and patterns of collaboration, relationship, and interdependence.

An example of the integration of TEK with the scientific method comes out of the Wild Rice and Ethnobiology Lab at Ferris State University, where Dr Scott Herron has worked in controlled lab settings with students to better understand the physiology, ecology, and pathogen responses of wild rice, *Zizania palustris*, an annual grass of the Great Lakes region (Herron *et al.*, 2010; Mitten and Herron, 2010). This research has been shared both in academic venues like the ESA meeting (Herron and Byers, 2011) and at multicultural public outreach wild rice conferences and camps (Robinson *et al.*, 2009). At the camps, the TEK of veteran wild rice harvesters and processers is taught by Anishinaabe experts, passing on to another generation the knowledge and skills of how to sustainably hand-build the tools, and then use those tools to harvest this native grain for a storable surplus food. To understand this plant, it should be understood that *manoomin*, as the Anishinaabe call their sacred relative, was part of their ancient prophecies that led them to the upper Great Lakes wild rice beds or marshes. When they reached the place where the food grows on the water, near the Kakagon

Sloughs of Bad River, Wisconsin, their several hundred year eastward journey from the Atlantic coast was complete (Minnesota Department of Natural Resources, 2008). Today, this ecosystem is recognized as a Ramsar site, and is the first and only owned by a Native American tribe, The Bad River Band of Lake Superior Chippewa (Wiggins, 2012). To the Anishinaabe, and to Herron, wild rice is a sacred relative, one to be honored throughout the year with ceremony, offerings, restoration plantings, and sustainable harvest during *manoominike giizis*, the wild rice harvest moon (August through September).

As discussed in Chapter 2, beginning with the agricultural revolution about 10,000 years ago, some humans began to disconnect from their earlier traditions of working with the environment to sustain their smaller family units. Cultural imperialism and the Industrial Revolution furthered the distance between much of the human race and TEK.

A similar concept is that of traditional indigenous knowledge, and local and indigenous knowledge (LINK). According to the United Nations Educational, Scientific and

Cultural Organization, LINK refers to 'the understandings, skills and philosophies developed by societies with long histories of interaction with their natural surroundings' (UNESCO, n.d.). Like TEK, LINK is of cultural origin, and implicates language, spirituality, ritual, and systems of practice. Its primary difference from TEK lies in the way that it informs day-to-day life and decision making in local places, as opposed to TEK's emphasis on the more practical aspects of resource management. In conclusion, to truly understand TEK, one must start with understanding the indigenous people where TEK originated as part of a system in an organic and holistic manner.

Friluftsliv

When Henrik Ibsen presented friluftsliv in print 150 years ago it was mystical and captured the hearts of the Norwegian populace. Roughly translated, the word means 'free air life', and now describes a philosophical lifestyle, or a way of life, embedded in Norwegian culture. It is characterized by a sense of freedom in nature and a spiritual connectedness with the landscape that goes beyond the superficial visits to nature or sporadic wilderness trips that more often make up today's nature experiences (Gelter, 2000).

At its very root, friluftsliv is a cultural phenomenon that defines a way of life and that implicates both identity and self-image. At the beginning of the 18th century Norway had been struggling for freedom from Denmark and Sweden for almost 500 years. A creative middle class decided to give becoming a free nation a try. They shunned Bacon's slogan for the paradigm of modernity, 'Knowledge is power over Nature', deciding instead to use the Swiss philosopher, Jean-Jacques Rousseau's terminology 'noble savages' for the inhabitants of the alpine lakes regions. Originating as a protest movement of European artists and philosophers against the reductionist, natural science mentality of the early 19th century, a unique Norwegian tradition developed 'identified with free nature—nature with undisturbed seasons, diurnal rhythms and growth rhythms' (Faarlund, 2010, p. 3).

By the 20th century, with the redefinition of a national culture at home in a sublime mountain and fjord landscape and with polar explorers returning as heroes, Norway had done what no other European nation had been able to do; it achieved freedom without militant nationalism. It had done so in large part by identifying with nature. Culturally one wonders if this was made even more possible because of biophilia, indigenous consciousness, or natural intelligence.

Historically, in the 18th century, friluftsliv was a means of realizing romantic ideals, especially for the bourgeoisie whose changing lifestyles had resulted in a disconnection from traditional ways of life. The concept has changed and grown with society, and this change can be aligned with other nature-based movements. For example, in the mid-1900s it was connected with the ideals of making strong and resilient citizens for the war effort, similar to the Outward Bound movement taking shape in England (Gelter, 2000). Then in 1966 under the leadership of rock climbing instructor and professor of philosophy, Arne Naess, the concept of friluftsliv evolved again. At the Stetind seminar friluftsliv was used as a foundation as the participants grappled with integrating ecology with the idea of 'why meet the dangers of the mountains' (Faarlund, 2010). Today Naess is credited with taking the emerging eco-philosophy into deep ecology and did so using the concept of friluftsliv.

More recently, a corresponding trend with the commercialization of the outdoors can be seen in friluftsliv. This trend involves both expensive gear and specialized training, and positions the environment as a testing ground, rather than the restorative environment originally conceptualized with the term. While often called friluftsliv, Gelter argues that this more modern application is not genuine in the philosophical sense. Still, for people from the city coming to natural areas for adventure and pleasure, friluftsliv has perhaps become a term for this certain kind of outdoor recreation. According to Vaagboe (1993) nine out of ten Norwegians enthusiastically stated that they actively took park in friluftsliv. Friluftsliv became a tradition with health benefits. It has evolved into understanding that free nature has intrinsic value.

In Norway, friluftsliv is taught in a variety of ways, including as a year of friluftsliv study for students prior to entering the workforce or college, or even as a college major. Friluftsliv also enters educational pedagogy as friluftsliv kindergartens, where the philosophy is that the children will learn better through having this phase of their educational experience take place primarily outdoors. By including children in early childhood, friluftsliv has been passed on to generations and is well established as part of Norwegian culture.

That there exists a cultural phenomenon of friluftsliv, and that so many people feel that immense draw to be free in nature, is both evidence of a connection and a cause to research this connection. While friluftsliv is tangible in that people live it, it is also intangible in that it is hard to define. Friluftsliv may be genetic, learned as a cultural practice, or both; it may have been born from a felt need to connect with indigenous consciousness or to express biophilia or naturalistic intelligence. The concept of friluftsliv provides identity as well as the opportunity to be in nature receiving many health benefits.

Ecofeminism

Eco, from the Greek word *oikos*, denotes the whole household of life; the term ecofeminism, therefore, means to care for all life. Ecofeminism is the merging of feminist and ecological principles for the purpose of mediating humanity's relationship with nature (Kelly, 1988). Ecofeminism fosters a sense of our belonging within, rather than being in control of, the community of life. Ecofeminism is directly related to human health and well-being because, as an adaptation, it makes possible a society where human health is situated at the forefront along with nature; though for some people it requires an attitude shift and acceptance of beliefs concerning systems thinking, connection, and healing. Ecofeminists have helped get human health issues on the mainstream environmental movement agenda. As Vandana Shiva (1994) articulated, 'Women's involvement in the environmental movement has started

with their lives and the severe threat to the health of their families. ... The "environment" is not an external, distant category. ... The "environment" for women ... is the place we live in and that means everything that affects our lives' (p. 2).

Francoise d'Eaubonne, a feminist activist, coined the term ecofeminism in her 1974 book, *Le féminisme ou la mort* (*Feminism or Death*). In 1978 she founded an ecology and feminism movement launched with her book *Écologie-Féminisme: Révolution ou Mutation?* (*Ecology-Feminism: Revolution or Mutation?*). This social movement was picked up primarily in Australia (see Val Plumwood's 1993 work) and the US. d'Eaubonne's premise was that society's disregard for women is comparable to its degradation of nature. While to some, d'Eaubonne appeared radical, her logic was simple. She understood how impacts on the biosphere, questions of energy choices, genetic engineering, and women's reproductive rights were concrete manifestations of the intersections of feminism and ecology. She said 'a complete feminist consciousness, since it takes all these disparate strands of thought (i.e., feminist ideas about patriarchy, economic systems, the multiple forms of oppression against specific types of humans and nature as a whole) and incorporates them all into its awareness, can only lead to the synthesis of feminism and ecology' (d'Eaubonne, 1978, p. 190). Linking the historical oppression of women to the oppression of the environment is echoed by both Carolyn Merchant (1980) and Vandana Shiva (1994).

Ecofeminism has been dismissed in the same way that feminism has been dismissed as unnecessary or portrayed as anti-men, elitist, angry, and irrelevant to women of color and poor people. Some people have trivialized it by associating it with earth-loving hippie women or spirituality. However, ecofeminism is about life and includes the care of all people. And it is about spirituality, though in the last few decades spirituality was more or less taboo in many spheres, including academic. Ecofeminists work for environmental justice for those statistically most at risk from, and powerless to prevent, environmental degradation: the poor, women, children, and indigenous peoples.

d'Eaubonne worked hard to help readers see that feminism is not the antithesis of patriarchy, putting women in charge. According to d'Eaubonne (1978), 'Patriarchy is immediately a society of adults against another, of one class against another, of one nation against another, and ultimately of all men against each other: the war of all against all' (p. 134). In contrast, ecofeminism offers a potentially transformative philosophy of the self and of society. Influenced by systems thinking and Gaia science, every entity is seen as internally related to all aspects of its environment, with that relationship as essential as itself. This awareness of ecological interdependence points to the need for an ethic of care within societies, including care for the fundamental elements of life and the recognition of limits, including the carrying capacities of land.

Ecofeminism teaches us to learn to live with and embrace our differences, both within our species and with other species. Ecofeminism works to heal the wounds caused by domination exemplified in the last 5000 plus years. The power-over or domination of one person or species over another is a philosophical stance, which has led to massive human-induced species extinctions and a huge amount of environmental destruction in the world. All domination is fundamentally connected, linking race, class, gender, and nature. Expressing this connection, Sheila Collins (1974) wrote: 'Racism, sexism, class exploitation, and ecological destruction are four interlocking pillars upon which the structure of the patriarchy rests' (p. 161). Through ecofeminism, the interconnectedness of all types of domination can be seen and begun to be dismantled. It offers and asks for a critique of culture and nature; it identifies dualism as an ideology and mechanism used to degrade the 'other'.

Many ecofeminists (e.g. Vandana Shiva) work worldwide, focusing on relationships between global economic policies and global ecological crises, arguing that a radical transformation of capitalist production, from a competitive system to a cooperative one, will benefit the global environment. Ecofeminism addresses issues of famine, deforestation, environmental degradation and pollution, and destruction of soils from monocultural agriculture and chemical fertilizers as global issues impacting human health.

Ecofeminism is a synthesis of feminist socialism and ecology, understanding that the health of the earth and the survival of humans are tightly bound with economics, reproductive rights, and women's liberation. It recognizes and works to change the subordination of nature, poor people, children, indigenous people, and women. The goal of ecofeminism is to increase the health and welfare of humans and nature.

Socio-ecological approach to human health

The socio-ecological approach to human health represents a way to enhance human connection to nature within the structure of today's society, and thus gain the health benefits from this association. As early as 1986 the World Health Organization proclaimed that healthcare is not separate from caring for the environment. The Ottawa Charter for Public Health Promotion, created at an international conference on health promotion in Ontario, Canada, calls for a socio-ecological approach to health management, including environmental protection in the name of health reform (Public Health Agency of Canada, 2008). The theory behind the socio-ecological approach to health management is that human behavior is a consequence of transactions at multiple levels, and ensuring health and well-being requires political commitment and a multidisciplinary approach. As disciplines begin to overlap and policy makers from environmental, public health, psychology, landscape architecture, medical, and urban planning backgrounds collaborate, a truer understanding of the need for, potential, and practice of a socio-ecological approach to health management and promotion will result.

A socio-ecological approach to human health rests on the belief that health results from an interwoven relationship between people and their environment. Human health is influenced by intrapersonal, interpersonal, organizational, community, and environmental factors, including the natural environment, with policy then influencing and creating further interdependencies that impact health (McLeroy *et al.*, 1988). Improvements in health may require interventions and changes, such as changes in policies at each or some of the governmental levels, or changes in personal health management. While similar to ecofeminism, a socio-ecological approach works to connect public policy (and public policy makers), community structures, organizations, individuals, and nature. Since environmental factors, including the natural environment, are components in the socio-ecological approach, this model necessarily looks at the relationship between people and the natural environment. Natural spaces, including public-owned parks, play a key role in a socio-ecological approach to health because these environments encourage and enable people to relate to each other and to the natural world (Maller *et al.*, 2006). This notion is also advocated by Dustin *et al.* (2010) in their call for an ecological model of health promotion. Drawing on research similar to that presented in this book, they emphasize the role of parks, recreation, and tourism entities in promoting the combined health of individuals and communities through a systems approach.

Incorporating natural areas into an overall health plan for people was the approach that Olmsted and others used in park development discussed in Chapter 3. All people, but especially those in urban environments, need natural areas maintained in close proximity to their living spaces in order to reap the benefits of being in nature. In the past, developers and community planners have not realized that research shows that benefits such as greenspace around housing units can increase social bonds, decrease family violence, and help immigrants transition to a new area (Wong in Rhode and Kendle, 1997; Kuo *et al.*, 1998). The socio-ecological approach to human health is an attempt at a true integration of science, society, and ideology. This major shift in beliefs about integrating nature into urban settings creates a needed underpinning to the concept of nature and well-being. This adaptation helps humans in more urban environments affiliate with nature.

Applications

This section focuses on the ways people have incorporated a relationship with the environment into health-enhancing practices. Spiritual connections, ecopsychology, conservation and development of a land ethic, medicine from nature and conservation medicine, and citizen science are ways to involve people in active interactions with natural environments. With this in mind, and within today's milieu, we can see a number of applications of the ongoing and emerging information we have about the natural world. Some of these applications come as a response to our sense of disconnection and a deep need or desire to reconnect to the natural world. Ranging from spiritual practice to the reconceptualization of the practice of medicine, these applications place greater focus on process and connectedness.

Spirituality: connecting directly with all that is

Spirituality is included as an application because most cultures incorporate spirituality into their lives as part of their worldview. Spiritually is a major aspect of individual health and the lack of spiritual health can be a major source of stress for humans (Blonna, 2011). For many people their spiritual health intertwines with time in nature. It may be as simple as watching a sunset or witnessing a birth.

The definition we use is that spirituality is a sense of connection, and more specifically interconnection, with something or someone beyond oneself; it is more about an emotional connection than an intellectual connection. Relating to interconnectedness, many spiritual teachers encourage being present or mindfulness—'the energy of being aware and awake to the present' (Thich Nhat Hanh, 2009). Faith can be an ingredient of spirituality. For some people that includes a supreme being or a god or gods. For other people there is no supreme being involved in their spirituality; it is based on experience, such as in Buddhism (Thich Nhat Hanh, 2009).

Spiritual ecology is an area that brings a spiritual sense to ecological concerns. Leaders in this area may include Rudolph Steiner, Wangari Maathai, Joanna Macy, Thomas Berry, Thich Nhat Hanh, and shamanic practitioner Sandra Ingerman. The idea is to bring environmentalism in sync with an awareness of the sacredness of life and creation. In *Green Sisters: A Spiritual Ecology*, Sarah McFarland Taylor (2009) identifies and ethnographically describes Roman Catholic nuns who she says respond to the call of the earth (p. 5). This is exemplified in seven areas: greening the religious vows, ecologically sustainable living as daily spiritual practice, re-inhabiting Western monasticism, ecological food choice and contemplative cooking, sacred agriculture, seed saving, and the greening of prayer and liturgy. Taylor explained that the sisters seek to harmonize the vibrant associations between the spiritual and the biophysical landscapes, calling this phenomenon 'ecospiritual mimetics' (p. 20). Essentially the sisters' ecological ideals reflect their spiritual values, which are reflected in green practices ranging from turning their lawns into community-supported gardens, to adopting green technology (e.g. hybrid vehicles, solar panels, and composting toilets), to converting their community properties into land trusts and wildlife sanctuaries.

The natural world complements, reinforces, and inspires a sense of connection. In nature many people are captivated by the wonders of life. For whatever reason, watching nature creates awe for many people (see Louie Schwartzberg's time-lapse photography films). Theists, atheists, and agnostic people can share in this nature-related spirituality. Time in nature can inspire the following feelings: (i) we are connected to life—we share DNA and common characteristics with other beings; (ii) we share a connection to everything in the cosmos because all elements came from the heat of a supernova, the only thing hot enough to merge atoms; (iii) we can better understand our place in the universe or cosmos and be humble and let go of needing to control because we are very small in relationship to the universe—we are at the mercy of the seeming randomness of accidents and the like; (iv) we all die and our elements will return to the earth and the Gaian system no matter what our beliefs; and (v) we can control our attitudes and beliefs and are free to live and explore our dreams.

In Elaine Howard Ecklund's (with Elizabeth Long, 2011) sociology of religion research, she found, not surprisingly, that many scientists were atheists and had no belief in a supreme being; what was surprising to her was that about 22% of the 60% of scientists who described themselves as either atheist or agnostic said they were spiritual and named being in nature as the primary place where they felt spiritual. They talked about something larger than themselves and found awe and beauty in nature as they did with the birth of their children, and in their work as scientists.

More than once, the Templeton Prize that rewards positive contributions to spiritual progress has been awarded to an atheist. Britain's Astronomer Royal Martin Rees received the 2011 prize for his idea of a multiverse. This concept, combined with the emergence of the cosmos, and the size of physical reality has the potential to reshape the philosophical and theological considerations that strike at the core of life. Rees says he is an atheist, profoundly spiritual, and has no religious beliefs.

Even though connection with the whole is a spiritual concept, it contains pieces from disciplines that include human development, psychology, and emotional aspects of life, among others. The view that spirituality and nature are connected is evident in the writing of Scandinavian American, Sigurd F. Olson, an environmental educator whose view of wilderness is antithetical to the historically typical US masculine view. His writings echo the essence of friluftsliv in his recognition of wilderness not only as a place to recreate but as a spiritual space. In the wake of World War II, Sigurd F. Olson said that wilderness is 'a spiritual backlog in the high speed mechanical world in which we live' (Olson, 1946). For Olson, the primary draw of the wilderness was perspective; he writes, 'They go to the wilderness for the good of their souls. These people know that wilderness to them is a necessity if they are to keep their balance' (Olson, 1946, pp. 62–63). People experience the outdoors in different ways; however, in the literature and as reported by outdoor leaders anecdotally, there is a prevalent thread that people feel at home and at one with the world or universe while in the outdoors. People feel spiritual and spiritually alive. The soul is nourished in nature.

Both the whys and hows of spirituality and consciousness continue to be mysteries. However, a number of references can be found that indicate that the natural world may act as a bridge between the physical world and the spiritual world. Doctrine in many religious areas such as process theology and feminist theology as well as non-traditional spiritualities and the spiritualities of indigenous communities understand the ecological interdependence and the value of biodiversity relating to nature as sacred or divine. It makes sense that the natural world helps humans develop consciousness and spiritual awakening. The forms, sounds, and smells of nature provide for many a window to understand the spirit. Tolle (2005) speculates that flowers may have been one of the first non-utilitarian things humans were attracted to. Flowers, as well as gems, crystals, some birds, and precious stones, have an ethereal quality that extends beyond form, thus aiding in transforming human behavior and conscious development.

In the gospel of Mary, discovered around 1945 in Egypt, it is written 'Be of good courage and if you are discouraged be encouraged in the presence of the different forms of nature' (Lee, 2009). This intrinsic value of natural spaces for spiritual development and attentiveness helps people be able to implement some of the great wisdom teachings that ask humans to rise above identification with form and be able to be aware of the formless. Being in nature can help people see the divine life force and recognize it as their own essence. Seeing beauty in flowers can awaken in humans their inner beauty. The evolution of consciousness is connected to nature and evidenced in the widening embrace of the land ethic, friluftsliv, ecopsychology, and, at least in Canada and Australia, the socio-ecological approach to human health.

Ecopsychology, ecotherapy, nature therapy, and terrapsychology

Ecopsychology is a blending of environmental philosophy, ecology, and psychology that has evolved into a growing body of knowledge that explores how our psychological health is related to the ecological health of planet Earth. It is believed that the mind can be comforted and healed through time in natural environments. Therefore the destruction of the natural environment negatively affects the physical and mental health of humans. The paradigm of ecopsychology supports the healing that occurs for people when they are in nature, including through the work of outdoor professionals in environmental education, organized camping, adventure education, wilderness and adventure therapy, and outdoor behavioral healthcare. Some people call the actual applied practice of ecopsychology ecotherapy or nature therapy, while others look at these as three names for the same practice, or as three different practices. Demonstrating the variability, Chalquist (2009) used ecotherapy as an umbrella that includes horticulture therapy (discussed in Chapter 3), wilderness excursion work (discussed in Chapter 9), time stress management, and certain kinds of animal-assisted therapy—more inclusive than ecopsychology typically is.

These therapies connect the health, especially the mental and emotional health, of humans with the health of the environment, and connect ill health to the destruction and dominance over the environment. Similarly, terrapsychology is a multidisciplinary set of approaches for investigating the deep connections between humans and the earth (Chalquist, 2007). These connections, mostly unconscious, are about connecting with the soul of a place. Terrapsychology helps make these connections known and seeks to begin cycles of mutual healing through helping humans understand the earth's story. Similar to gaining a sense of place, as humans know the earth better they create a bond and a sense of care for the earth and that care extends back to the individual. Ecopsychologists and terrapsychologists help people recognize, value, and act on a healthy connection between planetary health and human health.

In its early conceptions ecopsychology was informed by ecofeminism in that the domination of primarily Western people over nature caused humans to separate their identity from the rest of the natural world (this history is traced in Chapter 2). Both work to unravel the history of a mutualistic relationship with nature having become one of fear and mistrust of humans toward nature. Both agree that societal ills result from a culture based on scarcity and capitalism, which causes power-over relationships with others. Both seek to help people discover healthy alternatives to materialism and consumerism and to develop an ecological self (see Chapter 4 for a discussion about ecological identity).

Ecopsychology now specifically focuses on the interconnectedness of our psyches and the natural world and the ways in which our disconnection has resulted in destructive behavior toward ecosystems, spotlighting the environmental crisis as a psychological crisis. The premise is if the bonds between humans and the environment are recognized and felt by people, then people can psychologically heal individually, embed their psyches back into the natural world, and eventually the illness entrenched in Western culture can be healed. Individual healing often occurs in small groups through therapy using experiential activities including earth-based spiritual practices.

Robert Greenway (1999) coined the term ecopsychology in his 1963 essay. The modern wilderness immersion lineage was pioneered by Greenway at Sonoma State University in the early 1970s; his way with wilderness was an ecopsychological exploration, entering wilderness with a small community and leaving behind the distractions of the frontcountry in order to deepen relationship with place (Greenway, 1996). Greenway kept data for over 30 years as he took students on outdoor trips as part of his university classes. He found that 90% of the students described an increased sense of aliveness, well-being, and energy coming from their time in nature, and that 90% said the experience had allowed them to break an addiction (defined broadly) such as nicotine or chocolate (Greenway, 1996).

Ecopsychology and ecofeminism have long histories, as described in women's outdoor travel literature for over 100 years, and have been largely practiced by women's outdoor programs in the US and in other countries. The dominant culture's attitude toward nature as something to be used or conquered was destructive and belittling to both women and the land. In her introduction to *Wilderness Therapy for Women*, Cole noted 'the process by which individuals are encouraged to spend healing time in nature and as nature heals the individual, the individual also nurtures and heals the planet' (Cole *et al.*, 1994, p. 3). Women have found nature to be healing, were prone to find a sense of place, and felt spiritually connected to the land. As a result, women tended to have a mutual healing relationship with nature, notably exemplified by Julia Butterfly Hill who spent over 700 days living in a California redwood tree to prevent loggers from the Pacific Lumber Company from cutting it down. Hill later was inducted into the Ecology Hall of Fame for those efforts and her continuing work to heal the rift between humans and nature (Weiss, 2003). Susan Griffin (1978) in *Woman and Nature* wrote about the connections women have with the earth and how these connections have been eroded over the years. She says that our reconnection to the earth will heal and increase our capacity to love ourselves.

Edith Cobb's (1895–1977) work is an example of early research supporting ecopsychology and demonstrating this connection between healthy human development and nature. Trained in social work in the 1950s with an interest in the natural world, child development, and adult psychology, Cobb undertook a massive research project wherein she collected and analyzed more than 250 autobiographies. In *The Ecology of Imagination in Childhood* (Cobb and Mead, 1977), she establishes the importance of children's deep experience of the natural world to their adult cognition and psychological well-being.

By the 1990s, the concept of ecopsychology and our need to heal and be healed by the planet had become more widespread. Today ecopsychology has developed into a discipline with college textbooks, such as Deborah Du Nann Winter's (1996) *Ecological*

Psychology: Healing the Split between Planet and Self, and degrees offered at universities such as Naropa University in Colorado and Prescott College in Arizona. From Naropa University's web page, ecopsychology is based on the belief that 'human health, identity and sanity are intimately linked to the health of the earth and must include sustainable and mutually enhancing relationships between humans and the nonhuman world' (Naropa University, 2009).

Ecopsychology focuses on undoing the consequences of the extinguishment of humans' shared and reciprocal relationship with nature, and works to heal the wounds in humans and nature that have been caused by domination and over-indulgence or, as Pirages and Ehrlich (1974) termed it, the dominant social paradigm.

Conservation and the development of a land ethic

Another important application of our increased need and desire to connect to nature is found in various efforts to care for and conserve land, and in the development of a land ethic. Published posthumously, *Sand County Almanac* (1949) contains Leopold's now famous land ethic: 'A thing is right when it tends to preserve the integrity, stability, and beauty of the biotic community. It is wrong when it tends otherwise'. The land ethic helps take land from an economic realm where humans reaped goods but did not have obligation toward it for nurturance and protection to a place where humans ought to care for the land because they are part of the natural community. Friluftsliv and TEK, both originating before Leopold's writings, reflect this land ethic, lending support to the strong connections among these concepts.

But even beyond the benefits recognized by Leopold and imbued by the land itself, the act of conservation may also have health implications. For instance, Farmer *et al.* (2011) report that one reason why individuals adopt conservation easements is for a health-related purpose of engaging in restoration activities, such as tree planting, soil stabilization, and trail building. Similarly, Frumkin and Louv (2007) suggest that the benefits of experiencing

nature are so significant that practicing effective land conservation techniques and behaviors can be considered a strategy for improving public health. Perhaps Partridge (1993) said it most succinctly in his statement:

> But why should the systemic 'health' of the ecosystem be of interest to human beings? Because, of course, our personal health is inextricably tied in with the health of the ecosystem. We'd better take care of the ecosystem if we know what's good for us. (p. 6)

Thus, from a biocentric perspective the land ethic becomes entwined with the concept of human health. That is, caring for the land is valued because of the intrinsic values and opportunities it provides to health. Borrowing from authors such as Heft (2010) and Thompson and Aspinall (2011), healthy landscapes provide 'affordances' to humans by allowing for human development and health-based actions. Whether the landscape provides clean air or water, places for recreation, or simply a beautiful vista for reflection, healthy ecosystems afford humans a variety of health-promoting opportunities.

While not as overtly connected, the sustainability movement is also arguably linked to human health (Norton, 1992). In this case, sustainability is defined as development that meets the needs of the present without compromising the ability of future generations to meet their own needs. Whether ecosystem and, ultimately, human sustainability is based on the production of goods such as timber, water, and wildlife, or services such as water and air purification, or the recycling of nutrients, caring for the land is inextricably bound to variables associated with human health.

Medicine from nature and conservation medicine

Conservation medicine, also called ecological medicine and environmental medicine, links human and animal health with the environment and considers the health consequences of people's interactions with their environments. Environmental health problems are complex and their solutions involve a systems approach to health. Often teams include physicians, veterinarians, toxicologists, epidemiologists, microbiologists, and even political scientists. Carolyn Raffensperger uses the term ecological medicine (Raffensperger and Tickner, 1999), which acknowledges the role of nature in health and healing, while emphasizing prevention over cure.

Our bodies, like all objects we see as solids, are more space than matter. This leads some to say that a body is not a structure; it is a process. Furthermore, human bodies, the ecosystem, and the universe are all *one process* (Chopra, 2001). Most of us would agree that we are part of the ecosystem or, as some of us learned in school 40 years ago, part of the web of life. No process or being is separate from any other process. With good intent, Western medicine has taken a mechanistic mindset and tried to break down life functions into discrete structures with their own processes, then finding the process that appears to be broken, and fixing it. While this has seemingly worked in many cases, now we find that we actually spend more on healthcare each year, our life expectancy has not risen appreciably for 30 or more years, and younger generations may even have a shorter lifespan than baby boomers. The belief system behind that healthcare model does not account for the magnitude of interrelationship or connection of all life, including what we tend to see as inanimate objects.

The concept that the universe is a process and that humans are part of that process implies, as shaman medical systems such as Tibetan, Ayurvedic, Native American, and Traditional Chinese say, that we must stay in balance or return to balance with the other processes, including the natural world. Therefore instead of relying on a belief that the allopathic medical system can provide a pill to *cure* lifestyle-related diseases, including obesity, diabetes, and heart disease, individual responsibility ought to be encouraged to maintain a healthy life, including spending time in natural areas and being in harmony with nature. This can be explained in terms of process and connection to the whole and a need to stay in balance or in harmony.

In the past scientific uncertainty has been used to allow potentially harmful, but not yet proven harmful drugs, chemicals, and extraction practices to be used and implemented. However, many people have called for a precautionary approach, including Rachel Carson, other ecofeminists, and pediatricians, among others. In 1998 at a conference that included healthcare workers, environmentalists, ecofeminists, lawyers, and ethicists the Wingspread Statement on the Precautionary Principle (Science & Environmental Health Network, 1998) was drafted. The precautionary principle shifts the responsibility of proving the absence of harm from drugs, chemicals, and environmental changes to *before* they are enacted or utilized. The goal is to prevent harm to health and the environment before it occurs, rather than clean up the environment or deal with health concerns after an action with negative health impacts is approved. This upstream intervention or prevention approach discussed by Pryor (2009) is described as a measure to promote the health of whole populations and has greater long-term benefits than treating targeted or vulnerable populations after the fact. This upstream action can be inaction, e.g. not using the chemical, or it can include action, e.g. having barges drain their ballast water in the Atlantic Ocean before entering the Great Lakes area in order to prevent transporting disruptive species. Raffensperger and Tickner (1999) discussed the precautionary principle, stating that the threat of harm ought to be enough to warrant precautionary principles because establishing cause and effect can be time consuming and sometimes not possible with scientific limitations.

Citizen science[*]

Citizen science holds potential to positively affect individuals, their communities, and greater societal goals, while accomplishing scientific work and getting people outside in natural areas. Citizen science enlists members of the public to make and record useful observations, such as tracking the blooming dates of wildflowers in spring or counting aquatic

---

[*] This section was written with extensive input from Janet Ady.

invertebrates in local streams. Large numbers of volunteers can collect valuable research data, which are then aggregated to create an enormous body of scientific data on a vast geographic scale. In return, such projects increase participants' connections to science, place, and nature, while supporting science literacy and environmental stewardship (Dickinson & Bonney, 2012; Dickinson *et al.*, 2012). While performing this public service people are gaining the physical, psychological, spiritual, cognitive, and social health benefits of being in nature. In Chapter 7 and elsewhere in this book the specific benefits are named. Because people are learning about nature, developing a sense of place, becoming more attached to place, and providing a service, which helps develop an ethic of care, these volunteers are engaged in community development in a holistic manner that includes the environment. It will be easier for them to now see themselves as part of the environment and understand the systems approach to thinking about humans' connections to all else in the cosmos. The Muskegon River Watershed Assembly in Michigan uses citizen scientists for water quality and aquatic invasive species monitoring. Recently its board has sponsored kayak trips down the Muskegon River and its tributaries called 'Voyage of Discovery' in order to combine fun, learning, and opportunities to attach to place along with scientific monitoring.

Citizen science is sometimes described as 'public participation in scientific research (PPSR)' (Bonney *et al.*, 2009; Hand, 2010). Formally, citizen science has been defined as 'the systematic collection and analysis of data; development of technology; testing of natural phenomena; and the dissemination of these activities by researchers on a primarily avocational basis' (OpenScientist, 2011).

Citizen science is a fairly new term, yet it has been practiced for centuries. In fact, prior to the 20th century, amateur, self-funded researchers (among them notables such as Sir Isaac Newton, Benjamin Franklin, and Charles Darwin) actively participated in scientific research. Before that indigenous people and women healers of the 10th to 14th centuries conducted and applied their observations of the natural world. The relatively recent professionalization of science shifted the citizen role so that now researchers employed by universities and government research laboratories dominate ecological and other scientific research. Within this context, citizen science has evolved over the past two decades. Recent projects place more emphasis on scientifically sound practices and measurable goals for public education (Silvertown, 2009). Modern citizen science provides increasing access for broader public participation and technology is credited as one of the main drivers of the recent explosion of citizen science activity. Citizens can conveniently send in data through computers and phones. More and more, biologists are using lesser known datasets collected by citizen scientists to understand long-term changes in the environment and their causes and consequences, particularly as they study climate change and the timing of the biological events in plants and animals such as flowering, leafing, hibernation, reproduction, and migration (known as phenology) (Miller-Rushing *et al.*, 2012).

Educators and others interested in connecting young people with nature see that citizen science projects can serve as a model for youth engagement in scientific research goals. Projects may be designed for a formal classroom environment or an informal education environment such as museums, nature centers, parks, and wildlife refuges. In combining research with public education, citizen science adds to broader societal benefits by engaging members of the public in authentic research experiences at various stages in the scientific process and using modern communications tools to recruit and retain participants (Bauer *et al.*, 2000; Bonney and LaBranche, 2004; Brossard *et al.*, 2005).

Future success for the emerging field of citizen science in conservation depends on adopting a spirit of inclusion by embracing both centralized national programs and decentralized, local efforts, as well as managers and staff fostering a cooperative and supportive environment for all programs, practitioners, and participants and recognizing the value of each to the advancement of the field (Newman *et al.*, 2012). It is important to clearly define the desired balance between learning goals and scientific goals.

Scientists and education specialists working together can create and implement successful citizen science projects that use datasets gathered by citizen scientists to better understand the impact of environmental change. Training and practice in planning, designing, and implementing comprehensive education, citizen conservation science, and service-learning stewardship can encourage biologists and land managers to 'include young community members into their work, creating a lasting effect not only on the landscape, but also on the health and well being of the people who will live on and interact with that landscape for the generations to come' (Cramer, 2008).

Conclusion

Regardless of the level at which we are able to see it or understand it, as humans we are always dependent on the natural environment. Culture and worldview tend to dictate whether this interdependence is primarily implicit or explicit. Many of the adaptations and applications discussed in this chapter have positive implications for health and well-being.

As discussed in earlier chapters, our relationship with the natural environment has a long lineage of both connection and perceived disconnection. Cultural understanding of the natural environment, of course, varies, with some cultures valuing it more than or differently from others. In recent years, some elements of modern Western cultures have experienced a back-to-nature movement, in a sense uncovering and redefining the importance of the natural world. Chapters 7, 8, and 9 go into more detail about the empirical research about benefits, the role of education, and the use of adventure, respectively, in humans' relationship with the natural world.

References

Atran, S. (1998) Folk biology and the anthropology of science: cognitive universals and cultural particulars. *Behavioral and Brain Sciences* 21(4), 547–569.

Atran, S. (1999) Itzaj Maya folkbiological taxonomy: cognitive universals and cultural particulars. In: Medin, D.L. and Atran, S. (eds) *Folkbiology*. MIT Press, Cambridge, Massachusetts, pp. 119–213.

Bauer, M.W., Petkova, K. and Boyadjieva, P. (2000) Public knowledge of and attitudes to science: alternative measures that may end the 'science war'. *Science, Technology & Human Values* 25, 30–51.

Blonna, R. (2011) *Coping with Stress in a Changing World*, 5th edn. McGraw-Hill, Boston, Massachusetts.

Bonney, R. and LaBranche, M. (2004) Citizen science: involving the public in research. *ASTC Dimensions* May/June issue, 13.

Bonney, R., Ballard, H., Jordan, R., McCallie, E., Phillips, T., et al. (2009) *Public Participation in Scientific Research: Defining the Field and Assessing its Potential for Informal Science Education. A CAISE Inquiry Group Report*. Center for Advancement of Informal Science Education (CAISE), Washington, DC.

Brossard, D., Lewenstein, B. and Bonney, R. (2005) Scientific knowledge and attitude change: the impact of a citizen science project. *International Journal of Science Education* 27(9), 1099–1121.

Chalquist, C. (2007) *Terrapsychology: Re-engaging the Soul of Place*. Spring Journal Books, New Orleans, Louisiana.

Chalquist, C. (2009) A look at the ecotherapy research evidence. *Ecopsychology* 1(2), 64–74.

Chopra, D. (2001) *How to Know God: The Soul's Journey into the Mystery of Mysteries*. Random House, New York.

Cobb, E. and Mead, M. (1977) *The Ecology of Imagination in Childhood*. Columbia University Press, New York.

Cole, E., Erdman, E. and Rothblum, E.D. (eds) (1994) *Wilderness Therapy for Women: The Power of Adventure*. Harrington Press, New York.

Coley, J., Solomon, G. and Shafto, P. (2002) The development of folkbiology: a cognitive science perspective on children's understanding of the biological world. In: Kahn P.H. Jr and Kellert, S.R. (eds) *Children and Nature: Psychological, Sociocultural, and Evolutionary Investigations*. MIT Press, Cambridge, Massachusetts, pp. 65–89.

Collins, S.D. (1974) *A Different Heaven and Earth*. Judson Press, Valley Forge, Pennsylvania.

Cramer, J.R. (2008) Reviving the connection between children and nature through service-learning restoration partnerships. *Native Plants Journal* 9(3), 278–286.

d'Eaubonne, F. (1974) Feminism or death. In: Marks, E. and de Courtivron, I. (eds) *New French Feminisms: An Anthology*. University of Massachusetts Press, Amherst, Massachusetts, pp. 64–67.

d'Eaubonne, F. (1974) *Le féminisme ou la mort (Feminism or death)*, vol. 2. P. Horay, Paris.

d'Eaubonne, F. (1978) *Écologie-Féminisme: Révolution ou Mutation? (Ecology-Feminism: Revolution or Mutation?)*. Éditions ATP, Paris.

Dickinson, J.L. and Bonney, R. (2012) Introduction: why citizen science? In: Dickinson, J.L. and Bonney, R. (eds) *Citizen Science: Public Participation in Environmental Research*. Cornell University, Ithaca, New York, pp. 1–14.

Dickinson, J.L., Shirk, J., Bonter, D., Bonney, R., Crain, R.L., *et al.* (2012) The current state of citizen science as a tool for ecological research and public engagement. *Frontiers in Ecology & the Environment* 10(6), 291–297.

Dustin, D., Bricker, K. and Schwab, K. (2010) People and nature: toward an ecological model of health promotion. *Leisure Sciences* 32(1), 3–14.

Ecklund, E.H. and Long, E. (2011) Scientists and spirituality. *Sociology of Religion* 72(3), 253–274.

Faarlund, N. (2010) Challenging modernity by way of the Norwegian friluftsliv tradition. *Norwegian Journal of Friluftsliv*.

Farmer, J., Knapp, D., Meretsky, V.J., Chancellor, C. and Fischer, B.C. (2011) Motivations influencing the adoption of conservation easements. *Conservation Biology* 25(4), 827–834.

Feinstein, B.C. (2004) Learning and transformation in the context of Hawaiian traditional ecological knowledge. *Adult Education Quarterly* 54(2), 105–120.

Frumkin, H. and Louv, R. (2007) *The Powerful Link between Conserving Land and Preserving Health*. Land Trust Alliance, Washington, DC.

Gelter, H. (2000) Friluftsliv: the Scandinavian philosophy of outdoor life. *Canadian Journal of Environmental Education* 5(1), 77–92.

Greenway, R. (1996) Wilderness experience and ecopsychology. *International Journal of Wilderness* 2(1), 26–30.

Greenway, R. (1999) Ecopsychology: a personal history. *Gatherings: Journal of the International Community for Ecopsychology* 1(1). Available at: http://www.ecopsychology.org/journal/gatherings/personal.htm (accessed 15 September 2013).

Griffin, S. (1978) *Woman and Nature: The Roaring Inside Her*. Harper & Row, New York.

Hand, E. (2010) Citizen science: people power. *Nature* 466(7307), 685–687.

Heft, H. (2010) Affordances and the perception of landscape: an inquiry into environmental perception and aesthetics. In: Ward Thompson, C., Aspinall, P. and Bell, S. (eds) *Innovative Approaches to Researching Landscape and Health. Open Space: People Space 2*. Routledge, Abingdon, UK, pp. 9–32.

Herron, S. and Byers, J. (2011) TEK combined with laboratory research on wild rice is driving the restoration of the wild rice culture and ecosystems. Presented at *96th Annual Meeting: The Ecological Society of America*, Austin, Texas, 7–12 August 2011.

Herron, S., Byers, J. and Mitten, L. (2010) Reflections on a multi-year study of stratification needs and germination rates in wild rice. Presented at *95th Annual Meeting: The Ecological Society of America*, Pittsburgh, Pennsylvania, 1–6 August 2010.

Kelly, P. (1988) Linking arms, dear sisters, brings hope! In: Plant, J. (ed.) *Healing the Wounds*. New Society Publishers, Philadelphia, Pennsylvania, pp. ix–xi.

Kuo, F.E., Bacaicoa, M. and Sullivan, W.C. (1998) Transforming inner-city neighborhoods: trees, sense of safety, and preference. *Environment and Behavior* 30(1), 28–59.

Lee, J. (2009) The secret place of Thunder. *The Nag Hammadi Library*. Available at: http://www.youtube.com/watch?v=39U_zBOVbF0 (accessed 31 July 2009).

Leopold, A. (1949) *A Sand County Almanac*. Oxford University Press, London.

Maller, C., Townsend, M., Pryor, A., Brown, P. and St. Leger, L. (2006) Healthy nature healthy people: 'contact with nature' as an upstream health promotion intervention for populations. *Health Promotion International* 21(1), 45–54.

McGregor, D. (2004) Indigenous knowledge, environment, and our future. *The American Indian Quarterly* 28(3&4) 385–410.

McLeroy, K.R., Bibeau, D., Steckler, A. and Glanz, K. (1988) An ecological perspective on health promotion programs. *Health Education Quarterly* 15(4), 351–377.

Medin, D.L. and Atran, S. (eds) (1999) *Folkbiology*. MIT Press, Cambridge, Massachusetts.

Merchant, C. (1980) *The Death of Nature: Women, Ecology, and the Scientific Revolution*. Harper & Row, San Francisco, California.

Miller-Rushing, A., Primack, R. and Bonney, R. (2012) The history of public participation in ecological research. *Frontiers in Ecology & the Environment* 10(6), 285–290.

Minnesota Department of Natural Resources (2008) Natural Wild Rice in Minnesota. A Wild Rice Study document submitted to the Minnesota Legislature by the Minnesota Department of Natural Resources, February 15, 2008. Available at: http://files.dnr.state.mn.us/fish_wildlife/wildlife/shallowlakes/natural-wild-rice-in-minnesota.pdf (accessed 15 September 2013).

Mitten, D. (2012) Systems theory and thinking [course handout]. Master of Arts Program, Prescott College, Prescott, Arizona.

Mitten, L. and Herron, S.M. (2010) Elucidating the life cycle of *Claviceps zizaniae*, a fungal smut pathogen of wild rice. Presented at *95th Annual Meeting: The Ecological Society of America*, Pittsburgh, Pennsylvania, 1–6 August 2010.

Naropa University (2009) What is ecopsychology? Available at: http://www.naropa.edu/academics/gsp/grad/ecopsychology-ma/what-is-ecopsychology.php (accessed 6 April 2013).

Nemeroff, C. and Rozin, P. (2000) 'You are what you eat': applying the demand-free 'impressions' technique to an unacknowledged belief. In: Rosengren, K.S., Johnson, C.N. and Harris, P.L. (eds) *Imagining the Impossible: Magical, Scientific and Religious Thinking in Children*. Cambridge University Press, New York, pp. 1–33.

Newman, G., Wiggins, A., Crall, A., Graham, E., Newman, S., *et al.* (2012) The future of citizen science: emerging technologies and shifting paradigms. *Frontiers in Ecology & the Environment* 10(6), 298–304.

Norton, B. (1992) Sustainability, human welfare, and ecosystem health. *Environmental Values* 1(2), 97–111.

Olson, S. (1946) We need wilderness. *National Parks Magazine* January–March issue. Available at: http://www4.uwm.edu/letsci/research/sigurd_olson/articles/1940s/1946-01-00--We%20Need%20Wilderness--National%20Parks%20Mag.htm (accessed 15 September 2013).

OpenScientist (2011) Finalizing a definition of 'citizen science' and 'citizen scientists'. Available at: http://www.openscientist.org/2011/09/finalizing-definition-of-citizen.html (accessed 20 December 2012).

Partridge, E. (1993) The philosophical foundations of Aldo Leopold's 'Land Ethic'. Available at: http://gadfly.igc.org/papers/leopold.htm (accessed 3 December 2012).

Pirages, D. and Ehrlich, P. (1974) *Ark II: Social Response to Environmental Imperatives*. Viking Press, New York.

Plumwood, V. (1993) *Feminism and the Mastery of Nature*. Routledge, New York.

Pryor, A. (2009) Does adventure therapy have wings? In: Mitten, D. and Itin, C.M. (eds) *Connecting with the Essence of Adventure Therapy: Proceedings from the 4th International Adventure Therapy Conference (2006)*. Association for Experiential Education, Boulder, Colorado, pp. 46–71.

Public Health Agency of Canada (2008) Ottawa charter for health promotion. Available at: http://www.phac-aspc.gc.ca/ph-sp/docs/charter-chartre/index-eng.php (accessed 2 March 2009).

Raffensperger, C. and Tickner, J.A. (eds) (1999) *Protecting Public Health and the Environment: Implementing the Precautionary Principle*. Island Press, Washington, DC.

Rhode, C.L.E. and Kendle, A.D. (1997) Nature for people. In: Kendle, A.D. and Forbes, S. (eds) *Urban Nature Conservation – Landscape Management in the Urban Countryside*. E & FN Spon, London, pp. 319–335.

Robinson, P., Herron, S., Power, R. and Zak, D. (2009) A regional multicultural approach to sustaining wild rice. *Journal of Extension* 47(6), 1–5.

Science & Environmental Health Network (1998) Wingspread Statement on the Precautionary Principle. Available at: http://www.sehn.org/wing.html (accessed 15 April 2013).

Shiva, V. (1994) *Close to Home: Women Reconnect Ecology, Health, and Development Worldwide*. New Society, Philadelphia, Pennsylvania.

Silvertown, J. (2009) A new dawn for citizen science. *Trends in Ecology & Evolution* 24(9), 467–471.

Simpson, L.R. (2004) Anticolonial strategies for the recovery and maintenance of Indigenous knowledge. *The American Indian Quarterly* 28(3), 373–384.

Taylor, S.M. (2009) *Green Sisters: A Spiritual Ecology*. Harvard University Press, Cambridge, Massachusetts.

Thich Nhat Hanh (2009) *Happiness: Essential Mindfulness Practices*. Parallax Press, Berkley, California.

Thompson, C.W. and Aspinall, P.A. (2011) Natural environments and their impact on activity, health, and quality of life. *Applied Psychology: Health and Well-Being* 3(3), 230–260.

Tolle, E. (2005) *A New Earth: Awakening to Your Life's Purpose*. Penguin Books, New York.

UNESCO (n.d.) Local and Indigenous Knowledge System. Available at: http://www.unesco.org/new/en/natural-sciences/priority-areas/sids/enabling-environments/local-and-indigenous-knowledge-systems/ (accessed 29 January 2013).

Vaagboe, O. (1993) De forskjelige naturbrukeres verdipreferanser (The values and preferences of Norwegians who spend their time off in nature). In: FRIFO (ed.) *Frisk i Friluft* (*Fresh in the Open Air*). FRIFO, Oslo, pp. 29–38.

Weiss, D. (2003) Ecology hall of fame: Julia Butterfly Hill. Available at: http://ecotopia.org/ecology-hall-of-fame/julia-butterfly-hill/ (accessed 15 September 2013).

Wenzel, G.W. (1999) Traditional ecological knowledge and Inuit: reflections on TEK research and ethics. *Arctic* 52(2), 113–124.

Wenzel, G.W. (2004) From TEK to IQ: Inuit Qaujimajatuqangit and Inuit cultural ecology. *Arctic Anthropology* 41(2), 238–250.

Wiggins, M. Jr (2012) *Kakagon and Bad River Sloughs Recognized as a Wetland of International Importance.* The Bad River Band of Lake Superior Chippewa, Odanah, Wisconsin.

Winter, D. (1996) *Ecological Psychology: Healing the Split between Planet and Self.* HarperCollins College Publishers, New York.

7

Outcomes and Benefits

*Human-centredness is a complex syndrome
which includes the hyperseraparation of
humans as a special species and the
reduction of non-humans to their usefulness
to humans, or instrumentalism. Many have
claimed that this is the only prudent, rational
or possible course. I argue contrary to this
that human-centredness is not in the interests
of either humans or non-humans, that it is
even dangerous and irrational.*

Val Plumwood in 'Nature in
the active voice' (2009)

A substantial body of literature is now
available that specifically looks at the effect
of natural environments on health. In this
chapter specific outcomes and benefits from
time in natural surroundings are reviewed.
The kind of research that has been done
about time in nature, the settings in which
the research was done, and the category of
benefits, such as physical, mental, emotional,
spiritual, or social, are discussed. Previous
authors (Louv, 2005; Mitten, 2010; Mitten *et al.*,
2013) have discussed research in over 30 fields
that has contributed to the current knowledge
and literature about the natural environment and human health. In this chapter the
research is identified and reviewed in a comprehensive manner.

For example, research supports social
benefits of natural areas near communities by
showing that time in natural spaces strengthens
neighborhood ties, reduces crime, stimulates
social interactions among children, strengthens family connections and decreases domestic violence, assists new immigrants in coping
with transition, and is cost effective for health
benefits (Kweon *et al.*, 1998; Herzog *et al.*,
1997; Armstrong, 2000; Booth *et al.*, 2000; Kuo
and Sullivan, 2001; Bixler *et al.*, 2002; Leyden,
2003; Milligan *et al.*, 2004; Seeland and
Ballesteros, 2004; Rishbeth and Finney, 2006).

Research Background

Although the growing body of research showing the positive impacts of nature is exciting,
the research quality must be measured and
the methods and protocols must continually be
examined to be sure the research is sound, reli-
able, and valid. At the same time there are many
valid ways to do research, which is a gathering
of observations, data, and other information in
order to inform or create knowledge. The term
research may come from a French term in the
1500s meaning to search carefully, or it may
mean to search again and again. Our point is
that one research study should not be thought
of as conclusive and should be used as a catalyst to provide direction for further study. In
our desire to embrace research that shows
health benefits from being in nature we may

© CAB International 2014. *Natural Environments and Human Health*
(A.W. Ewert, D.S. Mitten and J.R. Overholt)

want to believe the results of every study show-ing positive correlations and not dig as deeply into the authentication or replication of the research as is needed. Research papers need to be scrutinized for the question that is asked, if appropriate methodology is used, and for the logic of the conclusions. At the same time, drawing from research across fields will help educate researchers about the kind of method-ologies and methods used in research about the effects of time in nature, and help spark new ideas. Research by the medical community might look different from research in sociology or education; however, all of the projects may contribute to the creation of more knowledge.

Some of the research about time in nature is exploratory and may have used small populations; other studies are correla-tional, experimental, or quasi-experimental. Some use qualitative methods, some quanti-tative, and many use mixed methods. While research, especially empirical research, is useful in proving the efficacy of the healing power of nature, we should not become too wedded only to empirical research. We also need to identify research gaps and replicate research using a variety of methods with larger populations and different populations. In general the studies found often used col-lege students as the study participants, fol-lowed by physically active individuals such as runners, backpackers, or athletes. More focus needs to be given to individuals not in these groups, such as people of color, elders, preschool children, middle-aged adults, and people who are ill in some manner.

In recent years, people in many different disciplines have engaged in research about the human–nature connection. In addition to outdoor-related fields (natural resources, rec-reation and leisure, environmental studies, environmental education and interpretation, organized camping, adventure education, wil-derness therapy, etc.), other fields that were examined as part of Mitten's (2010) research included biology, business, cognitive science, developmental psychology, ecology, ecopsy-chology, education, environmental psychology, evolutionary psychology, health, landscape architecture, medicine, nursing, political sci-ence, psychiatry, psychology, public health, recreation, religion, social psychology, social work, sociology, therapy, tourism, and urban planning (see Fig. 7.1). This variety of fields is encouraging because researchers from differ-ent disciplines approach questions about time in nature from different viewpoints and use different research methods. This variety of methods and approaches strengthens the overall body of research regarding the import-ance of humans spending time in nature. Additionally, professionals in these fields can work together both on research and in practice. For example, the research in environmental education and public health might be used to inform housing or community design. By looking at these different fields we remind readers to find and consider research in fields beyond their own discipline.

Research about the health effects of spend-ing time in nature varies given the particular question asked and the field of the researcher. A combination of theoretical research (testing actual theories) and empirical research (quan-tifying direct or indirect observation of inter-actions, usually through experimentation) as well as conceptual research (looking at the underlying theories and concepts guiding research or practice) and practice-based and anecdotal evidence is being reported in the literature. For example, outdoor-related fields often report conceptual beliefs supported by anecdotal evidence. Robert Greenway conceptually believed that taking college stu-dents into the wilderness would have bene-fits; he observed benefits and heard students anecdotally reporting them. From this base, Greenway began his research regime of col-lecting data from students after their class wilderness trips. As mentioned in Chapter 3, his 30 years of data now supports the anec-dotal evidence about the benefits of students spending time in the wilderness (Greenway, 1996). Supported by this sort of research, ecopsychology has grown into a field with both practitioners and academicians con-tributing to research and application. While there is a need for evidence-based research, anecdotal evidence often is crucial in help-ing to form the basis for evolving research agendas.

Another way to capitalize on existing data or research is to encourage meta-analyses of the outcomes of many studies. The results

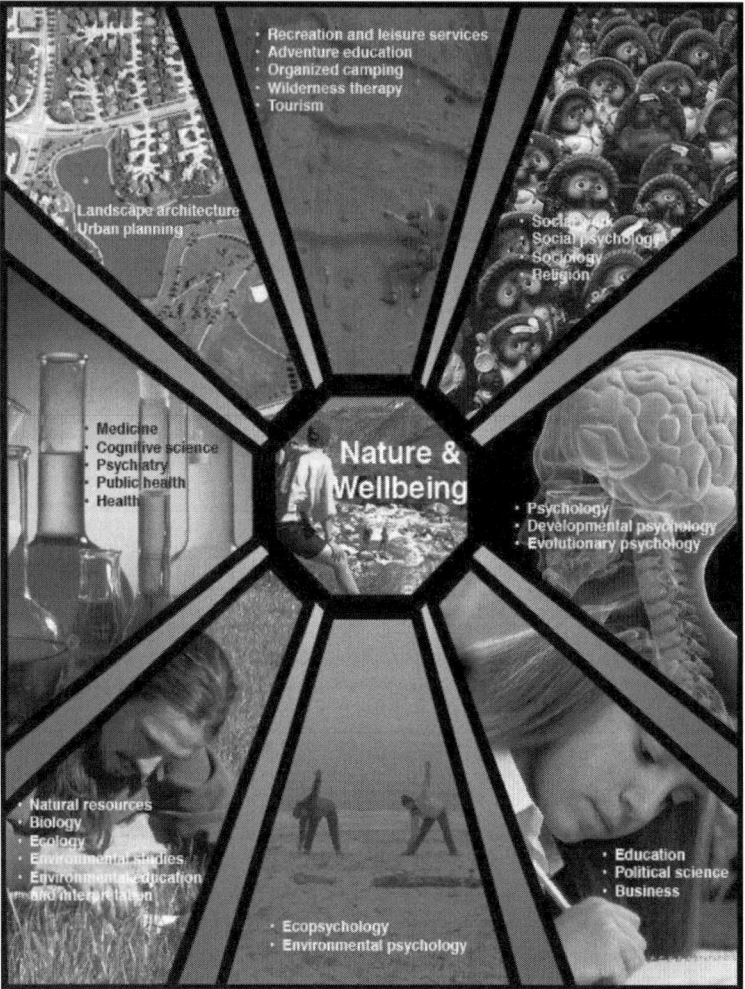

Fig. 7.1. Disciplines engaging in research about connections between human health, development and well-being, and nature. (Adapted from Mitten, 2010.)

of studies using similar or comparable hypotheses, variables, and methodologies can be combined in a statistical analysis in order to add reliability, to disprove the outcomes described, or to show that further research is needed. In wilderness therapy and adventure education research, Hans (2000) and Wilson and Lipsey (2000) used a meta-analysis to show the efficacy of these therapeutic outdoor modalities. In a meta-analysis of 24 articles addressing human health and nature, Bowler *et al.* (2010) found beneficial changes in energy, anxiety and anger reduction, as

well as improvements in fatigue and sadness upon exposure to natural environments.

Also strengthening theory and helping to identify research needs though combining studies are comprehensive reports and analyses of previous projects. Velarde *et al.* (2007) analyzed the health effects related to viewing a range of landscapes used in environmental psychology studies by using visual displays of the information. They reported on the evidence of health effects, the methods and measures used, and the groups studied. They found that broad categories of 'natural' or 'urban' mostly have been used in these studies, though a few studies used subcategories. Studies commonly looked at short-term recovery from stress or mental fatigue, faster physical recovery from illness, and long-term overall improvement in people's health and well-being.

In designing, evaluating, and applying research, keep in mind an underlying assumption in this book that our world is a system of which humans are a part, and that the interconnectedness and interdependence of all systems in the universe (and perhaps beyond) affects everything else. This cautions us not to become too mechanistic in our application of individual research projects and to look at the research in many fields from many projects as we use research for program design, community design, and the like. For example, wilderness therapy providers believe participants leave the programs 'better' somehow; however, providers acknowledge that they lack a clear idea of what specifically works to benefit the participants. In trying to gain legitimacy for wilderness therapy as a therapeutic intervention, practitioners and researchers have tried to isolate specifics about these endeavors to understand what makes them work, using mostly a reductionist and mechanistic mindset to determine efficacy. In the US the medical research model has in part driven this mechanistic mindset with the need to *prove* mechanistically how programs *work*. Using only mechanistic thinking and research to assess outcomes has pitfalls, as sometimes it is the holistic system that creates the healing or change (Mitten, 2004); for example, some combination of effects from the natural setting, the therapeutic relationship, and the therapeutic activities creates positive change and an increase in health. In these cases process research that examines systems may be an appropriate option.

Research on the health impacts of nature exposure has been conducted by many different fields and therefore in many different contexts. These include, but are not limited to, clinical settings, general home environments, sitting or walking in nature at parks or other natural areas, playing, gardening, specific nature activities, outdoor pursuits, and wilderness trips (see Fig. 7.2). Further, different populations (students, patients, children, etc.)

have been studied in these diverse settings and activities. Taken together these various research project results combine to add strength to the growing picture that spending time in and around natural environments benefits human health and well-being. Research has found that nature experiences can: contribute to physical health and psychosocial well-being (Ulrich *et al.*, 1991; Burdette and Whitaker, 2005; Norman *et al.*, 2010); foster a sense of community (Breunig *et al.*, 2010); increase spiritual awareness and well-being (Cosgriff *et al.*, 2010); and aid in recovery from mental fatigue and restore attention (Kaplan and Kaplan, 2005). For example, a study on the effects of 'greenness' on children's cognitive functioning found that proximity to, views of, and daily exposure to natural settings increases children's ability to focus and therefore enhances cognitive capabilities (Wells, 2000). Gardening benefits have been noted from prison projects as well as fewer medical requests from prisoners whose cell blocks have views of farmland rather than other buildings (Moore, 1981). Cimprich (1990) reported that breast cancer patients who engaged in outdoor restorative activities reported quality of life improvement and tended to initiate more projects than those who did not. Canin (1991) studied AIDS caregivers finding that those who passed time in quiet activities, especially nature activities that seemed particularly conducive to reflection, appeared to have less burnout and functioned more competently than those who passed time by watching television or by playing or watching organized sports.

Health and ways to increase health benefits in different settings have been researched with the majority of these studies typically occurring in parks, nature or wildlife preserves, wilderness areas, forests, or gardens (Bowler *et al.*, 2010). Built environments such as hospitals, nursing homes, playgrounds (both with and without greenspace), schools, and housing complexes (see Fig. 7.3) have also been researched. Much of what is known as 'green' building is done primarily to save energy and cut costs; however, the education sector also looks at the impact of the built environment on the health and well-being of its students. In 2012, Bernstein *et al.* (2013) estimated that 45% of the total construction starts

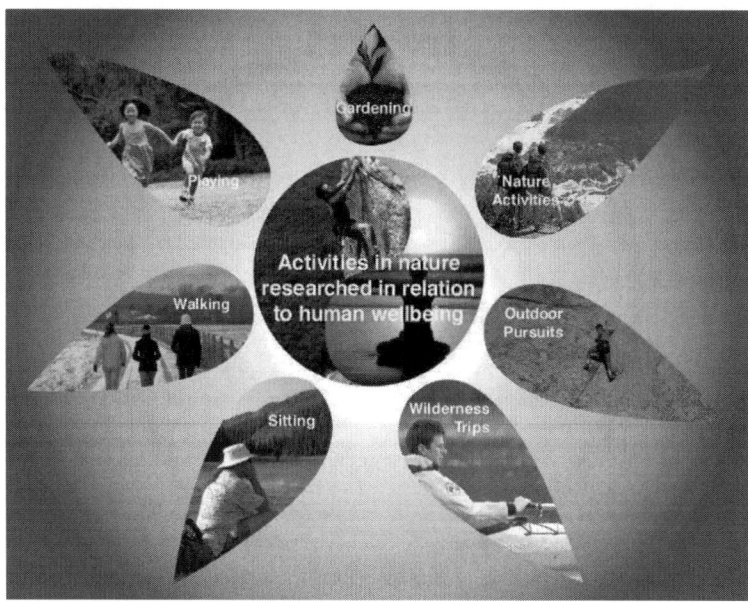

Fig. 7.2. Activities in nature researched in relation to human well-being. (Adapted from Mitten, 2010.)

in the educational sector were green, up from 15% in 2008. Both kindergarten to grade 12 schools (91%) and higher education (87%) report that green built buildings improve health and well-being as well as improve student productivity (74% and 63%, respectively). Integrating natural lighting, storm-water run-off strategies that integrate greenspaces with native landscaping to provide natural areas and less rapid runoff, and outdoor gardens and play spaces for educational use and the encouragement of physical activity are some of the design aspects of green schools. Some schools incorporate outdoor dining; many promote the natural environment and local natural resources. Tennessen and Cimprich (1995) looked at college students' capacity to direct attention with the variable being the view from their room. Using both objective and subjective measures the dormitory students with natural views scored higher on their attention measures.

Some hospitals are renewing their belief in the efficacy of nature in healing and are embracing therapeutic landscapes again (Therapeutic Landscapes Resource Center, Inc., 2000). As hospitals, nursing homes, primary care clinics, and rehabilitation facilities face high construction

and renovation costs, the economic benefits in terms of the healing impact of nature are being figured into the mix. The UK has spent at least $US 4 billion on new hospital construction since 2002, the Texas Medical Center in Houston about $US 1.8 billion. The US has averaged about $US 15 billion annually during the last decade for new hospitals (Ulrich, 2001). If incorporating gardens, such as the play-garden at the Rusk Institute of Rehabilitative Medicine in New York City, help children with brain injuries recover mobility faster than only exercising in a rehabilitation room, then costs are cut and healing enhanced. When burn patients are wheeled outside on paths designed to be wide enough for beds, and when the patients practice walking on different grades and experience different textures, they heal more quickly. The benefits are multiplied when their families can be part of their time outdoors, there is a relaxed setting for patient–visitor interaction away from the hospital interior, and staff are provided with needed retreat from the stress of clinical work. Both economic and satisfaction benefits are achieved.

Different population density settings have also been researched including urban, semi-urban, rural, and wilderness environments (see Fig. 7.3). In general, natural systems are fewer as population density increases. Therefore, natural systems typically need to be integrated into high population density areas in order to increase health benefits. For example, in the urban environment certain insects are helpful in creating a healthy human environment. Collaboration among social scientists, conservation biologists, community design practitioners, and the medical community might further understanding about how to integrate gardens and landscape into the built environment. MaryCarol Hunter, a landscape architect and ecologist, teamed with an entomologist to offer suggestions about how to design built projects that would encourage useful insects (Hunter and Hunter, 2008). They point out that cities, in turn, may offer viable environments for insects needed for sustaining life that otherwise might not be able to survive. Unfortunately the prevailing attitude about insects is mostly negative; however, maintaining biodiversity of insects is necessary for human health.

An often overlooked benefit of nature is the impact of landforms and how they influence human health and well-being. People in more urbanized and built communities are more vulnerable to negative impacts when natural landforms are altered. A number of recent disasters have provoked more research in the area of land forms and disaster prevention, warning, mitigation, and recovery. All natural disasters and hazards cannot be avoided, but appropriate use of nature and natural landforms can lessen human vulnerability. Two brief examples follow. Two hotels on a beach in Sri Lanka were in the path of the 2004 tsunami. One hotel had removed the natural sand barrier in order to have a better view and access to the water, and over 100 people died in this hotel. A second hotel was behind a natural sand barrier and lost only a

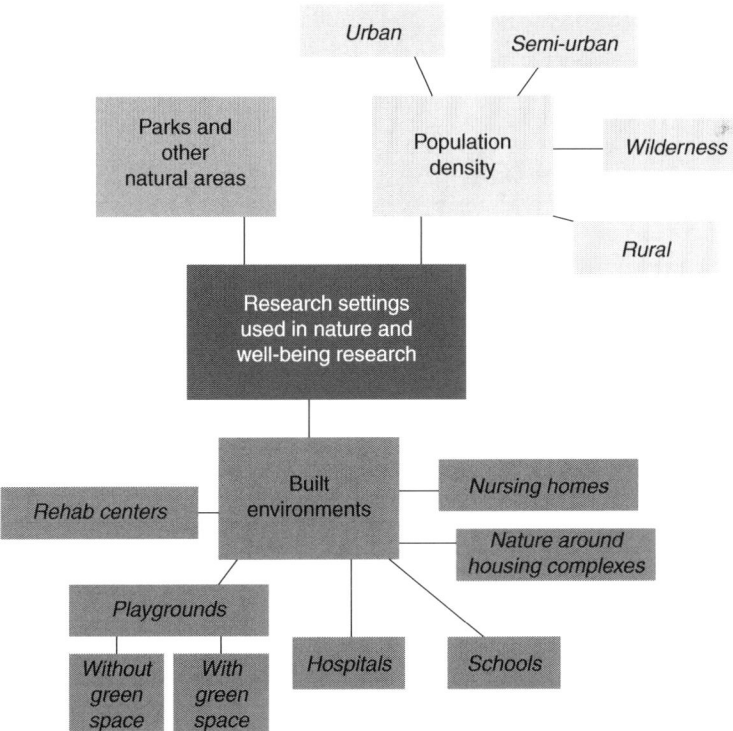

Fig. 7.3. Many different settings have been researched in relation to nature and human well-being. (Adapted from Mitten, 2010.)

few people. As a result of a 2004 hurricane that hit neighboring countries, Haiti and the Dominican Republic, massive amounts of water were released on the land at one time. In Haiti, deforested by mining, this excessive water caused massive landslides and over 6000 deaths. Neighboring Dominican Republic, still largely forested, benefited from the ability of the vegetation to stabilize the land and limit the flash flood problem, resulting in relatively few deaths during the aftermath of the storm (Ingram, 2009). Trees not only add in-soil stabilization during storms, they also help store water. Because of deforestation on Bali, water runs off into the ocean instead of being stored in large underground basins to be used year-round by residents. There, a water shortage is expected within 10 years. Allowing mangrove to remain in the saline coastal areas creates protection from storm damage and provides vital habitat for fish used in human consumption. Understanding how mutually dependent humans are with nature for our well-being

might lessen risks of drought, food shortages, and direct human injury.

Many different outcome measures have been used for research about nature and well-being. Data can be in the form of: (i) observable clinical signs or medical measures such as hormone levels, blood pressure, and intake of pain drugs; (ii) subjective measures such as reported satisfaction, mood, or pain level and developmental observations; (iii) estimated measures, which may include crime rate or domestic abuse rate changes in part due to natural areas, and preventive and general wellness benefits from being in nature; and (iv) economic measures such as rehabilitation or staff costs due to turnover (see Fig. 7.4). Of course the lines between these outcome measures are not exact. For example, a clinical outcome often leads to an economic outcome. As an overall example, Ulrich (2001) has shown that providing a hospital garden or plants indoors and outdoors improves clinical indicators, economic outcomes, and

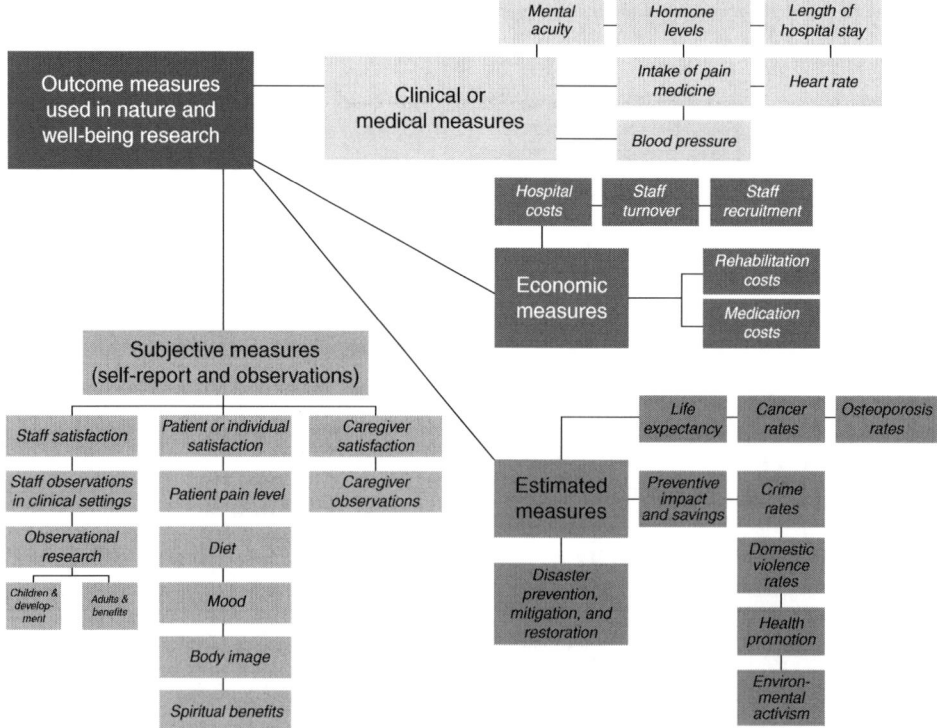

Fig. 7.4. Outcome measures used in nature and well-being research. (Adapted from Mitten, 2010.)

patient and staff satisfaction reports. In schools, absenteeism, student test scores, average yearly progress, student and staff satisfaction surveys, asthma incidence, nurse/health center visits, and student attentiveness are outcomes commonly measured and that commonly are positively affected by green building techniques that work in concert with the natural environment (Bernstein *et al.*, 2013).

Evidence from clinical research seems to be the most persuasive for physicians and other healthcare administrators as well as for federal funding requirements. Kreitzer *et al.* (2002) found that medical, nursing, and pharmacy faculty were most comfortable when presented with an evidence-based approach for understanding complementary and alternative medical treatments; this includes nature healing. Therefore, measuring health outcomes due to interaction with nature using clinical measures combined with other measures may be most promising for professional acceptance. Achieving health outcomes would be an indicator of healthcare quality or the efficacy of the treatment, which in this case would be interaction with the natural environment. Though clinical and economic outcome data are historically more influential in decision making, patient satisfaction measures along with staff and caregiver benefits are becoming better recognized as valid outcomes.

Research about the effect of nature on health and well-being usually comprises only a small part of the research that takes place in the variety of disciplines studied. Viewing the collective results, however, reveals fresh insights and significant findings within the combined body of research. For example, viewing the comprehensive body of knowledge about wilderness and adventure therapy and related fields and the research about nature and human health led to Mitten's (2004) proposal that intentional healthcare work by clinically trained outdoor leaders done in natural areas can provide such medically significant help that wilderness therapy and adventure therapy ought to be classified as a complementary modality by the US National Institutes of Health (NIH), National Center for Complementary and Alternative Medicine (NCCAM). Within their classifications, wilderness therapy and adventure

therapy most logically fits in the mind–body interventions, biologically based therapies, and the energy therapies domains.

Benefits from Exposure to Nature

This section explores the documented benefits from engagement in natural environments and categorizes these benefits into physiological, psychological, spiritual, and social domains. It should be recognized that all of the variables related to each of these areas play an important role in human health. All areas need more research but, to date, spirituality has received less research attention than the other variables.

Physiological benefits

Natural environments can promote human health in a multitude of ways. Perhaps most obviously, the natural environment provides a place for physical exercise or activity. This is particularly important because a number of authors suggest that, despite widespread attention toward increasing levels of physical activity, the overwhelming trend is toward a continuing decline in physical activity (Van Sluijs *et al.*, 2005; Beaton and Funk, 2008). For example, it is estimated that in Europe the decrease in energy expenditure for the average adult over the past 50 years is roughly equivalent to that of running a marathon each week (Pretty *et al.*, 2003). To a large extent, as a physical setting, natural environments provide an 'affordance' (Heft, 2010) or an opportunity to take action or practice a healthy activity (Ward Thompson, 2011). When combined with the substantial evidence that physical activity alone can promote positive health benefits (Bauman, 2004), the presence of natural environments provides both an affordance and behavioral attraction for activity and subsequent health benefits.

It has been suggested that physical exercise that takes place in natural environments may actually carry health-promoting benefits beyond that of similar types of exercise and energy expenditure in indoor environments (Pretty *et al.*, 2003, 2005). Coined 'green exercise' by Pretty and his colleagues, this idea capitalizes on existing restorative environment theories and proposes a synergistic effect where exercising in natural spaces is more beneficial than either exercise or time spent in natural environments alone.

A recent meta-analysis showed that green exercise improved both self-esteem and mood, with short engagements in both intensity and time yielding positive results (Barton and Pretty, 2010). While positive improvements were indicated for all environments and all subpopulations within the study, the greatest effects were linked to the presence of water, and for those who were diagnosed with a mental illness. Another meta-analysis, which looked at clinical trials involving subjects who exercised indoors on one occasion and outdoors on another, showed improvements in greater feelings of revitalization, positive engagement, decreases in tension, confusion, anger, and depression, and increased energy for exercise that occurred in natural environments (Thompson Coon et al., 2011). Additionally, and perhaps more importantly, subjects reported greater enjoyment and satisfaction and greater intent to repeat the activity in the future. While more research is needed in this area, existing research points to the natural environment as a place that not only promotes physical activity and facilitates enhanced wellness over traditional forms of exercise, but also that is enjoyable and therefore more likely to be adopted as part of one's lifestyle.

One specific example of the physical benefit of time spent outdoors is that sufficient time outdoors reduces the risk and diagnosis of myopia in children from across ethnic groups (McBrien et al., 2008). The critical factor seems to be total daylight hours outdoors, with both active and more passive outdoor activities being protective. Several studies undertaken in different cultures are in agreement that two to three hours per day outdoors seems to be sufficient to markedly lower the risk of myopia.

In addition, there is a growing body of knowledge that suggests that physical activity may be related to improved cognitive performance and subsequent academic achievement (Sibley and Etnier, 2003). While less developed at this time, some evidence supports the premise that teaching classes in a natural setting increases levels of physical activity (Mygind, 2007). This supports the belief that education in outdoor settings can facilitate the development of higher levels of physical activity and be facilitative in subsequent cognitive performance. O'Brien et al. (2011) expand on this connection between natural settings and education by pointing out that learning activities in the outdoors come in many forms and types. Structured learning activities that have specific connections to physical activity and other types of physicality include the following:

- school grounds/forests/gardens;
- guided walks and excursions;
- environmental education and naturalist events;
- structured or non-structured outdoor activities such as play or specific games such as orienteering or geo-caching; and
- adventure or outdoor recreation-based activities.

In another approach, Li et al. (2008) capitalized on the concept of just feeling 'good' in a forest environment by investigating 'forest bathing'. In this case, forest bathing is conceptually related to natural aromatherapy, and involves visiting a forest or wooded natural area for the purpose of relaxation and breathing in volatile substances called phytoncides. Phytoncides consist of antimicrobial volatile organic compounds produced principally by trees such as pines and citric species. Li et al. surmised that forest bathing trips of this type would increase natural killer cell activity to fight cancer in addition to positively affecting human endocrine and immune systems. Results of several studies (Li et al., 2007, 2008) indicated that the forest experience was associated with increased levels of natural killer activity among other indicators of anticancer proteins in the participants' blood. The researchers concluded that both the relaxation offered by the natural forest setting and the presence of phytoncides produced this beneficial physiological effect.

Finally, a correlation has been observed between life expectancy and natural environments. After controlling for sociodemographic and economic factors, Poudyal et al.

(2009) demonstrated a longer life expectancy for people living in counties that had larger areas of natural space such as farmlands or forest, longer daylight hours, and milder climates. Similar associations were found for the presence of parks, outdoor recreation amenities, and wilderness areas.

Psychological benefits

Increased levels of activity can also provide positive benefits related to mental health (Bowler *et al.*, 2010). In addition to health issues related to the cardiovascular system, physical activity is also associated with emotional and psychological outcomes such as improving mood and protecting against mild forms of depression (Cavill *et al.*, 2006; Bowler *et al.*, 2010). The mechanism by which this occurs can be represented as in Fig. 7.5. Further, natural environments provide a setting for engagement in a variety of physical activities wherein participation is often in concert with significant others (Street *et al.*, 2007). As a result, the effect of natural settings can be augmented by both heightened levels of physical activity and by engagement with social networks of friends, family, or other like-minded people. Further discussion of social benefits follows later in this chapter.

Natural landscapes also can promote mental well-being through mechanisms such as attention restoration, stress reduction, and the development of positive emotions (Abraham *et al.*, 2010). As described in Chapter 3, identifying the causal pathways for these benefits can be complex and overlapping. By that we mean that attention restoration may be related to the already discussed attention

Human beings like being active
↓
Using one's body in pursuit of a meaningful activity provides a personal reward
↓
Physiological effects are combined with the social effects of participating in social activities

Fig. 7.5. Benefits of physical activity

restoration theory (ART), but may also have connections with psycho-evolutionary theory (PET) and stress reduction theory. Similarly, benefits related to mental well-being may occur due to the synergistic effect of physical activity undertaken while in a natural environment, the social context present in that environment, and independent physiological and psychological responses occurring as a result of being in a natural environment (de Vries, 2010). Thus, as is true for mental well-being, the integrative nature of these effects can occur for other benefits acquired from natural environments. For example, natural environments may initially be attractive to individuals for mental restoration, and then the nature of the place may inculcate the desire to be more physically active. Hartig (2008) suggests that this intertwining appears at the core of physical activity and psychological restoration. That is, parks and other natural environments may engender certain types of physical activity and other behaviors which, in turn, bring about specific benefits to human health such as psychological changes or social networking, both of which are elements of enhanced human health.

While depression has been the variable most often linked to psychological well-being, this rubric also includes issues such as anxiety, social isolation, and an overall disconnect from the community or family. Processes that create these health issues include high levels of urbanization and crowding, a sense of social isolation, violence, and substance abuse. Just as with health in general, mental well-being involves more than just the absence of mental illness but rather includes the realization of individual potential and realized capacity for optimal development. Clearly, natural environments can play an important role in the enhancement of mental well-being; the benefits have clear implications for both the individual and society as a whole.

The psychological benefits of time in nature are increasingly clear. Chief among these benefits have been psychological healing, decreased anxiety and anger, enhanced mood, stress reduction, and increased coping. Further, much of the available literature demonstrates that psychological effects often lead to or are intertwined with physiological effects.

For example, viewing natural environments can be particularly valuable in the healing process for those confined to hospitals or nursing homes. In a landmark study, Ulrich (1984) found that patients in rooms with a view of nature had faster recovery rates and fewer complications. Later work by Ulrich *et al.* (1991) studied stress by showing subjects pictures of different natural and urban scenes. Using a variety of physiological measures including heart rate, skin conductance, muscle tension, and pulse transit time, they found that natural scenes were associated with faster and more comprehensive restoration.

Along similar lines, Bossen (2010) points to the healing properties of nature, particularly for individuals suffering from dementia. Of primary interest is the role that natural environments can play in providing an individual sense of well-being, natural multisensory stimulation, a sense of dignity, and an increase in the ability to focus, all of which are often degraded with aging or in situations involving dementia. As the nation prepares for an increasing proportion of older adults within the general population, research on the benefits of natural environment interaction for older adults becomes increasingly important. Researchers are working to emphasize the role of the natural environment in the broader healthcare system as an effective treatment that is accessible, affordable, and carries few if any side-effects. Park-based leisure experiences have been identified as a means of connecting older adults with nature and with each other in a way that reduces stress and impacts physiological variables such as blood pressure and body mass index (Orsega-Smith *et al.*, 2004). While the findings regarding physiological variables in this study were only modest, overall study findings indicate the importance of park-based leisure experiences to enhance mental health and create social support networks.

Another psychological variable that appears susceptible to beneficial effects of nature is mood enhancement. Generally implying a relaxed or peaceful state, natural environments often serve as the antithesis of many of the relatively unpleasant psychosomatic symptoms of urbanized and other built environments, such as tension, deadlines, low-grade conflict, and irritability (Furnass, 1979). Kaplan and Kaplan (1989) suggest that much of the pleasure and mood enhancement from exposure to natural settings is a result of directly viewing natural scenes. This does not require directed or focused attention—only undirected or effortless attention, which is restorative.

Social well-being

The positive benefits associated with the development of an individual's social network are well documented. The strength of an individual's social network has been linked to a wide variety of issues pertaining to health and well-being, ranging from likelihood to commit suicide (Pescosolido and Georgianna, 1989) to participation in leisure activities (Stokowski, 1990). Social networks are particularly important when considering health behaviors because individuals are socialized into various behaviors through interactions with those around them. While the research on social networks and time spent in natural environments is limited, existing studies demonstrate an important link between leisure, social networks and social capital, and appreciation of the natural world (Burch, 1969; Harshaw and Tindall, 2005).

Natural environments have generally been associated with social well-being through the development of factors such as social integration, social engagement, and social support (Abraham *et al.*, 2010). The types of natural environments most cited in these works and in relation to social well-being are parks and gardens, often rich in vegetation, or trails for walking (Frumkin, 2001). While physical activity and restorative effects interact to assist in social well-being, another, perhaps more interesting variable that impacts the development of social well-being is collectively experiencing nature. Thus, in addition to the effects of nature, the interaction between the collective (i.e. two or more people) and the natural environment appears to offer another avenue towards social well-being. For example, Townsend (2006) found that participation in forest/woodland management practices proved to be effective in reducing symptoms related to depression. This was especially true if an individual was participating with a small group of family or friends.

Different populations have been studied to see if camping in natural areas or attending camps in natural settings impacts family bonds. Mactavish and Schleien (1998) showed that from a parental perspective family recreation such as a camping vacation was viewed as a beneficial catalyst for skill, interest, and self-development, as well as being a potentially accepting and enduring social and recreation outlet for children with a developmental disability. Family camps in natural settings help family members strengthen their relationships and create more meaningful bonds with one another (Lewicki et al., 1996; Taylor, 2006). These findings are consistent with family leisure studies showing compelling connections between family functioning and family leisure (Zabriskie and McCormick, 2001).

Degenhardt and Buchecker (2012) offer another perspective through their concept of nearby outdoor recreation behaviors (NORBs). Their study links outdoor recreation behavior with the explicit purpose to reduce loads and strains. Loads are defined as demands that affect the individual and must be dealt with, such as work-related problems. Strains refer to the subjective consequences of dealing with loads and include psychological or physiological impacts (e.g. stress, anxiety, anger, etc.). Degenhardt and Buchecker found that recreation behavior in terms of visitation to a nearby natural setting was used to accommodate personal goals, social-based goals, or environmental-based goals, thus providing further evidence of how natural environments such as NORBs can be useful in effecting positive health benefits.

Spiritual benefits of nature

Nature affords myriad spiritual benefits; the awe and wonder inspired by nature (Fox, 1997; Heintzman, 2000; Loeffler, 2004;

Grafanaki et al., 2005; Schmidt and Little, 2007) may even be viewed as the genesis of spirituality (as well as, perhaps paradoxically, empirical science). Childhood experiences in nature help children develop their sense of wonder, creativity, and imaginative thinking (Cobb 1977/1998; Crain, 2001; Taylor et al., 2001; Atchley et al., 2012). Spirituality is considered by Kellert and Wilson (1993) to be an innate human biophilic response to nature, inspiring connections of self with a wider world or a larger 'other', such as a higher power; and sense of oneness with the world (Heintzman, 2000, 2007a,b, 2008; Crain, 2001; Loeffler, 2004; Livengood, 2009; Kellert, 2012). Kaplan's (1995) restorative environments theory suggests nature often offers captivating settings and aesthetic vistas that stimulate exploration and contemplation. These environments provide tranquility and calm (Stringer and McAvoy, 1992; Fox, 1997; Loeffler, 2004; Grafanaki et al., 2005; Heintzman, 2007a), allowing for spiritual reflection (Bobilya et al., 2009). Further, contact with nature fosters expansion of human connection from self to others, nurturing empathy, generosity, and altruism (Weinstein et al., 2009; Mathur et al., 2010).

Summary of Research on Benefits

The connection between natural environments and human health issues is of current interest for a number of organizations and researchers across a variety of fields. Not surprisingly, the search for verifiable evidence of outcome efficacy is of prime importance. While the previous section points out some findings about positive health impacts with respect to physical, psychological, spiritual, and social variables from engagement in natural environments, this section summarizes the overall findings of natural settings and health research. Note that although Chapter 4 discusses theoretical constructs describing reasons why natural environments may create positive health outcomes, the pathways in which these outcomes are realized are still not fully understood (Bowler et al., 2010; Stigsdotter et al., 2011). Bowler et al. (2010) and others report characterizations of research on the outcomes of interaction with natural environments and human health as presented in Table 7.1.

Clearly, Table 7.1 represents only a sample of items depicted. However, it shows the wide range of factors that exist, such as locations, samples, methods, types of analyses, and outcomes. Moreover, many of the studies investigating the effects of natural environments upon human health have demonstrated a positive relationship—many but not all! Some of the research conducted on natural settings and health has resulted in little or no change or effect. As was discussed earlier in this chapter, there are issues connected to these research efforts and confirming research projects is necessary.

Future Research

Careful consideration of the previously described issues regarding the investigation of the effects of natural environments upon human health is critical given the importance to humanity of the credibility and validity of the research findings. Three questions emerge:

1. Does exposure to and engagement in natural environments aid in the development of beneficial health outcomes?
2. How does this exposure influence health outcomes?

3. What influence do mediating variables such as length of exposure (i.e. dosage), type of exposure, and the underlying context of the exposure have in the positive health outcomes?

The first question presents several issues for consideration. Among these is the question of whether or not the sample represents a larger population. As previously indicated, much research done on outcomes involved college students and/or physically active individuals. Moreover, many of these studies use self-report instrumentation, which by its very nature encourages people to respond in socially and environmentally desirable manners (Ewert and Galloway, 2009). Additionally, Bowler *et al.* (2010) report that many studies have focused on very short-term experiences in natural environments, and have focused only on immediate impacts rather than long-term implications.

Relative to the second question, much existing research posits that natural environments influence the creation of beneficial outcomes but often fail to explain *how* this effect is produced. This deficiency is partially accounted for by the use of underlying explanatory theories such as ART and PET (see Chapter 4). Another conceptual model that provides an explanatory framework includes individual factors (e.g. personal characteristics, age, sex, ethnicity, and life events) that flow into environmental factors including physical changes to the environment (e.g. improved access to paths and trails) and social

Table 7.1. Characterizations of health-related studies and natural environments.

Variable	Findings
Most common locations	Parks, university campuses, nature reserves, wilderness areas, forests, gardens
Comparisons	Built environments, indoor gyms, shopping centres
Duration of exposure	One hour or less, one day
Study participants	College/university students, physically active, specific health, physical or cognitive issues
Methods	Self-report, randomized, counterbalanced, pre/post tests, observational study, interviews
Analysis	Comparison tests, physical measurements, association measures, effect sizes
Outcome variables	Emotions (e.g. anger, tension, energy, etc.), level of attention/focus/concentration/memory, cardiovascular (e.g. blood pressure, heart rate), endocrine (e.g. levels of cortisol), immune functions

changes (e.g. walking programs). These factors lead to proximal outcomes such as enhanced environments (e.g. feeling safer) and behavior outcomes (e.g. more walking) that in turn lead to health outcomes (e.g. lower stress levels) and subsequent health behaviors (increased walking and physical activity) (Ward Thompson and Aspinall, 2011).

Elsewhere, O'Brien *et al.* (2011) describe the DPSEEA (drivers, pressures, state, exposure, effects, and actions) model. In one example from this model, pressures (e.g. lack of access to greenspace) are influenced by the state or situation (e.g. local environment does not encourage walking), which leads to exposure (e.g. lack of walking leads to a sedentary lifestyle), which in turn leads to effects (e.g. greater incidence of cardiovascular disease, type 2 diabetes, and mental ill health). Thus, while an absolute description of how natural environments influence human health remains elusive, there are a number of models and theories that elucidate the process.

Regarding the mediating or confounding variables, a number of issues arise that can influence or cloud understanding of the relationship between natural environments and human health. For example, an individual's receptiveness to engage in a natural landscape and thus experience improved health may be related to the ways and the extent that she/he experienced natural environments earlier in life (Ewert *et al.*, 2005; Cheng and Monroe, 2012).

Similarly, context often plays an important role in realized outcomes. Measurements taken after the respondent already experienced mental fatigue or had a stressful day or event can impact results. For example, Felsten (2009) reports that murals or scenes of nature were the most effective in generating a restorative environment for college students experiencing stress or attention fatigue. Moreover, it is unclear if the actual types of activities or experiences within a natural environment play differential roles in developing health-related outcomes.

In conclusion, numerous positive outcomes have been associated with engagement with natural environments. Much of the literature supports the contention that natural environments can serve an effective function in enhancing human health. A number of theories inform our understanding about human health and natural environments, many of which have been previously discussed. Moreover, a number of instruments and research approaches exist to assist in ascertaining the effectiveness of natural environments and health. As the complexity of the available instruments, theories, and models grows, doubtless the explanations of the outcomes related to the health–environment interface will become more sophisticated and complex. In the final analysis (using Occum's razor principle that the simplest explanation is usually the correct one), perhaps Hartig's (2008) point regarding 'intertwining of the mechanisms' is most useful: the extent to which people engage in natural environments is simply related to the feelings of restoration they experience in nature.

References

Abraham, A., Sommerhalder, K. and Abel, T. (2010) Landscape and well-being: a scoping study on the health-promoting impact of outdoor environment. *International Journal of Public Health* 55(5), 59–69.

Armstrong, D.A. (2000) Survey of community gardens in upstate New York: implications for health promotion and community development. *Health & Place* 6(4), 319–327.

Atchley, R.A., Strayer, D.L. and Atchley, P. (2012) Creativity in the wild: improving creative reasoning through immersion in natural settings. *PLoS ONE* 7(12), e51474.

Barton, J. and Pretty, J. (2010) What is the best dose of nature and green exercise for improving mental health? A multi-study analysis. *Environmental Science & Technology* 44(10), 3947–3955.

Bauman, A.E. (2004) Updating the evidence that physical activity is good for health: an epidemiological review 2000–2003. *Journal of Science and Medicine in Sport* 7(1), 6–19.

Beaton, A.A. and Funk, D.C. (2008) An evaluation of theoretical frameworks for studying physically active leisure. *Leisure Sciences* 30(1), 53–70.

Bernstein, H.M., Russo, M.A. and Laquidara-Carr, D. (2013) *New and Retrofit Green Schools: The Cost Benefits and Influence of a Green School on its Occupants*. McGraw-Hill Construction, Bedford, Massachusetts.

Bixler, R.D., Floyd, M.F. and Hammitt, W.E. (2002) Environmental socialization: quantitative tests of the childhood play hypothesis. *Environment and Behavior* 34(6), 795–818.

Bobilya, A.J., Akey, L. and Mitchell, D. Jr (2009) Outcomes of a spiritually focused wilderness orientation program. *Journal of Experiential Education* 31(3), 440–443.

Booth, M.L., Owen, N., Bauman, A., Clavisi, O. and Leslie, E. (2000) Social-cognitive and perceived environment influences associated with physical activity in older Australians. *Preventive Medicine* 31(1), 15–22.

Bossen, A. (2010) The importance of getting back to nature for people with dementia. *Journal of Gerontological Nursing* 36(2), 17–22.

Bowler, D.E., Buyung-Ali, L.M., Knight, T.M. and Pullin, A.S. (2010) A systematic review of evidence for the added benefits to health of exposure to natural environments. *BMC Public Health* 10, 456.

Breunig, M., O'Connell, T., Todd, S., Anderson, L. and Young, A. (2010) The impact of outdoor pursuits on college students' perceived sense of community. *Journal of Leisure Research* 42(4), 551–572.

Burch, W.R. Jr (1969) The social circles of leisure: competing explanations. *Journal of Leisure Research* 1(2), 125–147.

Burdette, H. and Whitaker, R. (2005) Resurrecting free play in young children: looking beyond fitness and fatness to attention, affiliation, and affect. *Archives of Pediatrics & Adolescent Medicine* 159(1), 46–50.

Canin, L.H. (1991) Psychological restoration among AIDS caregivers: maintaining self care. Doctoral dissertation, University of Michigan, Ann Arbor, Michigan.

Cavill, N., Kahlmeier, S. and Racioppi, F. (eds) (2006) *Physical Activity and Health in Europe: Evidence for Action*. World Health Organization, Geneva, Switzerland.

Cheng, J. and Monroe, M.C. (2012) Connection to nature: children's affective attitude toward nature. *Environment and Behavior* 44(1), 31–49.

Cimprich, B.E. (1990) Attentional fatigue and restoration in individuals with cancer. Doctoral dissertation, University of Michigan, Ann Arbor, Michigan. *Dissertation Abstracts International* 51(4): 1740B.

Cobb, E. (1977/1998) *The Ecology of Imagination in Childhood*. John Wiley, New York.

Cosgriff, M., Little, D. and Wilson, E. (2010) The nature of nature: how New Zealand women in middle to later life experience nature-based leisure. *Leisure Sciences* 32(1), 15–32.

Crain, W. (2001) How nature helps children develop. *Montessori Life* 9(2), 41–43.

Degenhardt, B. and Buchecker, M. (2012) Exploring everyday self-regulation in nearby nature: determinants, patterns, and a framework of nearby outdoor recreation behavior. *Leisure Sciences* 34(5), 450–469.

de Vries, S. (2010) Nearby nature and human health: looking at the mechanisms and their implications. In: Ward Thompson, C., Aspinall, P. and Bell, S. (eds) *Innovative Approaches to Researching Landscape and Health. Open Space: People Space 2*. Routledge, Abingdon, UK, pp. 77–96.

Ewert, A. and Galloway, G. (2009) Socially desirable responding in an environmental context: development of a domain specific scale. *Environmental Education Research* 15(1), 55–70.

Ewert, A., Place, G. and Sibthorp, J. (2005) Early-life outdoor experiences and an individual's environmental attitudes. *Leisure Sciences* 27(3), 225–239.

Felsten, G. (2009) Where to take a study break on the college campus: an attention restoration theory perspective. *Journal of Environmental Psychology* 29(1), 160–167.

Fox, R.J. (1997) Women, nature and spirituality: a qualitative study exploring women's wilderness experience. In: Rowe, D. and Brown, P. (eds) *Proceedings of ANZALS Conference 1997*. Australian and New Zealand Association for Leisure Studies and Department of Leisure and Tourism Studies, University of Newcastle, Newcastle, Australia, pp. 59–64.

Frumkin, H. (2001) Beyond toxicity: human health and the natural environment. *American Journal of Preventive Medicine* 20(3), 234–240.

Furnass, B. (1979) Health values. In: Messer, J. and Mossley, J.G. (eds) *The Value of National Parks to the Community: Values and Ways of Improving the Contribution of Australian National Parks to the Community*. Australian Conservation Foundation, University of Sydney, Sydney, Australia, pp. 60–69.

Grafanaki, S., Pearson, D., Cini, F., Godula, B., McKenzie, B., *et al.* (2005) Sources of renewal: a qualitative study of the experience and role of leisure in the life of counselors and psychologists. *Counselling Psychology Quarterly* 18(1), 31–40.

Greenway, R. (1996) Wilderness experience and ecopsychology. *International Journal of Wilderness* 2(1), 26–30.

Hans, T. (2000) A meta-analysis of the effects of adventure programming on locus of control. *Journal of Contemporary Psychotherapy* 30(1), 33–60.

Harshaw, H.W. and Tindall, D.B. (2005) Social structure, identities, and values: a network approach to understanding people's relationships to forests. *Journal of Leisure Research* 37(4), 426–449.

Hartig, T. (2008) Green space, psychological restoration, and health inequality. *Lancet* 372(9650), 1614–1615.

Heft, H. (2010) Affordances and the perception of landscape: an inquiry into environmental perception and aesthetic. In: Ward Thompson, C., Aspinall, P. and Bell, S. (eds) *Innovative Approaches to Researching Landscape and Health: Open Space: People Space 2*. Routledge, Abingdon, UK, pp. 9–32.

Heintzman, P. (2000) Leisure and spiritual well-being relationships: a qualitative study. *Society and Leisure* 23(1), 41–69.

Heintzman, P. (2007a) Men's wilderness experience and spirituality: a qualitative study. In: Burns, R. and Robinson, K. (comps) *Proceedings of the 2006 Northeastern Recreation Research Symposium. General Technical Report NRS-P-14*. US Department of Agriculture, Forest Services, Northern Research Station, Newton Square, Pennsylvania, pp. 432–439.

Heintzman, P. (2007b) Rowing, sailing, reading, discussing, praying: the spiritual and lifestyle impact of an experientially based, graduate, environmental education course. Paper presented at the *Trails to Sustainability Conference*, Kananaskis, Canada, 24–27 May 2007.

Heintzman, P. (2008) Men's wilderness experience and spirituality: further explorations. In: LeBlanc, C. and Vogt, C. (eds) *Proceedings of the 2007 Northeastern Recreation Research Symposium. General Technical Report NRS-P-23*. US Department of Agriculture, Forest Services, Northern Research Station, Newton Square, Pennsylvania, pp. 55–59.

Herzog, T.R., Black, A.M., Fountaine, K.A. and Knotts, D.J. (1997) Reflection and attentional recovery as distinctive benefits of restorative environments. *Journal of Environmental Psychology* 17(2), 165–170.

Hunter, M.C. and Hunter, M.D. (2008) Designing for conservation of insects in the built environment. *Insect Conservation and Diversity* 1(4), 189–196.

Ingram, J. (2009) The role of ecological systems and natural resource management in reducing vulnerability to hazards. Presented at *Ecological Society of America 94th Annual Meeting*, Albuquerque, New Mexico, 2–7 August 2009.

Kaplan, R. and Kaplan, S. (1989) *The Experience of Nature: A Psychological Perspective*. Cambridge University Press, Cambridge, UK.

Kaplan, R. and Kaplan, S. (2005) Preference, restoration, and meaningful action in the context of nearby nature. In: Barlett, P.F. (ed.) *Urban Place: Reconnecting with the Natural World*. MIT Press, Cambridge, Massachusetts, pp. 271–298.

Kaplan, S. (1995) The restorative benefits of nature: toward an integrative framework. *Journal of Environmental Psychology* 15(3), 169–182.

Kellert, S.R. (2012) *Birthright: People and Nature in the Modern World*. Yale University Press, New Haven, Connecticut.

Kellert, S.R. and Wilson, E.O. (eds) (1993) *The Biophilia Hypothesis*. Island Press, Washington, DC.

Kreitzer, M.J., Mitten, D., Harris, I. and Shandeling, J. (2002) Attitudes toward CAM among medical, nursing, and pharmacy faculty and students: a comparative analysis. *Alternative Therapies in Health and Medicine* 8(6), 44–47.

Kuo, F.E. and Sullivan, W.C. (2001) Environment and crime in the inner city: does vegetation reduce crime? *Environment and Behavior* 33(3), 343–367.

Kweon, B.S., Sullivan, W.C. and Wiley, A.R. (1998) Green common spaces and the social integration of inner-city older adults. *Environment and Behavior* 30(6), 832–858.

Lewicki, J., Goyette, A. and Marr, K. (1996) Family camp: a multimodal treatment strategy for linking process and content. *Journal of Child and Youth Care* 10(4), 51–66.

Leyden, K.M. (2003) Social capital and the built environment: the importance of walkable neighborhoods. *American Journal of Public Health* 93(9), 1546–1551.

Li, Q., Morimoto, K., Nakadai, A., Inagaki, H., Katsumata, M., et al. (2007) Forest bathing enhances human natural killer activity and expression of anti-cancer proteins. *International Journal of Immunopathology and Pharmacology* 20(2 Suppl. 2), 3–8.

Li, Q., Morimoto, K., Kobayashi, M., Inagaki, H., Katsumata, M., et al. (2008) A forest bathing trip increases human natural killer activity and expression of anti-cancer proteins in female subject. *Journal of Biological Regulators and Homeostatic Agents* 22(1), 45–55.

Livengood, J. (2009) The role of leisure in the spirituality of New Paradigm Christians. *Leisure/Loisir* 33(1), 389–417.

Loeffler, T.A. (2004) A photo elicitation study of the meanings of outdoor adventure experiences. *Journal of Leisure Research* 36(4), 536–556.

Louv, R. (2005) *Last Child in the Woods: Saving our Children from Nature-Deficit Disorder.* Algonquin Books of Chapel Hill, Chapel Hill, North Carolina.

Mactavish, J.B. and Schleien, S.J. (1998) Playing together growing together: parents' perspectives on the benefits of family recreation in families that include children with a developmental disability. *Therapeutic Recreation Journal* 32(3), 207–230.

Mathur, V.A., Harada, T., Lipke, T. and Chiao, J.Y. (2010) Neural basis of extraordinary empathy and altruistic motivation. *Neuroimage* 51(4), 1468–1475.

McBrien, N.A., Morgan, I.G. and Mutti, D.O. (2009) What's hot in myopia research – The 12th International Myopia Conference, Australia, July 2008. *Optometry & Vision Science* 86(1), 2–3.

Milligan, C., Gatrell, A. and Bingley, A. (2004) 'Cultivating health': therapeutic landscapes and older people in northern England. *Social Science & Medicine* 58(9), 1781–1793.

Mitten, D. (2004) Adventure therapy as complementary and alternative therapy. In: Bandoroff, S. and Newes, S. (eds) *Coming of Age: The Evolving Field of Adventure Therapy.* Association of Experiential Education, Boulder, Colorado, pp. 240–257.

Mitten, D. (2010) Friluftsliv and the healing power of nature: the need for nature for human health, development, and wellbeing. In: *Henrik Ibsen: The Birth of 'Friluftsliv'. A 150 Year International Dialogue Conference Jubilee Celebration.* North Troendelag University College, Levanger, Norway, Mountains of Norwegian/Swedish Border, 14–19 September 2009. Available at: http://norwegianjournaloffriluftsliv.com/doc/122010.pdf (accessed 8 September 2013).

Mitten, D., Ady, J.C. and D'Amore, C. (2013) The healing power of nature: the need for human health, development, and well being. Poster session presented at the *42nd Annual North American Association of Environmental Education Conference*, Baltimore, Maryland, 9–12 October 2013.

Moore, E.O. (1981) A prison environment's effect on health care service demands. *Journal of Environmental Systems* 11(1), 17–34.

Mygind, E. (2007) A comparison between children's physical activity levels at school and learning in an outdoor environment. *Journal of Adventure Education & Outdoor Learning* 7(2), 161–176.

Norman, J., Annerstedt, M., Boman, M. and Mattson, L. (2010) Influence of outdoor recreation on self-rated health: comparing three categories of Swedish recreationists. *Scandinavian Journal of Forest Research* 25(3), 234–244.

O'Brien, L., Burls, A., Bentsen, P., Hilmo, I., Holter, K., *et al.* (2011) Outdoor education, lifelong learning and skills development in woodlands and green spaces: the potential links to health and well-being. In: Nilsson, K., Sangster, M., Gallis, C., Hartig, T., de Vries, S., *et al.* (eds) *Forests, Trees and Human Health.* Springer, New York, pp. 343–372.

Orsega-Smith, E., Mowen, A.J., Payne, L.L. and Godbey, G. (2004) The interaction of stress and park use on psycho-physiological health in older adults. *Journal of Leisure Research* 36(2), 232–257.

Pescosolido, B.A. and Georgianna, S. (1989) Durkheim, suicide, and religion: toward a network theory of suicide. *American Sociological Review* 54(1), 33–48.

Plumwood, V. (2009) Nature in the active voice. *Australian Humanities Review* 46(May). Available at: http://www.australianhumanitiesreview.org/archive/Issue-May-2009/plumwood.html (accessed 16 September 2013).

Poudyal, N.C., Hodges, D.G., Bowker, J.M. and Cordell, H.K. (2009) Evaluating natural resource amenities in a human life expectancy production function. *Forest Policy and Economics* 11(4), 253–259.

Pretty, J., Griffin, M., Sellens, M. and Pretty, C. (2003) *Green Exercise: Complementary Roles of Nature, Exercise and Diet in Physical and Emotional Well-being and Implications for Public Health Policy. CES Occasional Paper 2003-1.* University of Essex, Colchester, UK.

Pretty, J., Peacock, J., Sellens, M. and Griffin, M. (2005) The mental and physical health outcomes of green exercise. *International Journal of Environmental Health Research* 15(5), 319–337.

Rishbeth, C. and Finney, N. (2006) Novelty and nostalgia in urban greenspace: refugee perspectives. *Tijdschrift voor Economische en Sociale Geografie* 97(3), 281–295.

Schmidt, C. and Little, D.E. (2007) Qualitative insights into leisure as a spiritual experience. *Journal of Leisure Research* 39(2), 222–247.

Seeland, K. and Ballesteros, N. (2004) *Cultural Comparative Studies of the Potential for Social Inclusion within Green Spaces in the Agglomerations of Geneva, Lugano and Zurich. Forstwissenschaftliche Beiträge No. 31.* ETH Zürich, Zürich, Switzerland.

Sibley, B.A. and Etnier, J.L. (2003) The relationship between physical activity and cognition in children: a meta-analysis. *Pediatric Exercise Science* 15(3), 243–256.

Stigsdotter, U.K., Palsdottir, A.M., Burls, A., Chermaz, A., Ferrini, F., *et al.* (2011) Nature-based therapeutic interventions. In: Nilsson, K., Sangster, M., Gallis, C., Hartig, T., de Vries, S., *et al.* (eds) *Forests, Trees and Human Health*. Springer, New York, pp. 309–342.

Stokowski, P.A. (1990) Extending the social groups model: social network analysis in recreation research. *Leisure Sciences* 12(3), 251–263.

Street, G., James, R. and Cutt, H. (2007) The relationship between organized physical recreation and mental health. *Health Promotion Journal of Australia* 18(3), 236–239.

Stringer, L.A, and McAvoy, L.H. (1992) The need for something different: spirituality and wilderness adventure. *Journal of Experiential Education*, 15(1), 13–20.

Taylor, A.F., Kuo, F.E. and Sullivan, W.C. (2001) Coping with ADD: the surprising connection to green play settings. *Environment and Behavior* 33(1), 54–77.

Taylor, S. (2006) Family camps: strengthening family relationships at camp and at home. *Camping Magazine* 79(4), 54–59.

Tennessen, C.M. and Cimprich, B. (1995) Views to nature: effects on attention. *Journal of Environmental Psychology* 15(1), 77–85.

Thompson Coon, J., Boddy, K., Stein, K., Whear, R., Barton, J., *et al.* (2011) Does participating in physical activity in outdoor natural environments have a greater effect on physical and mental wellbeing than physical activity indoors? A systematic review. *Environmental Science & Technology* 45(5), 1761–1772.

Therapeutic Landscapes Resource Center, Inc. (2000) Therapeutic Landscapes Database. Available at: http://www.healinglandscapes.org/sites.html (accessed 1 May 2009).

Townsend, M. (2006) Feel blue? Touch green! Participation in forest/woodland management as a treatment for depression. *Urban Forestry & Urban Greening* 5(3), 111–120.

Ulrich, R.S. (1984) View through a window may influence recovery from surgery. *Science* 224(4647), 420–421.

Ulrich, R.S. (2001) Effects of healthcare environmental design on medical outcomes. In: Dilani, A. (ed.) *Design and Health*. Svensk Byggtjanst, Stockholm, pp. 49–59.

Ulrich, R.S., Simons, R.F., Losito, B.D., Fiorito, E., Miles, M.A., *et al.* (1991) Stress recovery during exposure to natural and urban environment. *Journal of Environmental Psychology* 11(3), 201–230.

Van Sluijs, E.M.F., Van Poppel, M.N., Twisk, J.W., Brug, J. and Van Mechelen, W. (2005) The positive effect on determinants of physical activity of a tailored, general practice-based physical activity intervention. *Health Education Research* 20(3), 345–356.

Velarde, Ma.D., Fry, G. and Tveit, M. (2007) Health effects of viewing landscapes-landscape types in environmental psychology. *Urban Forestry & Urban Greening* 6(4), 199–212.

Ward Thompson, C. (2011) Linking landscape and health: the recurring theme. *Landscape and Urban Planning* 99(3), 187–195.

Ward Thompson, C. and Aspinall, P.A. (2011) Natural environments and their impact on activity, health, and quality of life. *Applied Psychology: Health and Well-Being* 3(3), 230–260.

Weinstein, N, Przybylski, A.K. and Ryan, R.M. (2009) Can nature make us more caring? Effects of immersion in nature on intrinsic aspirations and generosity. *Personality and Social Psychology Bulletin* 35(10), 1315–1329.

Wells, N.M. (2000) At home with nature: effects of 'greenness' on children's cognitive functioning. *Environment and Behavior* 32(6), 775–795.

Wilson, S.J. and Lipsey, M.W. (2000) Wilderness challenge programs for delinquent youth: a meta-analysis of outcome evaluations. *Evaluation and Program Planning* 23(1), 1–12.

Zabriskie, R.B. and McCormick, B.P. (2001) The influences of family leisure patterns on perceptions of family functioning. *Family Relations* 50(3), 281–289.

8

Sense of Place and the Role of Education

_____

*To understand such a language of the land
requires a deep acquaintance with some
place, or perhaps a group of places.*
Val Plumwood in *Environmental Culture: The
Ecological Crisis of Reason* (2002, p. 231)

In order to protect and preserve the natural environment, people need to feel a sense of connection or attachment to the environment, or in some way assign importance to its existence. For most residents of the Western world, it is now easier to be disconnected from the environment than it is to be connected—physically or emotionally. We know that in the developed world both children and adults are spending less time outdoors than in previous generations. By the 1980s people in the industrialized nations spent over 90% of their time indoors (National Research Council, 1981). An extensive survey done in England found that less than 10% of children reported playing in natural places, whereas 40% of their parents reported playing in natural places when they were children (Natural England, 2009). The same survey showed that the largest percentage of children (41%) reported 'indoors' as their favorite place to play, whereas previous generations reported neighborhood streets and gardens. It is clear that many of us must purposefully seek out experiences in the natural environment to work and to play. But we will only do so if those experiences are both meaningful and accessible.

As discussed in Chapter 2, human behaviors and ideals stem from our worldviews. Thus, environmentally sustainable behaviors are dependent on culture and, in turn, on political and educational institutions. This chapter explores connectedness to the natural environment through the development of a land ethic, as well as through an understanding of the concepts of place and sense of place. An exploration of ways that we can educate and manage for sense of place, and the role of education in reconnecting children to the natural environment follow. Finally, we explore the idea that the development of a sense of place accompanied by an ethic of care, critical thinking, and environmental literacy may in turn lead to positive action through environmental stewardship.

Why We Need a Land Ethic

*A thing is right when it tends to preserve
the integrity, stability, and beauty of the
biotic community. It is wrong when it tends
otherwise.*

Aldo Leopold

Aldo Leopold coined the term 'land ethic' in his famous work, *A Sand County Almanac* (1949). According to Leopold, a land ethic is a philosophy that guides decisions and actions about land use and management.

© CAB International 2014. *Natural Environments and Human Health*
(A.W. Ewert, D.S. Mitten and J.R. Overholt)

Leopold saw land as an enlargement of the community boundaries, such that the soil, plants, animals, and water are part of the human community rather than separate from it: 'In short, a land ethic changes the role of *Homo sapiens* from conqueror of the land-community to plain member and citizen of it. It implies respect for his [sic] fellow-members, and also respect for the community as such' (Leopold, 1949). As a proponent of humans-as-citizens rather than humans-as-conquerors of the land, Leopold was farsighted in his ability to recognize that a conquering mentality was ultimately self-defeating and that we were better off learning to work with the natural environment rather than trying to impose our will upon it. Thus, a land ethic meant learning to live in harmony with the land, as well as a willingness to sacrifice short-term economic gain for long-term environmental good.

Now, more than ever, we need a land ethic to guide our decision making and interaction with the land around us. Individuals spend increasingly less time outdoors, ever increasing the sense of disconnection, yet human survival is as dependent upon the natural world as ever. A land ethic, or a philosophy that underscores our connectedness, may be the missing piece to help lift the veil of disconnection and aid in reawakening our sense of interdependence with the surrounding ecosystems.

A land ethic is an attempt to instill an ethic of care—to encourage people to act out of a sense of compassion and connectedness—rather than setting minimum environmental regulations and incentives based on rewards or punishments to incentivize people. Leopold himself recognized the futility of that approach: 'we asked the farmer to do what he [sic] conveniently could to save his [sic] soil, and he has done just that and only that' (Leopold, 1949, p. 204). Environmental regulation and enforcement—although well-intentioned and factual—can drive people further from the cause at hand. This behavior has been attributed to emotion-management designed to reduce anxiety (Crompton and Kasser, 2009). In other words, people employ behavioral strategies to reduce anxiety created by hearing claims of environmental destruction. They may limit their exposure to upsetting information, focus on present gratification, or only engage in small actions (such as replacing light bulbs) instead of focusing on larger-scale systemic change.

Crompton and Kasser (2009) suggested that engaging people's identities is more effective than engaging behaviors or organizations. Behaviors, such as those advocated through green consumerism, do little to actually address the root causes and while organizations can create important policy, individuals must have personal buy-in to support actual change. In this sense, identities are implicated in an individual's sense of connectedness to nature (see Chapter 4 for more on environmental identity-related concepts). Even in the early 1900s, Leopold recognized that this sentiment was missing from US society. This sense of connection, or more simply a love of the land, is a key component of a land ethic. Love of the land forms the basis for ethics, which ultimately underlie environmental behavior. Furthermore, Leopold saw a connection between a land ethic and the existence of an environmental conscience, which 'reflects a conviction of individual responsibility for the health of the land' (1949, p. 221).

The study of ethics evolved as a study of human interests; to include the land in ethics was a radical move on Leopold's behalf, but one that has caught on in recent years. According to Rolston (2000), the development of environmental ethics in Western philosophy did not occur until the mid-1970s, so for much of recent history, nature was considered value-less, morally speaking. The dominant Western worldview assigned value to land only insofar as it was useful to humans, as opposed to being valuable in its mere existence. In contrast, many ancient civilizations and religions, such as Chinese Taoism, Native Americans, or Pagan spirit traditions (see Chapters 2 and 3 for more on this topic), include environmental ethics as an important part of daily life and an essential component of their dominant philosophies and worldviews. Perhaps the emergence only relatively recently of environmental ethics as a distinct sub-field of philosophy is because, until recently, the natural environment was not considered to be separate and therefore did not necessitate a separate philosophy.

Western ethics are prevailingly anthropocentric and utilitarian. Thus, moral duty or obligation to the natural environment is often dismissed and protection of nature is driven by its perceived benefit to humans. For example, rainforests are saved because of potential undiscovered medications, air pollution is reduced once it is linked to lung disease, and exotic habitats are protected as vacation destinations. It has been argued, however, that in order to fully develop an ethic towards nature, humans must include themselves in their conceptualization of nature (Rolston, 2000). Only then does environmental stewardship equate to care of ourselves and our communities.

Here, an understanding of ecology and ecosystems, as well as systems theory (see the description of systems thinking in Chapter 6), is essential in order to perceive ecological connectedness, rather than looking at the natural environment as aggregated separate parts. While modern thought tends toward reductionist notions of separate phenomena, it is likely that the problems facing today's world are highly correlated and interconnected. An environmental ethic encompassing human and ecological systems interdependence could motivate environmental protection. In fact, it can be argued that concern for the environment is interconnected with many major global societal issues:

> Environmental ethics, started by a forester spending his weekends in a shack in the rural sand countries, will be taken by some, even yet, to be peripheral concern about chipmunks and daisies, extrapolated to rocks and dirt. But not so. The four most critical issues that humans currently face are peace, population, development, and environment. All are entwined. (Rolston, 2000, p. 1056)

Biophilia and a land ethic

The biophilia hypothesis suggests that humans are innately affiliated with the natural environment in ways that transcend many of our typical conceptualizations of our relationship to the natural world. Kellert (1993) writes that the biophilia hypothesis 'proclaims a human dependence on nature that extends far beyond the simple issues of material and physical sustenance to encompass as well the human craving for aesthetic, intellectual, cognitive, and even spiritual meaning and satisfaction' (p. 20). Precisely because of this deep relationship, Kellert suggests the biophilia hypothesis accounts for the development of a land ethic and of environmentally responsible behaviors. This is due to a combination of self-interest in preserving our own habitat and the deeply ingrained learning rules that E.O. Wilson (1984) hypothesizes to exist (even if weakly) from millennia of human inhabitance of the natural world. It is these learning rules that guide our thinking about the natural environment and therefore also influence the development of environmental ethics, and of our personal connections and convictions regarding the land around us.

Wilson takes a decidedly anthropocentric approach to the ethics of conservation, sidestepping the issue of whether a given species has an innate right to exist. Instead, his approach looks at the utilitarian, aesthetic, and spiritual values of species diversity as the impetus for conservation. This is also reflected in Kellert's typology of land valuation, in which he identifies nine basic values that appear to exist across cultures (see Chapter 4 for more on this topic). In contrast, Knapp (1999) wrote that a biophilic worldview would be guided by reverence and affinity for ecological life. His notion is that an anthropocentric worldview allows nature to be exploited and for people to disregard their responsibilities as members of a biotic community, thus propagating the current ecological predicament. As population continues to increase and development advances, we are increasingly forced to make difficult decisions about the ways the Earth and its resources are utilized. We have a long history of dealing reactively with consequences of poor land-use decisions, as opposed to preemptively accounting for potential negative impacts. The precautionary principle, developed by

Carloyn Raffensperger, lends guidance to situations where clear scientific guidance does not yet exist regarding the environmental impacts of a certain policy decision or practice:

> When an activity raises threats of harm to human health or the environment, precautionary measures should be taken even if some cause and effect relationships are not fully established scientifically. In this context the proponent of an activity, rather than the public, should bear the burden of proof. The process of applying the precautionary principle must be open, informed and democratic and must include potentially affected parties. It must also involve an examination of the full range of alternatives, including no action. (Wingspread Statement on the Precautionary Principle, Science & Environmental Health Network, 1998)

This principle places the burden of proof on the entity proposing the action, rather than on those who oppose it on environmental grounds. Lately, this principle has been adopted as is or used to formulate policy by corporations, school districts, municipalities, intergovernmental treaties, and the European Union. For example, early pesticide bans in Canada cited this principle, the Maastricht Treaty on the European Union dictates community environmental policy be based on this principle, and cities in the US such as San Francisco and Berkeley have created city ordinances utilizing this principle (Science & Environmental Health Network, 1998).

Place

Leopold's use of an old farmstead in Wisconsin as the setting for his writing and development of his land ethic is instructive in what it tells us about the innate value of land. He was not writing about an isolated wilderness, a dramatic mountain range, or an exotic rainforest—he was writing about the lands he knew best: his home in Wisconsin and the places he had visited over the years. It was in the act of stewarding this land, hunting, and farming that he developed his philosophy. It is clear in the way he wrote that the land was essential to who he was, and that he considered it as much a part of him as he was of it. This is the epitome of sense of place—another essential component to the development of a land ethic.

An important aspect of identity is place—that is, part of 'knowing what one is' is 'knowing where one is' (Roberts, 1996, p. 66). While place is a foundational concept in the study of geography as well as other academic fields, its meaning and understanding have changed over the years (Relph, 2001). One current understanding of place refers to the interactions between cultural and physical aspects of setting and personality and behavior of individuals, and depends on experience to acquire meaning (Greene, 1996). Place has also been described as being politically charged and connected to global developments (Gruenewald, 2008); the combination of physical environment, human behavior, and social/psychological processes (Stedman, 2003); the meaning of a landscape (Fishwick and Vining, 1992); and a field of care (Tuan, 2001). Most salient to the concept of place, however, is that it is differentiated from physical space by its meaning or value (Relph, 1976; Tuan, 2001). In other words, a location becomes a place based on our personal interactions, experiences, culture, beliefs, and values.

Experiences occur in places. But our experiences of place don't just shape the way we understand and view a given location, they also shape the way we view the world. Places are ultimately the environments in which people live, work, go to school, relax, and vacation; they play a fundamental role in shaping the way we understand, think about, and experience the Earth. In other words, places are our immediately accessible window to the larger world—where we see beauty, where we understand degradation, where we feel big or small, and ultimately where we come to understand our relationship to the surrounding physical environment. If we want to understand people's relationship to the natural environment we must first start with the natural environment as a place, and then consider the various ways people experience that place—a concept often referred to as 'sense of place'.

Sense of place

Place and space have received a great deal of scholarly attention in the past few decades, especially within the study of geography and landscapes. However, widespread agreement as to the use and understanding of these terms is still lacking. Sense of place refers specifically to the emotional, spiritual, and symbolic aspects of places (Kaltenborn, 1997) or to the meanings and emotional bonds people form with places (Williams and Stewart, 1998; Eisenhauer *et al.*, 2000). While lacking a clear definition among researchers, sense of place brings attention to the subjective nature and cognitive and perceptual aspects of experience (Foote and Azaryahu, 2009). Sense of place invokes the concept of human relationships with non-human nature and has been shown to be important to child and adolescent development, self-regulation, attention restoration, self-esteem, self-concept, and identity (Korpela *et al.*, 2001, 2002). Thus, sense of place speaks to the ways in which places are important and meaningful to individuals and how they are connected to sense of self. This relationship between humans and places allows us to better understand how humans connect to the natural environment.

Over the years a variety of place-related terms have been coined to further understand human experiences of place, including place attachment, place satisfaction, and place meaning. Place attachment speaks to the bond or connection that individuals form with meaningful places, while place satisfaction speaks more to an overall judgment of a place's quality or desirability. Place meaning, on the other hand, is a more global construct that refers to the cultural and societal norms that influence the ways people make sense of places and construct their meanings. The proliferation of these place-related terms demonstrates the intensely personal relationship people have with places.

Lately the connection between health and place has garnered more attention in the academic literature. This includes a discussion of therapeutic landscapes, which are 'places that have achieved lasting reputations for physical, mental and spiritual healing' (Kearns and Gesler, 1998, p. 8). Kearns and Gesler discuss common features of these places, including natural elements and outstanding scenery, contribution to sense of place and identity, and sensitivity to cultural values and beliefs.

Place and identity

Environmental issues that are personally relevant, invoke emotion, and are connected to individuals' lives are more important and salient than those issues that seem more disconnected, as might be the case with the deforestation of the Amazon or global warming (Clayton and Opotow, 2003; Crompton and Kasser, 2009). Place is connected to where people actually live and is grounded in global trends as they impact local places. Typically, one does not personally experience the extinction of the polar bears or the hole in the ozone layer. But individuals do experience changes that occur in the places that are important to them. This is important because it focuses attention on economic and political decisions as they impact a given location (Orr, 1992). For example, Sommer (2003) writes about the importance of neighborhood trees to the identity and well-being of local residents, pointing out how trees are woven through our lives physically and metaphorically, and citing a variety of behaviors to underscore this relationship such as city tree-planting programs, activism to save trees, home purchases specifically based on the presence of trees, etc. For example, in the 2013 Dosie fire on Granite Mountain in Arizona, fire crews were able to safely and successfully save an ancient alligator juniper tree. This particular tree had personal significance to many members of the local community; it was associated with people's identity.

Grounding sense of place literature in the existing tenets of social psychology, including beliefs, attitudes, and identity, Stedman (2002) suggests that place attachment is an analog to identity, whereas place satisfaction is an analog to attitudes. In other words, place attachment is aligned with how we view ourselves in relation to place, whereas place satisfaction is related to our positive or negative feelings about a place. This connection between identity and place attachment is further supported in the environmental identity literature (Korpela, 1989; Altman and Low, 1992; Clayton and Opotow, 2003). In a study of land owners near lakes in the North Country of Wisconsin, Stedman (2002) found that both place attachment (identity) and place satisfaction (attitudes) can lead to place-protective behaviors, but through different mechanisms. *High* place attachment fosters behaviors that are protective of places that are salient to our identities, and *low* place satisfaction leads people to improve on non-optimal conditions. In both cases, these behaviors are especially likely to occur when the meanings of a given setting are threatened. While acknowledging the critical nature of meaning in determining attachment, satisfaction, and behavior, Stedman cautions that the nature and origin of place meaning still remain a relative mystery and should be the focus of future investigations. In today's mobile society, individuals may attach to a number of places in their lifetime. It may behoove communities to have ongoing programs that help newcomers to experience, learn about, and value place. For example, numerous communities including Milford, Delaware, Big Rapids, Michigan, and Prescott, Arizona have plaques around town describing unique natural settings such as along river walks, and sometimes describing the interface of culture and the natural environment. Visitors and new residents alike can experience, learn about, and begin to appreciate the setting, which is a step in developing a sense of place.

Placelessness

Place can be both local and global, but if the local is not emphasized then people may experience a sense of deterritorialization or placelessness. Understanding the symbolic nature of place meaning involves acknowledging the social and cultural structures that influence and shape the structure of the landscape, the meanings that ensue, and the

valuation of experiences (Stedman, 2002). Urbanization and suburban development are two such structures that influence sense of place, or the lack thereof (also see the landless discussion in Chapter 2). Placelessness is a phenomenon that is often associated with such development. Relph (1976) describes placelessness as 'the casual eradication of distinctive places and the making of standardized landscapes that results from an insensitivity to the significance of place' (Preface). This is most evident in the overly generic built landscapes of the developed world—the sense of sameness forestalls connection. Relph's book *Place and Placelessness* (1976) takes a phenomenological approach to understanding place and space. Like many place researchers, Relph has come under fire for what some view as an overly simplistic and dichotomous theory of place. However, Seamon and Sowers (2008) argued that the framework espoused in *Place and Placelessness* is flexible and adaptable and provides the necessary language to further explore the concepts of place and space as they relate to human experience. One of the many arguments lodged against place theories is that they fail to take into account the globalizing nature of society and, as such, are somewhat quaint or antiquated. Those who take this standpoint believe that we are no longer connected to specific locales, due to the ease of movement around the globe. Massey (1991) identifies two differing views on place: the first involves a desire for rootedness and place-based identity as a response to globalization and the resulting feelings of vulnerability and unsettledness, and the second eschews the romanticized notion of place as reactionary and ultimately as something that will impede real progress. Massey herself takes a centrist approach, calling for a 'global sense of place' (p. 8). In her critique, she identifies several problems with traditional place theories—the idea that places have single identities, the introverted sense of history, the necessity of drawing geographical boundaries and the resulting distinction between inside and outside, and the perceived interchangeability between place and community. Massey does not attempt to deny the importance of sense of place, only the way in which it is conceptualized. Her view espouses place as a process of interactions, which is integrated with other places 'with a consciousness of the wider world' (p. 7). This viewpoint allows for the concept of multiple identities and for the inherent political and social aspects of place. Similarly, Heise (2008) advocates for an 'environmental world citizenship', which capitalizes or builds on the globalizing nature of society and the resulting 'deterritorialization' (p. 10). Deterritorialization refers to new forms of culture that are no longer anchored in place, forming the basis of Heise's argument that ecological advocacy should be connected to larger systems rather than to specific, local places.

Both Heise and Massey make an important point—environmental advocacy and education cannot afford to become lost in anachronistic longings that fail to take into account the current trajectory of society. Still, studies show that the 'gloom and doom' predictions of global environmental degradation often result in feelings of hopelessness and self-preservation strategies that may actually increase behaviors that are destructive to the natural environment (Crompton and Kasser, 2009). Crompton and Kasser demonstrate that these anxiety-inducing claims lead to a variety of reactions, including strategies for diversion (limiting exposure, focusing in the present, doing something, seeking pleasure), reinterpretation of the threat (e.g. denial of guilt, projection), indifference (apathy, resignation), orienting towards materialism, and denigration of the out-group. These negative effects are sometimes exacerbated by the tactics of environmental campaigns and are often enabled by Western society and culture. However, aspects of environmental identity, which are often tied to specific places, may serve as part of the antidote to some of these socially driven problems.

The Role of Education

Developing a sense of place, establishing healthy environmental identities, and formulating positive relationships with the natural world are arguably best accomplished in childhood. Connecting children with the natural environment increasingly falls on formal and non-formal educational programs, as young people now spend less time living and playing on farmland or undeveloped areas. The US Centers for Disease Control and Prevention recognizes the role of schools in promoting healthy behaviors and providing safe and supportive environments—a mission which should include learning about and spending time in the natural environment. Richard Louv and other leaders in the current movement to reconnect children with nature emphasize the importance of schools. According to the US Department of Education (2008), children spend an average of 6.6 hours each day in school. During this time, implicitly or explicitly, they learn how to view and interact with nature.

Simply being outside or having outdoor recreation experiences may not be enough to instill a sense of stewardship in young people. Rather, the combination of four things—a sense of place, an ethic of care, critical thinking skills, and ecoliteracy skills—fosters stewardship (Litz and Mitten, 2013). There is a recognized need for schools to educate for sense of place by creating innovative curriculum that allows students to create ties with their local environments; many call this place-based education. This personal connection, combined with an ethic of care, motivates positive environmental behavior. Critical thinking and ecoliteracy skills enable successful action. This chapter speaks to how an effective curriculum can address all four of these areas. This is followed by a discussion about how the natural environment can be utilized to create more effective and healthy school environments through the creation of greener school grounds, the act of educating in the outdoors, and the incorporation of the natural environment into classrooms.

Educating for sense of place

What is common to all human societies is their need for a sense of 'place'—a feeling of living in an environment which has boundaries and identity. (Marsh, 1988, p. 27)

Sense of place is especially important in today's globalized and fast-paced society, and concern for sense of place has risen in proportion to the rise in mass culture (Williams and Stewart, 1998). Hay (1998) analyzed sense of place in a developmental context, concluding that sense of place is a basic need—something that all humanity requires and desires; something that cannot be fulfilled by the generic places so prevalent in modern society. Emphasis of sense of place as part of formal and non-formal education programming in schools and in land management situations may enable people to develop a connection to nature and to certain locales that may not have otherwise come to fruition.

Edith Cobb's work (also see Chapters 4 and 6) suggested that a sense of place is vital to a child's evolving personality; for a child 'place' can be a tree, a stream, a knoll, or a hiding spot beneath a group of shrubs. Cobb also correlates adult creativity and happy childhood experiences with time in nature (Cobb and Mead, 1977). Chawla (2002) expanded on Cobb's work and reinforced the importance of childhood time spent in nature in an unthreatening way that encourages a bond based on connection rather than on fear. An overwhelming majority of the adults working in environmental positions whom Chawla (2002) interviewed experienced 'positive experiences of natural environments in childhood and adolescence, and family role models who demonstrated an attentive respect for the natural world' (p. 212).

Acknowledging that human society and culture are nested in ecological systems (Bowers, 2001), Gruenewald (2008) argued for a critical pedagogy of place. This perspective blends critical pedagogy and place-based education into a proposed approach to link the classroom, cultural politics, and local socio-ecological places. More specifically, Greunewald described the following:

> A politicized, multicultural, place-based education would explore how humanity's diverse cultures attempt to live well in the age of globalization, and what cultural patterns should be conserved or transformed to promote more ecologically sustainable communities. (Gruenewald, 2008, p. 318)

Critical pedagogy of place aims: (i) to identify and recover spaces and places that teach people to live well and steward their environments, a process Greunewald identifies as 'reinhabitation'; and (ii) to change ways of thinking that injure or exploit other people and places, or 'decolonization'. This philosophy has strong ties to environmental justice and is important in that it moves beyond simple notions of stewardship or preservation.

A critical pedagogy of place employs the concept of place in two ways—it helps to create a sense of place where one may not have previously existed, and it uses place to further student engagement and understanding. There is a careful balance which must be struck in order to not inundate students with ecological problems before they are ready to think critically about solutions. Sobel (1996) emphasizes the importance of allowing children to develop a relationship with the natural world and a sense of place within their environment before asking them to tackle the myriad of environmental crises. Focusing on place is a way to avoid the psychological shut-down in response to major environmental destruction described by Crompton and Kasser (2009).

Developing an ethic of care

The modern world attempts to solve the ecological crisis through objective rationalism using empirical science, economics, and objectivity to explain and resolve environmental issues. This, however, is insufficient. Such limited reasoning ignores cultural influences, de-emphasizes interpersonal communication, and disregards relationships (Russell and Bell, 1996; Caduto, 1998). In other words, our society lacks a modality dictated by care. Such a schema is necessary for helping people understand and respect themselves, other humans, other species, and the environment (Mitten, 1996; Noddings, 2005). An ethic of care encompasses love for oneself and the surrounding world. It incorporates love for friends, strangers, and distant others. It also includes love for biodiversity and its supporting environment (Noddings, 2005).

Implementing an ethic of care teaches people to accept and embrace all others, including the non-human world. Such admiration is conducive to building relationships and cultivating appreciation for the natural world (Russell and Bell, 1996; Knapp, 1999). This builds deep respect, which encourages an internal commitment to care for another's well-being, including that of the environment (Mitten, 1996). Such respect for the natural world is the first step toward promoting effective environmental stewardship. Knapp (1999) said that this respect translates into a life-loving worldview, which is an essential component of environmental education. Love of nature, love of one's community, and love of self is a prime motivating factor in personal and cultural transformation (Sobel, 2005).

Developing critical thinking skills

Critical thinking involves the ability to grapple with information, reflect, recognize bias, and come to an informed decision or judgment about the topic at hand. Critical thinking skills help students distinguish perceptions from facts, helping them to expand their worldviews and value systems. The Foundation for Critical Thinking (2009) explains that by teaching students critical thinking skills, educators also help them understand the interdependency of ecological, social, political, and cultural influences. Students gain an awareness of the complexity of environmental issues. The development of critical thinking skills is fostered through an education process where students learn how to think instead of what to think (Nosich, 2012). When applied to environmental issues and problems, critical thinking skills enable students to better understand the issue itself and consider various solutions while incorporating a wide range of information and sources. This is an essential corollary to the development of sense of place and an ethic care, because the inability to think critically about the problem precludes the ability to move towards action or solutions (Litz and Mitten, 2013).

Developing ecoliteracy skills

Finally, education for the environment requires fostering development of ecoliteracy skills in students. Ecoliteracy, or ecological literacy, is the ability to understand and apply ecological principles and processes to everyday life and decision making (Center for Ecoliteracy, 2000). Knowing the principles of ecology and conservation biology is essential to attaining ecoliteracy; however, appropriate action based on this knowledge rests on the development of sense of place and ethic of care. In other words, students must feel an emotional connection to a place before they will act on factual knowledge they have also gained (Litz and Mitten, 2013).

Combining skills, attitudes, and knowledge gained by nurturing a sense of place, an ethic of care, critical thinking skills, and ecoliteracy skills fosters stewardship and forms a larger educational whole. Students are then equipped to know and experience the natural environment, to care for both the environment and for those around them, to think critically about the problems they face in their communities and their places, and to skillfully act on those issues in informed and sustainable ways. Essentially they learn to live in accord with nature (Knapp, 1999), guided by an environmental ethic of care. Figure 8.1 demonstrates the relationship between these concepts.

'Green' schools

The value of school grounds for learning is often underestimated, especially in secondary schools. Additionally, maintaining green open space in schoolyards is usually a lower priority than building classrooms and sport facilities. A charity or non-profit organization in England called Learning through Landscapes aims to improve school grounds and researches the effects of green schoolyard projects on pupils. Its evaluative reports indicate that teachers whose schools have taken part in this transformation process report more creative learning and environmental awareness in their students. Teachers also

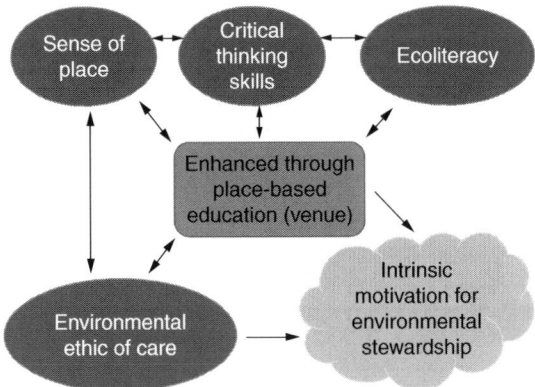

Fig. 8.1. Model of motivation for environmental stewardship (Adapted from Litz & Mitten, 2013.)

report improved learning attitudes, behavior, self-esteem, and social interaction, and decreased bullying when kids play outside in improved greenspaces.

Forest kindergartens

Ruth Wilson (2012) in her book, *Nature and Young Children: Encouraging Creative Play and Learning in Natural Environments*, explains the need to create many opportunities for children to feel nature and not just think about it. Evidence of a growing trend to capitalize on this notion is the increasing number of schools that take place primarily or entirely outdoors. For example, Germany, the country responsible for the idea of kindergarten, recently implemented waldkindergarten, or 'forest kindergartens'. These new schools are just as they sound—kindergarten programs for 3- to 6-year-olds that take place almost entirely outdoors. The more than 700 of these programs currently operating in Germany testify that connecting children with nature is essential, especially during formative years (Esterl, 2008). Children enrolled in these programs learn, grow, and explore in the natural environment, rain or shine. These programs are catching on in other developed countries, with a forest kindergarten in the US recently opened in Portland, Oregon. Friluftsliv also enters educational pedagogy, with friluftsliv kindergartens in Norway and other Scandinavian countries, where the philosophy is that the children will learn better through having this entire phase of their educational experience outdoors. Won Sop Shin, recently appointed Secretary of Forestry for the Republic of Korea and Vice President of the International Society of Nature and Forest Medicine (INFOM), announced that, as part of the 'From cradle to grave: Life with Forests' policy, a Forest Kindergarten Program was established that by 2012 included 3910 institutions and 420,000 people (Anon., 2013).

Natural playgrounds

Another recent trend in connecting children with nature is the development of 'natural playgrounds'. These playgrounds remove traditional play structures like swings and slides, which have prescribed uses and limit creativity, and replace them with natural elements like logs, stumps, rocks, and sand, which encourage creative interaction such as climbing and building. These types of playgrounds encourage cooperative play, healthy risk taking, and hands-on interaction with natural elements, as well as safe and comfortable transitions from indoor or constructed playground areas to wild natural areas. In addition to playgrounds, a variety of resources are increasingly available to schools, parks, and other children's venues to assist in the creation of natural play areas that are designed to promote exploration and free play. These include Nature Explore Classrooms (Arbor Day Foundation and Dimensions of Educational Research Foundation, 2008), nature discovery areas, and natural playscapes.

School environment and architecture

In addition to viewing schools as places to educate about and forge connections with the natural environment, researchers have also identified ways in which the natural environment enhances the educational experience. The idea of 'greening' schools, or making them more environmentally friendly, has been shown to be beneficial not only to the environment, but also to student and teacher health, academic performance, community health and well-being, as well as the overall financial health of the institutions themselves (Kats, 2006). For example, daylight exposure in the classroom has been linked with improved math and reading scores and overall improved health for students (Frumkin and Fox, 2011). Similarly, plants in classrooms and gardens on school grounds have been shown to have a positive effect on learning outcomes. In one study, plants were placed in college classrooms for the duration of the semester. Students' course performance and evaluations of their instructor were compared with those of students who took the same class in the same room, with the same professor during the same semester, but without plants (Doxey *et al.*, 2009). Plants were found to have a positive impact on students' performance and their feelings about the class and the instructor. This was especially true in classrooms without other natural elements such as windows.

In another study that examined 101 public high schools in Michigan, nearby nature was shown to have positive effects on test scores, graduation rates, plans to attend college, and reductions in criminal behavior (Matsuoka, 2010). Nearby nature includes views of trees and shrubs from classroom and cafeteria windows, and the ability to directly interact with natural elements on school grounds. Interestingly, the exposure to landscapes lacking in natural features had a negative relationship with tested items. In a report for a Canadian organization called Evergreen,

the greening of school grounds has been linked not just to a number of benefits for students, but also for teachers, the surrounding community, and the schools and curricula (Raffan, 2000). Benefits for students included increased meaningful play that is more inclusive and gender-neutral; a safer, less hostile, less toxic environment; and improved academic performance, greater creativity, enhanced sense of place, and more pride in learning than in less 'green' school grounds. Creating 'green' educational environments provides direct and tangible benefits to students and teachers, while also conveying the significance of coexisting with our natural surroundings.

The Role of Public Land Management

The connections between schools and greenspaces are being realized in multiple ways—from holding class outdoors, to greening school environments and creating more natural play spaces, to the maintenance and propagation of school forests. This ensures at least some form of connectedness to nature for children; however, exposure to nature at school is not enough. For the same reasons that necessitate greener school grounds, it is becoming increasingly important for individuals to have access to public lands and other open greenspaces. Without these managed and protected areas, many individuals would not have access to natural places. Managing natural areas with this in mind means actively creating spaces where people can connect to the natural environment in meaningful ways. It is this sense of connection, meaning, and value that will encourage regular visitation and ultimately stewardship and support.

Managing for sense of place

Public lands are another realm in which a sense of place can be developed. Williams and Stewart (1998) discuss the importance of sense of place in land management, noting that place meanings are not limited to residents of a given place. The increasing complexity of social, cultural, and political forces leads Williams and Stewart to describe place meanings as less stable than in the past and fraught with increasing complexity from competing land interests: 'Virtually any resource or land-use planning effort is really a public exercise in describing, contesting, and negotiating competing senses of place and ultimately working out a shared future sense of place' (p. 23). Despite these challenges, sense of place is an important aspect of public lands and should be taken into management consideration (Ewert, 1996; Putney, 2003). Williams and Stewart (1998) suggest several day-to-day management tactics that can help to engender a sense of place, including knowing and using local place names, communicating management plans in place-specific terms, understanding the politics of place, and paying special attention to places that have multiple meanings among user groups, including indigenous groups.

Ultimately, effective natural resource management strategies rely on understanding the forces that drive human behavior and understanding how to modify human behavior (Ewert, 1996). Managing for sense of place requires an understanding of the complex interactions between a place and the various groups of people who associate with it. The activities that people do in a given place are usually of less importance than the environmental features or the social interactions that take place there (Eisenhauer *et al.*, 2000). These connections are important because they often result in a higher sense of caring and stewardship of the land.

Stern *et al.* (1999) bring attention to the importance of individual environmental values and the sense that those values are threatened. As the population of the US becomes increasingly more urban and suburban, education and land management will play increasingly influential roles in fostering the valuation of nature. Additionally, these agencies will play a role in shaping peoples' perceptions and evaluations of environmental degradation. Sense of place is an important aspect of one's identity and results from the meaning and value of specific locales. Developing a sense of place, or a relationship with the natural environment, may be the key to creating an environmental identity and can mediate the relationship between global effects and personal relevance (Linneweber *et al.*, 2003).

Inspiring Action

We cannot win this battle to save species and environments without forging an emotional bond between ourselves and nature as well— for we will not fight to save what we do not love.

S.J. Gould

All of this is to say that nature is good for us, and that we, in turn, can be good for nature. It starts with the formation of a connection— something local, something small, and something personal and meaningful. We do not necessarily connect with the rainforests of the Amazon or with the mountains of the Himalayas (unless we live there, of course), but we can develop an appreciation for the beauty, grandeur, immensity, and importance of these places if we can first appreciate that which is most immediately around us. One connection leads to the next. The human species has demonstrated time and again its incredible capacity for creative problem solving, innovation, and compassion, so perhaps it is merely a matter of beginning with a sense of importance—a sense of place—and the rest will follow.

References

Altman, I. and Low, S.M. (eds) (1992) *Place Attachment*. Plenum, New York.

Anon. (2013) An Interview with Republic of Korea Secretary of Forestry Won Sop Shin posted July 21, 2013. Available at: http://hikingresearch.wordpress.com/2013/07/21/an-interview-with-republic-of-korea-secretary-of-forestry-won-sop-shin/ (accessed 23 July 2013).

Arbor Day Foundation and Dimensions of Educational Research Foundation (2008) Nature Explore. Available at: http://www.natureexplore.org/ (accessed 22 July 2013).

Bowers, C. (2001) *Educating for Eco-justice and Community*. University of Georgia Press, Athens, Georgia.

Caduto, M. (1998) Ecological education: a system rooted in diversity. *Journal of Environmental Education* 29(4), 11–16.

Center for Ecoliteracy (2000) *Ecoliteracy. Mapping the Terrain*. Center for Ecoliteracy, Berkeley, California.

Chawla, L. (2002) Spots of time: manifold ways of being in nature in childhood. In: Kahn, P.H. Jr and Kellert, S.R. (eds) *Children and Nature: Psychological, Sociocultural, and Evolutionary Investigations*. MIT Press, Cambridge, Massachusetts, pp. 199–225.

Clayton, S. and Opotow, S. (2003) *Identity and the Natural Environment: The Psychological Significance of Nature*. MIT Press, Cambridge, Massachusetts.

Cobb, E. and Mead, M. (1977) *The Ecology of Imagination in Childhood*. Columbia University Press, New York.

Crompton, T. and Kasser, T. (2009) *Meeting Environmental Challenges: The Role of Human Identity*. WWF-UK, Godalming, UK.

Doxey, J.S., Waliczek, T.M. and Zajicek, J.M. (2009) The impact of interior plants in university classrooms on student course performance and on student perceptions of the course and instructor. *HortScience* 44(2), 384–391.

Eisenhauer, B., Krannich, R. and Blahna, D. (2000) Attachments to special places on public lands: an analysis of activities, reason for attachments, and community connections. *Society & Natural Resources* 13(5), 421–441.

Esterl, M. (2008) German tots learn to answer call of nature. *The Wall Street Journal*, 14 April 2008. Available at: http://online.wsj.com/article/SB120813155330311577.html (accessed 16 September 2013).

Ewert, A. (ed.) (1996) *Natural Resource Management: The Human Dimension. Social Behavior and Natural Resources Series*. Westview Press, Boulder, Colorado.

Fishwick, L. and Vining, J. (1992) Toward a phenomenology of recreation place. *Journal of Environmental Psychology* 12(1), 57–63.

Foote, K.E. and Azaryahu, M. (2009) Sense of place. In: Rob, K. and Nigel, T. (eds) *International Encyclopedia of Human Geography*. Elsevier, Oxford, UK, pp. 96–100.

Foundation for Critical Thinking (2009) Our concept of critical thinking. Available at http://www.criticalthinking.org/aboutCT/ourConceptCT.cfm (accessed 26 September 2013).

Frumkin, H. and Fox, J. (2011) Healthy schools. In: Dannenberg, A., Frumkin, H. and Jackson, R. (eds) *Making Healthy Places: Designing and Building for Health, Well-being, and Sustainability*. Island Press, Washington, DC, pp. 216–228.

Greene, T. (1996) Cognition and the management of place. In: Driver, B.L., Dustin, D., Baltic, T., Elsner, G. and Peterson, G. (eds) *Nature and the Human Spirit: Toward an Expanded Land Management Ethic*. Venture Publishing, State College, Pennsylvania, pp. 301–310.

Gruenewald, D. (2008) The best of both worlds: a critical pedagogy of place. *Environmental Education Research* 14(3), 308–324.

Hay, R. (1998) Sense of place in developmental context. *Journal of Environmental Psychology* 18(1), 5–29.

Heise, U. (2008) *Sense of Place and Sense of Planet: The Environmental Imagination of the Global*. Oxford University Press, New York.

Kaltenborn, B. (1997) Nature of place attachment: a study among recreation homeowners in southern Norway. *Leisure Sciences* 19(3), 175–189.

Kats, G. (2006) *Greening America's Schools*. Capital E, Washington, DC.

Kearns, R.A. and Gesler, W.M. (1998) *Putting Health into Place: Landscape, Identity, and Well-being*. Syracuse University Press, Syracuse, New York.

Kellert, S.R. (1993) Introduction. In: Kellert, S.R. and Wilson, E.O. (eds) *The Biophilia Hypothesis*. Island Press, Washington, DC, pp. 20–31.

Knapp, C.E. (1999) *In Accord with Nature: Helping Students Form an Environmental Ethic Using Outdoor Experience and Reflection*. ERIC/CRESS, Appalachia Educational Laboratory, Charleston, West Virginia.

Korpela, K. (1989) Place-identity as a product of environmental self-regulation. *Journal of Environmental Psychology* 9(3), 241–256.

Korpela, K., Hartig, T., Kaiser, F. and Fuhrer, U. (2001) Restorative experience and self-regulation in favorite places. *Environment and Behavior* 33(4), 572–589.

Korpela, K., Kyttä, M. and Hartig, T. (2002) Restorative experience, self-regulation, and children's place preferences. *Journal of Environmental Psychology* 22(4), 387–398.

Learning through Landscapes (2013) Homepage. Available at: http://www.ltl.org.uk/index.php (accessed 2 September 2013).

Leopold, A. (1949) *A Sand County Almanac*. Oxford University Press, London.

Linneweber, V., Hartmuth, G. and Fritsche, I. (2003) Local environment threatened by global climate change: toward a contextualized analysis of environmental identity in a coastal area. In: Clayton, S. and Opotow, S. (eds) *Identity and the Natural Environment: The Psychological Significance of Nature*. MIT Press, Cambridge, Massachusetts, pp. 227–246.

Litz, K. and Mitten, D. (2013) Inspiring environmental stewardship: developing a sense of place, critical thinking skills, and ecoliteracy to establish an environmental ethic of care. *Pathways, The Ontario Journal of Outdoor Education* 25(2), 4–8.

Marsh, P. (1988) *Tribes*. Peregrine Smith Books, Salt Lake City, Utah.

Massey, D. (1991) A global sense of place. *Marxism Today* 35(6), 24–29.

Matsuoka, R.H. (2010) Student performance and high school landscapes: examining the links. *Landscape and Urban Planning* 97(4), 273–282.

Mitten, D. (1996) The value of feminist ethics in experiential education teaching and leadership. In: Warren, K. (ed.) *Women's Voices in Experiential Education*. Kendall/Hunt, Dubuque, Iowa, pp. 159–171.

National Research Council (1981) *Indoor Pollutants*. National Academy Press, Washington, DC, p. 537.

Natural England (2009) *Childhood and Nature: A Survey on Changing Relationships with Nature across Generations*. England Marketing, Warboys, UK.

Noddings, N. (2005) *The Challenge to Care in Schools: An Alternative Approach to Education*. Teachers College Press, New York.

Nosich, G. (2012) *Learning to Think Things Through: A Guide to Critical Thinking in the Curriculum*. Prentice Hall, New York.

Orr, D.W. (1992) *Ecological Literacy: Education and the Transition to a Postmodern World*. State University of New York Press, Albany, New York.

Plumwood, V. (2002) *Environmental Culture: The Ecological Crisis of Reason*. Routledge, New York.

Putney, A. (2003) Introduction. In: Putney, A. and Harmon, D. (eds) *The Full Value of Parks: From Economics to the Intangible*. Rowman & Littlefield Publishers, Lanham, Maryland, pp. 3–12.

Raffan, J. (2000) *Nature Nurtures: Investigating the Potential of School Grounds*. Evergreen, Toronto, Canada.

Relph, E. (1976) *Place and Placelessness*. Pion, London.

Relph, E. (2001) Place in geography. In: Neil, J.S. and Paul, B.B. (eds) *International Encyclopedia of the Social & Behavioral Sciences*. Pergamon, Oxford, UK, pp. 11448–11451.

Roberts, E. (1996) Place and spirit in public land management. In: Driver, B.L., Dustin, D., Baltic, T., Elsner, G. and Peterson, G. (eds) *Nature and the Human Spirit: Toward an Expanded Land Management Ethic*. Venture Publishing, State College, Pennsylvania, pp. 61–80.

Rolston, H. (2000) The land ethic at the turn of the millennium. *Biodiversity and Conservation* 9(8), 1045–1058.

Russell, C.L. and Bell, A.C. (1996) A politicized ethic of care: environmental education from an ecofeminist perspective. In: Warren, K. (ed.) *Women's Voices in Experiential Education*. Kendall/Hunt Publishing Co., Dubuque, Iowa, pp. 172–181.

Science & Environmental Health Network (1998) Wingspread Statement on the Precautionary Principle. Available at: http://www.sehn.org/wing.html (accessed 15 April 2013).

Seamon, D. and Sowers, J. (2008) Place and placelessness, Edward Relph. In: Hubbard, P., Kitchen, R. and Vallentine, G. (eds) *Key Texts in Human Geography*. Sage, London, pp. 43–51.

Sobel, D. (1996) *Beyond Ecophobia: Reclaiming the Heart in Nature Education*. The Orion Society and the Myrin Institute, Great Barrington, Massachusetts.

Sobel, D. (2005) *Place-based Education: Connecting Classrooms and Communities*, 2nd edn. The Orion Society, Great Barrington, Massachusetts.

Sommer, R. (2003) Trees and human identity. In: Clayton, S. and Opotow, S. (eds) *Identity and the Natural Environment: The Psychological Significance of Nature*. MIT Press, Cambridge, Massachusetts, pp. 179–204.

Stedman, R.C. (2002) Toward a social psychology of place: predicting behavior from place-based cognitions, attitude, and identity. *Environment and Behavior* 34(5), 561–581.

Stedman, R. C.(2003) Is it really just a social construction? The contribution of the physical environment to sense of place. *Society & Natural Resources* 16(8), 671–685.

Stern, P., Dietz, T., Abel, T., Guagnano, G.A. and Kalof, L. (1999) A value-belief-norm theory of support for social movements: the case of environmentalism. *Human Ecology Review* 6(2), 81–97.

Tuan, Y. (2001) *Space and Place: The Perspective of Experience*. University of Minnesota Press, Minneapolis, Minnesota.

US Department of Education (2008) Public School Data Files, 2003–04 and 2007–08. National Center for Education Statistics, Schools and Staffing Survey (SASS). Available at: http://nces.ed.gov/surveys/sass/tables_list.asp (accessed 16 September 2013).

Williams, D. and Stewart, S. (1998) Sense of place: an elusive concept that is finding a home in ecosystem management. *Journal of Forestry* 96(5), 18–23.

Wilson, E.O. (1984) *Biophilia: The Human Bond with Other Species*. Harvard University Press, Cambridge, Massachusetts.

Wilson, R. (2012) *Nature and Young Children: Encouraging Creative Play and Learning in Natural Environments*. Routledge, Abingdon, UK.

9

Innovative Approaches to Integrating Natural Environments and Health

---

We give the grass a name, and earth a name. We say earth and grass are separate. We know this because we can pull grass free of the earth and see its separate roots – but when the grass is free, it dies.

Susan Griffin (1980)

In this chapter we explore the use of adventure-based activities as an innovative way to integrate the concepts of natural environments and human health. In this case, adventure activities are defined as 'A variety of educational and recreational activities and experiences utilizing a close interaction with the natural environment, that contain elements of real or apparent risk, in which the outcome is uncertain, but can be influenced by the participant and/or circumstance' (Ewert *et al.*, 2006). From an individual health perspective, these types of activities can be most effective in promoting health in three major ways: (i) the activities themselves usually demand a relatively high level of physical activity; (ii) adventure activities can be powerful determinants of individual behaviors and attitudes; and (iii) adventure activities most often take place in natural environment settings, such as wilderness areas, and therefore provide an avenue for exposure to the natural world. We begin this chapter with an overview of the adventure experience, move to the therapeutic uses of adventure, discuss adventure as a health-transforming experience, and conclude with

the use of adventure as intentionally designed experiences that can work in concert with natural environments to promote human health.

An Overview of Adventure-Based Activities

While there is a long history of considering natural environments as places to experience health, in 1986, the President's Commission on Americans Outdoors formally stated that the outdoors was a 'giant health machine' that both enriched the lives of millions of citizens and promoted the development and practice of positive health-related behaviors. By 2010, participation data clearly indicated that what people do within a recreational framework in the outdoor environment can play a significant role in the health of specific individuals as well as the collective health of society. Consider the data generated from the *Outdoor Recreation Participation Report 2011*, regarding why people participate in outdoor activities for the first time (Outdoor Foundation, 2011). As presented in Table 9.1, the data convincingly point to an increasing number of people choosing to participate in outdoor recreation specifically to enhance their health from a personal, familial, and social perspective.

When considering health and wellness, it is of particular interest that for individuals over

Table 9.1. Reasons for first-time participation in outdoor activities. (Data are from the *Outdoor Recreation Participation Report 2011*; Outdoor Foundation, 2011.)

Reason	Percentage of respondents citing the reason	
	Age 6–24 years	Age 25+ years
My friends and/or family participate in outdoor activities	54	35
I wanted to try something new	38	39
Outdoor activities help me stay fit and healthy	25	39
Outdoor activities bring my family together and strengthen family ties	23	23
Exercising outdoors is more fun/motivating than exercise indoors	22	30
Outdoor activities are affordable	17	27
Outdoor activities are close to my home	15	21
I was introduced to outdoor activities at school	14	3
Outdoor activities help me relax and manage stress	11	31
Outdoor activities give me a chance to get back to nature	11	25
My kids are the right age now	8	17
I was pressured by others	5	7
I saw an article, show or video	2	3

the age of 25 years, the desire to stay fit and healthy ties with wanting to try something new as the number one reason for first-time participation in outdoor adventure activities (39%), followed by managing stress (31%) and trying to connect with nature (25%). Clearly, the data suggest that outdoor recreational activities can be an important variable in considering individual health, wellness, and quality of life. Educational and recreational outdoor activities that involve elements of potential risk, danger, and challenge are an important and growing component of the use of outdoor environments and an important consideration in terms of their relationship to human health and well-being.

There are a number of outdoor-related activities and venues that have documented connections with human health and have received varying degrees of research attention. A sampling of these activities would include gardening, walking, viewing the scenery, playing, and environmental educative endeavors. Of primary focus for this chapter, however, are activities related to adventure-based recreational and educational types of engagements such as wilderness trips, camps, and participation in rope courses. A sample of these types of adventure activities is illustrated in Table 9.2.

Despite the potential risk and danger attendant in these activities, their popularity can be ascertained both through participant numbers

Table 9.2. Sample of adventure-based activities.

Rafting
Rock climbing
Kayaking/canoeing
Challenge activities
Rope courses
Ice climbing
Backpacking
Wilderness/backcountry trekking

and types of activities. For example, since 2009, participation rates for adventure-based leisure pursuits such as kayaking and climbing have increased by over 25% (Outdoor Foundation, 2011). Outdoor adventure activities have seen dramatic growth in participation over the last few decades for a variety of reasons. The gear is increasingly accessible and affordable, it is easier to find a class, trip or program, and the visibility of adventure is heightened. But perhaps the underlying reason for all of this is more systemic—perhaps it is an artifact of an increasing discontent with the safety and confines of modern society. Adventure has been regarded by a number of scholars as a means of psychological escape from modernism (Mortlock, 2002; Lynch and Moore, 2004). Adventure is, after all, a journey into the unknown and represents the search for authenticity in the modern age.

Therapeutic and Educational Uses of Adventure

The lure of adventure and the accompanying risk and uncertainty make it ripe for personal growth and development. Over the last several decades in the US, practitioners have capitalized on this potential, utilizing adventure not just for recreation, but also for education and for therapy. When adventure activities are done within an educational structure, the term adventure education is generally used. Adventure education has been defined as 'a variety of self-initiated activities utilizing an interaction with the natural environment that contains elements of real or apparent danger, in which the outcome, while uncertain, can be influenced by the participant and the circumstance' (Ewert, 1989, p. 6). If, on the other hand, these activities are engaged in primarily for recreational purposes, the term adventure recreation or outdoor pursuits is often used. In a similar fashion, if the principal goal underlying participation is therapeutic or rehabilitative, a number of terms have emerged including wilderness therapy, adventure based-counseling, and adventure therapy (Gass et al., 2012).

According to Priest and Gass (1997), therapeutic adventure programming is 'aimed at changing dysfunctional behavior patterns, using adventure experiences as forms of habilitation and rehabilitation'. Using adventure as a therapeutic modality was popularized in the 1970s and continues to be an effective form of treatment for a variety of physical, behavioral, and psychological issues. Often inherent in these programs is the use of an underlying clinical approach, couched within an APIE (assessment, planning, implementation, and evaluation) process or structure (Bullock and Mahon, 2001). Usually each client or participant is assessed on an individual basis for the purpose of goal setting. The *assessment* of a client then becomes the foundation for treatment and the *planning* of an appropriate therapeutic experience, often in the form of an 'individual treatment plan'. The treatment plan directs the interventions or activities that are then *implemented* and the client is observed and assessed for improvement or, in some cases, reassessment. The outcomes of the treatment plans, or therapeutic interventions, are measured through *evaluation* at the end of participation in the adventure activity or program (Groff and Dattilo, 2000).

Adventure-based activities have served a therapeutic function for a broad spectrum of situations and populations. For example, adventure-based therapy has been used to treat overweight adolescents (Jelalian et al., 2006), as a form of marriage and family therapy (Gass, 2007), as a means of enhancing quality of life for adolescents with cancer (Stevens et al., 2004), as life skills development for women offenders (Mitten, 1998), and as a means of promoting behavioral health for at-risk youth (Russell, 2003). Recently, adventure programs have also been shown to be an effective treatment setting for military veterans. For example, Van Puymbroeck et al. (2012) found that an adventure-based program (Outward Bound) could be effective in enhancing levels of sense of coherence and subsequent health-related factors.

While there is currently a plethora of both adventure education programs and adventure therapy programs, Baldwin et al. (2004) posit that there are some essential commonalities across many of these programs. These commonalities involve the following:

1. planned and purposive use of adventure-based activities with specific goals;
2. real-life learning contexts;
3. participant becomes the agent of change;
4. goal-directed challenges necessitating the use of individual or group-generated solutions;
5. outdoor or natural environment setting;
6. small group (usually ten or less) context; and
7. structured facilitation.

Thus, the adventure education experience can be a systematic and structured event that can be both health-enhancing and health-promoting.

The Intersection of Adventure and Health

While it is increasingly clear that adventure is both attractive and desirable to many individuals, its relationship to health may be a little less clear. Beyond the more obvious uses of adventure toward a specific end such as education or therapy, how does the act of participating in adventure lead to enhanced health and well-being? Relative to this relationship, three primary questions need to be considered. First, is adventure beneficial? Second, in what specific ways? And third, how is adventure beneficial, or through what mechanisms? Following a discussion of these three questions we consider some of the specific components of adventure and their relationship to human health.

Is adventure beneficial to human health?

Beginning with Shore's (1977) overview of outdoor adventure research, a substantial body of research has accumulated revealing the effectiveness of adventure education activities and its relationship to human health. For example, participation in outdoor and adventure education has often been associated with developmental outcomes such as personal growth, enhanced interpersonal skills, and group development (Ewert and Garvey, 2007; Passarelli et al., 2010).

A substantial number of research studies now point to the development of variables that have a direct association with health. For example, resilience has been conceptualized as the ability to bounce back from a disruptive event (Flach, 1997) and can be thought of as reflecting the ability to maintain a stable equilibrium, often through protective systems such as a supportive group or improved self-efficacy beliefs. Thus, resiliency could be impacted through participation in training experiences where people experience challenge, learn to negotiate difficulty, and grow from the experience. This idea of growth and development through challenge is aligned with the philosophies of adventure education, and research demonstrates that adventure experiences can be effective in enhancing resilience scores. For example, Ewert and Yoshino (2011) found that relatively short-term adventure experiences can influence levels of resilience among college students, and the inclusion of adventure programming in an anti-bullying initiative

was also found to positively impact resilience in youth (Beightol et al., 2012).

Another variable that appears in much of the adventure literature and has connections to human health is self-efficacy (Bandura, 1986). Self-efficacy can be defined as an 'individual's beliefs in his or her abilities to execute necessary courses of action to satisfy situational demands' (McAuley et al., 2001, p. 236). Self-efficacy is hypothesized to influence behavior related to health because of several mechanisms: (i) through the adoption of health behaviors or the cessation of unhealthy behaviors; and (ii) by influencing biological and physiological factors such as stress and control over one's life (Maddux and Gosselin, 2003). Adventure activities are thought to be effective in enhancing variables such as self-efficacy because of the interaction between success and challenge that they provide the participant (Sibthorp, 2003; Sheard and Golby, 2006). In a similar fashion, Havitz et al. (2013) found a positive relationship between the development of self-efficacy and levels of physical activity.

While much of the existing research focuses on psychological outcomes, adventure is also beneficial through its promotion of physical activity, strengthening of familial bonds and social relationships, reduction of stress, and enhancement of spiritual well-being. Recent research indicates that adventure activities can be a viable means of promoting family togetherness and strengthening parent–child bonds (Hill et al., 2001; Ewert et al., 2011). This is partially because of the small-group atmosphere, intensive togetherness, and separation from technology and daily life. These same elements have been shown to enhance overall feelings of social support, another important element of personal well-being. Whether it is a part of a recreational endeavor or an educational experience, adventure appears to be effective in creating positive health outcomes through factors such as the demands for physical activity and the enhancement of variables that can lead to positive health behaviors and subsequent positive benefits. With these things in mind, this chapter proposes that five specific benefits arise from participation in adventure activities:

1. encouragement of physical activity (de Vries, 2010);
2. development of social contacts and relationships (McAvoy *et al.*, 1992);
3. encouragement of optimal development;
4. opportunities for developing a greater sense of personal awareness and a sense of achievement and empowerment (Mitten, 1992; Ewert *et al.*, 2010); and
5. recovery from stress and attention fatigue.

How is adventure beneficial?

In many ways, measuring the outcome of participation in an adventure experience is easier than understanding the process through which the outcome is created. McKenzie (2000) attributes these outcomes to four characteristics common to many adventure education experiences: (i) the unfamiliar nature of the physical environment; (ii) the incremental and progressive sequencing of the challenges presented through the adventure education experience; (iii) the 'processing' of the experience in order to identify and organize meaning for the participant; and (iv) the use of small groups to facilitate issues such as reciprocity, group cohesiveness, interpersonal relationships, and the balance between group belongingness and individual autonomy. In addition to the four components described by McKenzie, the adventure education experience also provides opportunities for outdoor physical activity, facing concrete challenges while being part of a small team, and being part of a shared experience that can facilitate future dialogue and discourse. Adventure scholars have produced a number of models that work to conceptualize this process.

Outward Bound process model

From a theoretical perspective, the most widely emulated process model used in adventure education is the Outward Bound process model (OBPM), advocated by Walsh and Golins (1976). As depicted in Fig. 9.1, this model describes a series of seven processes to explain the impact and effects of Outward Bound programs on participants.

As can be seen in Fig. 9.1, the OBPM suggests that upon conclusion of the experience, the learner will continue to be positively oriented to further learning and developmental experiences; a process known as transference. This is an important concept within the adventure field because most participants experience adventure for a relatively short duration before returning to their daily life. It is the process of transference that often captures the benefits of adventure participation.

Components of the adventure experience

The unique components of adventure experiences, such as interactions with the natural environment, elements of risk and challenge, and uncertainty of outcome, are often key elements in the contribution to personal transformation, as well as enhanced health and well-being. This section provides an overview of each these components while linking them to existing research.

The learner is placed into a

↓

Prescribed physical and social environment,
then given a

↓

Characteristic set of problem-solving tasks,
which creates a state of

↓

Adaptive dissonance, leading to

↓

A sense of mastery and competence,
which, in turn, leads to

↓

Re-organization of the meaning and direction
of the experience, which, in turn, leads to

↓

Transference of new skills and attitudes

Fig. 9.1. The Outward Bound process model (OBPM). (Adapted from Walsh and Golins, 1976.)

Interaction with the natural environment

A salient feature of many adventure education experiences is the expressed use of natural or outdoor environments. Numerous studies now provide support for the restorative value that outdoor and natural environments can have on an individual's health parameters, including stress reduction, activity levels, and quality of life (Ward Thompson and Aspinall, 2011).

There are a number of ways in which natural environments may promote human health by reducing stress. Natural environments can often provide the setting for physical activity, with numerous studies reporting the beneficial effects of 'green' exercise (Driskell *et al.*, 2001). Natural environments have been linked to the attention restoration theory (ART) in which these settings possess a particular set of properties that promote restoration from attention fatigue (Kaplan and Kaplan, 1989; Kaplan and Berman, 2010). Similar to ART, the psychoevolutionary theory (PET) posits that natural environments are effective at reducing levels of stress because they offer specific attributes that our species viewed as having inherent survival qualities, such as water and spatial openness (Ulrich, 1984). Hartig *et al.* (1991) integrate these two theories by suggesting that there is an 'intertwining of the mechanisms', whereby the extent to which people are attracted to and use a natural environment is dependent on how restorative that specific environment is to them. Finally, Degenhardt *et al.* (2011) identified a number of variables such as state of health, self-efficacy, and quality of the neighborhood that have a direct bearing on the frequency and type of use of natural environments.

Garst *et al.* (2010) found that restoration was a primary meaning associated with developed forest camping, and that participants viewed the natural environment as an opportunity for escape, rest, and recovery from their daily routines. The authors compared contemporary meanings with those from earlier research in the 1960s and found that while the meanings have not changed dramatically, the increasing societal disconnect from the natural environment is creating a 'greater urgency and a deeper need for nature in a world in which some of the experiences afforded by camping have become fewer'. Similarly, a number of authors have found links between participation in adventure programs and resistance to stress (Ewert, 1988; Bunting *et al.*, 2000; Pryor *et al.*, 2005).

Directly related to this is the work by Korpela *et al.* (2010) that demonstrated a link between the desire for restoration (e.g. freedom from worries and stress) and the use of 'environmental self-regulation strategies (ERS)'. Korpela and his colleagues define ERS as personal behaviors, such as visitation to selected locations (e.g. parks and other natural environments), for the specific acquisition of restorative outcomes. That is, individuals deliberately engage in selected activities for the specific purpose of achieving a particular goal such as reducing stress.

Elements of risk and challenge

As stated previously, adventure education activities often present structurally inherent possibilities for risk and challenge. This risk can be either apparent or real. Apparent risk is often contingent on the perceptions of the participant; that is, the activity may appear risky but in reality has relatively low levels of actual risk, or vice versa. On the other hand, real risk typically involves objective hazards that pose a potential for loss, such as avalanche slopes, rock falls, fast and deep water, or lightening.

What makes risk and challenge important in understanding the connection between adventure education and health is their ability to facilitate and create change in the individual. Through the mechanisms of risk and challenge, it has been proposed that adventure education activities can be effective in enhancing a number of variables often associated with health such as resilience (Ewert and Yoshino, 2011), empowerment (Autry, 2001), and body image (McDermott, 2004). The OBPM (Walsh and Golins, 1976) describes the process used by many adventure education programs for creating personal change through the purposive use of risk and challenge. Perhaps not surprisingly, the OBPM is similar to the psychosomatic model (Snegroff, 2012) depicted in Fig. 9.2.

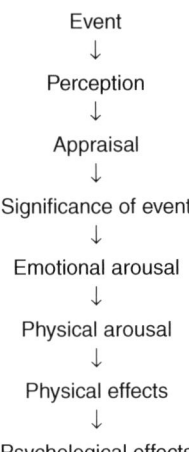

Event
↓
Perception
↓
Appraisal
↓
Significance of event
↓
Emotional arousal
↓
Physical arousal
↓
Physical effects
↓
Psychological effects

Fig. 9.2. The psychosomatic model. (Adapted from Snegroff, 2012.)

What is unique about the intersection of these two models is that instead of considering concepts such as risk and challenge as negative or disease-producing variables, adventure education utilizes these situations as positive and health-enhancing tools. That is, engaging with risk and challenge in a structured and purposive way can enhance individual attributes such as resilience that, in turn, can provide for positive enhancement or promotion of positive health-related behaviors, such as increased physical activity or smoking cessation.

Uncertain outcomes

Finally, one of the hallmarks of participation in adventure education activities is the ability to consistently enhance individual self-system outcomes (Sibthorp *et al.*, 2007). For example, the constructs of self-efficacy, self-confidence, and self-esteem are often altered and enhanced as a natural by-product of adventure participation, and may be a transitional mechanism through which a person can begin to change former, maladaptive coping behaviors (i.e. overeating, smoking, depression, etc.) and replace them with positive alternatives learned through personal testing, challenge, and high-risk experiences. When persons are engaged in adventure activities where they are encouraged to stretch

their usual comfort zone into circumstances of uncertainty (e.g. 'I have never done anything like this before', 'I'm not sure if I will make it through this'), and even into circumstances of fear (e.g. 'If I fall here, that's it', 'I am terrified of heights'), the opportunities for real transition and growth become possible. The foundational and philosophical elements for many organizations that deliver adventure education programs include the promotion of the development of self-systems. For example, Outward Bound's core principles espouse the goals of increasing self-confidence and self-actualization, being your best self, and living a healthy and balanced life (Outward Bound, n.d.).

Intentionally Designed Experiences (IDEs) Within an Adventure Context

One of the major issues that arise in terms of wilderness experiences involves the importance of simply being in the natural environment versus doing structured activities in that natural environment. Several researchers have proposed that a major factor in the efficacy of wilderness programs is simply being in contact with the natural environment itself (Bardwell, 1992; Mitten, 1994). It has also been suggested, however, that structured programs work to 'focus the power' of nature and that highlighting this relationship could work to further enhance health-related outcomes (Mitten, 2009).

More specifically, the benefits from both PET and ART are typically based on simply being exposed to the natural environment (e.g. wilderness). Thus, simply being in a natural setting will create the potential for a positive connection to human health through PET and/or ART. As such, the natural setting serves as a somewhat passive connection to PET and ART (see Fig. 4.1 in Chapter 4).

Another way of enhancing the relationship between humans and natural environments and creating subsequent health benefits, however, is through the active use of Intentionally Designed Experiences (IDEs) (Sheard and Golby, 2006). Similar to Mitten's (2009) idea that wilderness programs may serve to focus the benefits of nature, the idea behind IDEs is that programming in the natural environment can, and should, be purposeful in its planning and implementation in order to achieve specific benefits. In this way, wilderness-based programs become a type of vector for the benefits imbued by the more passive theories of the psychological and physical benefits of nature such as ART and PET.

The type of benefits resulting from participation in an IDE can be separated into two orders or levels. First-order benefits can be considered major outcome variables (often health-related) that occur from participation in an IDE and include achievement, restoration, empowerment, and resilience. Second-order benefits emerge from the occurrence of these first-order outcome variables and include benefits related to self-systems (e.g. esteem, concept, awareness, and efficacy), stress reduction, identity formation, and social support.

The idea of the IDE posits that the purpose, type, and specifics of the program, as well as the type of clients, all impact the outcome. The IDE is an active mechanism that, depending on its design, incorporates many of the theoretical underpinnings of the human–nature benefit interaction, involving constructs such as ART and PET, into a program or experience that ultimately contributes to health and/or quality of life.

Adventure Experiences as Health-Transforming Experiences

Adventure education experiences may be especially beneficial because they entail engaging with the landscape rather than merely viewing it (Frumkin, 2001). This immersion may lead to a sense of feelings of awe, wonder, humility, comfort in and connection to nature, increased appreciation of others, and a feeling of renewal and vigor (Cumes, 1998 in Frumkin, 2001). Indeed, a number of authors have written about spiritual connections that take place within wilderness environments, using terms such as awe, rapture, and transcendence. According to Williams and

Harvey (2001), characteristics of a transcendent experience may include aspects such as a strong positive affect, overcoming personal limits, a sense of union with the universe, absorption in the moment, and a sense of timelessness. Transcendence has also been identified as a limitless experience, one that goes beyond the 'essential being' (Fox, 1999). The idea of rapture is similar to the transcendent experience, and has been described as 'at-one-ment' with the wilderness or a transcending of the ego (Vest, 2008). Thus, rapture is an intensely religious or spiritual experience, which may arise through wilderness solitude or experience. Williams and Harvey (2001) ultimately identified six different types of forest experiences, two of which—diminutive and deep flow—can be classified as transcendent. The diminutive experience was characterized by high fascination, high novelty, low compatibility, and high levels of transcendence, and was likely to include feelings of insignificance and humility. The deep flow experience was different in that it included high compatibility and moderate novelty, and was likely to be described as relaxing or creating a sense of belonging. The other four types of experience—non-transcendent, aesthetic, restorative-familiar, and restorative-compatibility—contain similar characteristics as those deemed transcendent, but could not be distinctly categorized as such.

Williams and Harvey's (2001) work on the transcendent forest experience is important because it provides empirical data that links the spiritual experience with Kaplan's (1995) restorative environments work. Specifically, the authors posit that the concept of fascination may be influential in both restorative and transcendent experiences in nature. Thus, this study not only brings additional credence to the ART that has been utilized for a number of decades, but it adds much needed theoretical and empirical knowledge to the existing spirituality literature. Transcendent experiences have also been related to other theories currently used to describe the outdoor recreation experience, including flow (Csikszentmihalyi and Csikszentmihalyi, 1990) and peak experience (Maslow, 1962).

The idea of transcendence has often been thought of as a type of spiritual experience in a natural setting, but it may also be regarded as a transformational aspect. One way of conceptualizing this may be to examine the constructs of Fredrickson's (1998, 2001) broaden-and-build theory. Fredrickson's model posits that negative experiences narrow peoples' thought-action responses, while positive experiences broaden them. As such, new resources are built physically, intellectually, socially, and psychologically via these positive experiences. These new resources may facilitate coping with stress and adversity (Fredrickson and Joiner, 2002). As the idea of the transcendent experience is built around experiencing intensely positive emotions triggered in a wilderness setting, Fredrickson's theory helps to explain how these experiences are then translated into the benefits so often touted as being a result of time spent in the wilderness.

In a similar fashion, Roggenbuck and Driver's (2000) taxonomy of wilderness benefits includes developmental, therapeutic/mental health, physical health, self-sufficiency, independence, social identity, educational, spiritual, and aesthetic/creativity benefits. The literature generally tends to agree that wilderness experiences imbue benefits such as those listed in the above taxonomy (Miles, 1987; Scherl, 1989; Fox, 1999; Ewert and McAvoy, 2000; White and Hendee, 2000). However, the exact meaning and role of the wilderness setting with regard to the benefits and outcomes of adventure programs remain largely unknown. Fredrickson and Joiner (2002) propose an upward spiral of positive emotions that lead to emotional well-being for the individual, and that these feelings accumulate and compound over time. Thus, the wilderness experience could be a first step in creating these positive emotions. This is an area in need of future investigation by wilderness researchers.

Thus, it is very possible that many of the benefits that have resulted from adventure education-based IDEs are due, in part, to the natural environments in which they take place. Knowing that this is the case may allow us to more fully harness or focus the power of nature through intentional design.

For example, Stringer and McAvoy (1992) suggest that programs might set aside more structured reflection time, allow for group discussion on the spiritual nature of the experience, provide for emotional challenge and the testing of personal limits, teach more about the natural history of an area in which a group is travelling, and facilitate personal connections among group members. A greater understanding of how adventure education can promote the acquisition of specific types of benefits, such as the transcendent experience, stress reduction, or sense of well-being, coupled with a greater understanding of the ways in which IDEs can help to facilitate those benefits, may ultimately result in enhanced adventure experiences for those who participate.

Conclusion

Adventure education offers a relatively 'new' but potentially powerful avenue for enhancing components of both individual and public health. The popularity of these types of activities and experiences, with many of the subsequent benefits being linked to personal growth and development, strengthening family and social relationships, and increased physical activity, is demonstrated by documented increases in both participation rates and types of activities in which people are participating. In addition, when formally facilitated and structured, adventure education can play a vital role in the health of the individual, particularly through its therapeutic and rehabilitative properties. Therapeutic uses of adventure have led to the establishment of myriad adventure therapy programs to bring about personal change within an individual. Inherent in this process are the unique components of a natural setting, risk and challenge, and uncertain outcomes that can be influenced by the individual. The reason adventure education can be effective in promoting positive health-related attitudes and subsequent behaviors lies in the ability of these types of experiences to strengthen and build principally on first-order outcomes such as empowerment, resilience, sense of achievement, and restoration. As a result of these initial outcomes, second-order outcomes, or benefits, often occur which include self-system variables such as self-esteem, stress reduction, increased physical activity, and social support. Organizations such as Outward Bound have long espoused adventure activities and programs as a conduit for enhancing self-systems, creating personal change and developing greater resilience through the purposive use of challenge and risk. Far from being simply activities and experiences sought out by the daredevil or individual seeking an adrenaline rush, adventure education, through its use of IDEs, restorative environments, and structured activities, can enhance the wellness, health, and quality of life for many of our citizens.

References

Autry, C.E. (2001) Adventure therapy with girls at-risk: responses to outdoor experiential activities. *Therapeutic Recreation Journal* 35(4), 289–306.

Baldwin, C., Persing, J. and Magnuson, D. (2004) The role of theory, research, and evaluation in adventure education. *Journal of Experiential Education* 26(3), 167–183.

Bandura, A. (1986) *Social Functions of Thought and Action*. Prentice-Hall, Englewood Cliffs, New Jersey.

Bardwell, L. (1992) A bigger piece of the puzzle: the restorative experience and outdoor education. In: Henderson, K. (ed.) *Coalition for Education in the Outdoors Research Symposium Proceedings, Bradford Woods, Indiana, 17–19 January 1992*. Coalition for Education in the Outdoors, Cortland, New York, pp. 15–20.

Beightol, J., Jevertson, J., Carter, S., Gray, S. and Gass, M. (2012) Adventure education and resilience enhancement. *Journal of Experiential Education* 35(2), 307–325.

Bullock, C. and Mahon, M. (2001) *Introduction to Recreation Services for Persons with Disabilities: A Person-centered Approach*, 2nd ed. Sagamore, Champaign, Illinois.

Bunting, C., Tolson, H., Kuhn, C., Suarez, E. and Williams, R.B. (2000) Physiological stress response of the neuroendocrine system during outdoor adventure tasks. *Journal of Leisure Research* 32(2), 191–207.

Csikszentmihalyi, M. and Csikszentmihalyi, M. (1990) *Flow: The Psychology of Optimal Experience*. Harper & Row, New York.

Degenhardt, B., Frick, J., Buchecker, M. and Gutscher, H. (2011) Influences of personal, social, and environmental factors on workday use frequency of the nearby outdoor recreation areas by working people. *Leisure Sciences* 33(5), 420–440.

de Vries, S. (2010) Nearby nature and human health: looking at the mechanisms and their implications. In: Ward Thompson, C., Aspinall, P. and Bell, S. (eds) *Innovative Approaches to Researching Landscape and Health. Open Space: People Space 2*. Routledge, Abingdon, UK, pp. 77–96.

Driskell, J.E., Johnston, J.H. and Salas, E. (2001) Does stress training generalize to novel settings? *Human Factors: Journal of the Human Factors and Ergonomics Society* 43(1), 99–110.

Ewert, A. (1988) Reduction of trait anxiety through participation in Outward Bound. *Leisure Sciences* 10(2), 107–117.

Ewert, A. (1989) *Outdoor Adventure Pursuits: Foundations, Models, and Theories*. Publishing Horizons, Scottsdale, Arizona.

Ewert, A. and Garvey, D. (2007) Philosophy and theory of adventure education. In: Prouty, D., Panicucci, J. and Collinson, R. (eds) *Adventure Education: Theory and Applications*. Human Kinetics, Champaign, Illinois, pp. 19–32.

Ewert, A. and McAvoy, L. (2000) The effects of wilderness settings on organized groups: a state-of-knowledge paper. In: McCool, S.F., Cole, D.N., Borrie, W.T. and O'Loughlin, J. (comps) *Wilderness Science in a Time of Change Conference—Volume 3: Wilderness as a Place for Scientific Inquiry*, Missoula, Montana, 23–27 May 1999. *USDA Forest Service Proceedings RMRS-P-15-VOL-3*. US Department of Agriculture, Forest Service, Rocky Mountain Research Station, Ogden, Utah, pp. 13–26.

Ewert, A. and Yoshino, A. (2011) The influence of short-term adventure-based experiences on levels of resilience. *Journal of Adventure Education and Outdoor Learning* 11(1), 35–50.

Ewert, A., Attarian, A., Hollenhorst, S., Russell, K. and Voight, A. (2006) Evolving adventure pursuits on public lands: emerging challenges for management and public policy. *Journal of Park and Recreation Administration* 24(2), 125–140.

Ewert, A., Overholt, J., Voight, A. and Wang, C.C. (2010) Understanding the transformative aspects of the wilderness and protected lands experience upon human health. In: Watson, A., Murrieta-Saldivar, J. and McBride, B. (comps) *Science and Stewardship to Protect and Sustain Wilderness Values. Ninth World Wilderness Congress Symposium,* Merida, Yucatan, Mexico, 6–13 November 2009. *Proceedings RMRS-P-64.* US Department of Agriculture, Forest Service, Rocky Mountain Research Station, Fort Collins, Colorado, pp. 140–146.

Ewert, A., Davidson, C., Overholt, J., Luo, Y., Billingham, R. & Bishop, C. (2011) Retying the knot: enhancing father/son relationships through adventure education. Presented at *13th Canadian Conference on Leisure Research,* St. Catharines, Ontario, Canada, 18–21 May 2011.

Flach, F.F. (1997) *Resilience: How to Bounce Back when the Going Gets Tough!* Hatherleigh Press, New York.

Fox, R. (1999) Enhancing spiritual experience in adventure programs. In: Miles, J.C. and Priest, S. (eds) *Adventure Programming.* Venture Publishing, State College, Pennsylvania, pp. 455–461.

Fredrickson, B.L. (1998) What good are positive emotions? *Review of General Psychology* 2(3), 300–319.

Fredrickson, B.L. (2001) The role of positive emotions in positive psychology: the broaden-and-build theory of positive emotions. *American Psychologist* 56(3), 218–226.

Fredrickson, B.L. and Joiner, T. (2002) Positive emotions trigger upward spirals toward emotional well-being. *Psychological Science* 13(2), 172–175.

Frumkin, H. (2001) Beyond toxicity: human health and the natural environment. *American Journal of Preventive Medicine* 20(3), 234–240.

Garst, B.A., Williams, D.R. and Roggenbuck, J.W. (2010) Exploring early twenty-first century developed forest camping experiences and meanings. *Leisure Sciences* 32(1), 90–107.

Gass, M.A. (2007) Bringing adventure into marriage and family therapy: an innovative experiential approach. *Journal of Marital and Family Therapy* 19(3), 273–286.

Gass, M.A., Gillis, H.L. and Russell, K. (2012) *Adventure Therapy: Theory, Research, and Practice.* Routledge, New York.

Griffin, S. (1980) *Woman and Nature: The Roaring Inside Her.* Harper Colophon, New York.

Groff, D. and Dattilo, J. (2000) Adventure therapy. In: Dattilo, J. (ed.) *Facilitation Techniques in Therapeutic Recreation.* Venture Publishing, State College, Pennsylvania, pp. 13–37.

Hartig, T., Mang, M. and Evans, G. (1991) Restorative effects of natural environment experiences. *Environment and Behavior* 23(1), 3–26.

Havitz, M.E., Kaczynski, A.T. and Mannell, R.C. (2013) Exploring relationships between physical activity, leisure involvement, self-efficacy, and motivation via participant segmentation. *Leisure Sciences* 35(1), 45–62.

Hill, B.J., Freeman, P.A. and Huff, C. (2001) The influence of challenging family recreation on family functioning. In: Havitz, M.E. and Floyd, M.F. (eds) *Book of Abstracts: 2001 Leisure Research Symposium.* National Recreation and Park Association, Ashburn, Virginia, abstr. 65.

Jelalian, E., Mehlenbeck, R., Lloyd-Richardson, E.E., Birmaher, V. and Wing, R.R. (2006) 'Adventure therapy' combined with cognitive-behavioral treatment for overweight adolescents. *International Journal of Obesity* 30(1), 31–39.

Kaplan, R. and Kaplan, S. (1989) *The Experience of Nature: A Psychological Perspective.* Cambridge University Press, Cambridge, UK.

Kaplan, S. (1995) The restorative benefits of nature: toward an integrative framework. *Journal of Environmental Psychology* 15(3), 169–182.

Kaplan, S. and Berman, M.G. (2010) Directed attention as a common resource for executive functioning and self-regulation. *Perspectives on Psychological Science* 5(1), 43–57.

Korpela, K.M., Ylén, M., Tyrväinen, L. and Silvennoinen, H. (2010) Favorite green, waterside and urban environments, restorative experiences and perceived health in Finland. *Health Promotion International* 25(2), 200–209.

Lynch, P. and Moore, K. (2004) Adventures in paradox. *Australian Journal of Outdoor Education* 8(2), 3–12.

Maddux, J.E. and Gosselin, J.T. (2003) Self-efficacy. In: Leary, M.R. and Tangney, J.P. (eds) *Handbook of Self and Identity.* Guilford Press, New York, pp. 218–238.

Maslow, A. (1962) *Toward a Psychology of Being.* Van Nostrand, New York.

McAuley, E., Pena, M.M. and Jerome, G.J. (2001) Self-efficacy as a determinant and an outcome of exercise. In: Roberts, G.C. (ed.) *Advances in Motivation in Sport and Exercise.* Human Kinetics, Champaign, Illinois, pp. 235–261.

McAvoy, L.D., Mitten, J., Steckart, L. and Stringer, L. (1992) Research in outdoor education: group development and group dynamics. In: Henderson, K. (ed.) *Coalition for Education in the Outdoors Research*

Symposium Proceedings, Bradford Woods, Indiana, 17–19 January 1992. Coalition for Education in the Outdoors, Cortland, New York, pp. 23–34.

McDermott, L. (2004) Exploring intersections of physicality and female-only canoeing experiences. *Leisure Sciences* 23(3), 283–301.

McKenzie, M. (2000) How are adventure education program outcomes achieved? A review of the literature. *Australian Journal of Outdoor Education*, 5(1), 19–28.

Miles, J. (1987) Wilderness as healing place. *Journal of Experiential Education* 10(3), 4–10.

Mitten, D. (1992) Empowering girls and women in the outdoors. *Journal of Physical Education, Recreation, & Dance* 63(2), 56–60.

Mitten, D. (1994) Ethical considerations in adventure therapy: a feminist critique. In: Cole, E., Erdman, E. and Rothblum, E.D. (eds) *Wilderness Therapy for Women: The Power of Adventure*. The Haworth Press, Inc., Binghamton, New York, pp. 55–84.

Mitten, D. (1998) 'You ain't gonna get me on that rock'. *ZipLines: The Voice for Adventure Education* (34), 46–50.

Mitten, D. (2009) Under our noses: the healing power of nature. *Taproot Journal* 19(1), 20–26.

Mortlock, C. (2002) *Beyond Adventure: An Inner Journey*. Cicerone Press, Milnthorpe, UK.

Outdoor Foundation (2011) *Outdoor Recreation Participation Report 2011*. Outdoor Foundation, Boulder, Colorado.

Outward Bound (n.d.) Outward Bound philosophy. Available at: http://www.outwardbound.org/about-outward-bound/philosophy/ (accessed 17 September 2013).

Passarelli, A., Hall, E. and Anderson, M. (2010) A strength-based approach to outdoor and adventure education: possibilities for personal growth. *Journal of Experiential Education* 33(2), 120–135.

President's Commission on Americans Outdoors (1986) *Report and Recommendations to the President of the United States*. US Government Printing Office, Washington, DC.

Priest, S. and Gass, M. (1997) *Effective Leadership in Adventure Programming*. Human Kinetics, Champaign, Illinois.

Pryor, A., Carpenter, C. and Townsend, M. (2005) Outdoor education and bush adventure therapy: a social-ecological approach to health and wellbeing. *Australian Journal of Outdoor Education* 9(1), 3–13.

Roggenbuck, J. and Driver, B.L. (2000) Benefits of nonfacilitated uses of wilderness. In: McCool, S.F., Cole, D.N., Borrie, W.T. and O'Loughlin, J. (comps) *Wilderness Science in a Time of Change Conference—Volume 3: Wilderness as a Place for Scientific Inquiry*, Missoula, Montana, 23–27 May 1999. *USDA Forest Service Proceedings RMRS-P-15-VOL-3*. US Department of Agriculture, Forest Service, Rocky Mountain Research Station, Ogden, Utah, pp. 33–49.

Russell, K.C. (2003) An assessment of outcomes in outdoor behavioral healthcare treatment. *Child and Youth Care Forum* 32(6), 355–381.

Scherl, L. (1989) Self in wilderness: understanding the psychological benefits of individual–wilderness interaction through self-control. *Leisure Sciences* 11(2), 123–135.

Sheard, M. and Golby, J. (2006) The efficacy of an outdoor adventure education curriculum on selected aspects of positive psychological development. *Journal of Experiential Education* 29(2), 187–209.

Shore, A. (1977) *Outward Bound: A Reference Volume*. Topp Litho, New York.

Sibthorp, J. (2003) An empirical look at Walsh and Golins' adventure education process model: relationships between antecedent factors, perceptions of characteristics of an adventure education experience, and changes in self-efficacy. *Journal of Leisure Research* 35(1), 80–106.

Sibthorp, J., Paisley, K. and Gookin, J. (2007) Exploring participant development through adventure-based programming: a model from the National Outdoor Leadership School. *Leisure Sciences* 29(1), 1–18.

Snegroff, S. (2012) Psychoneuroimmunology and performance anxiety: mind and body health and adaptation. Presented at *Second International Conference on Health, Wellness, & Society*, University Center, Chicago, Illinois, 10–11 March 2012.

Stevens, B., Kagan, S., Yamada, J., Epstein, I., Beamer, M., *et al.* (2004) Adventure therapy for adolescents with cancer. *Pediatric Blood & Cancer* 43(3), 278–284.

Stringer, L. and McAvoy, L. (1992) The need for something different: spirituality and wilderness adventure. *Journal of Experiential Education* 15(1), 13–20.

Ulrich, R.S. (1984) View through a window may influence recovery from surgery. *Science* 224(4647), 420–421.

Van Puymbroeck, M., Ewert, A.W., Luo, Y. and Frankel, J. (2012) The influence of the Outward Bound Veterans Program on sense of coherence. *American Journal of Recreation Therapy* 11(3), 31–38.

Vest, J.H.C. (2008) The philosophical significance of wilderness solitude. *Environmental Ethics* 9(4), 303–330.

Walsh, V. and Golins, G. (1976) *The Exploration of the Outward Bound Process*. Outward Bound Publications, Denver, Colorado.

Ward Thompson, C. and Aspinall, P.A. (2011) Natural environments and their impact on activity, health, and quality of life. *Applied Psychology: Health and Well-Being* 3(3), 230–260.

White, D.D. and Hendee, J. (2000) Primal hypotheses: the relationship between naturalness, solitude, and the wilderness experience benefits of development of self, development of community, and spiritual development. In: McCool, S.F., Cole, D.N., Borrie, W.T. and O'Loughlin, J. (comps) *Wilderness Science in a Time of Change Conference—Volume 3: Wilderness as a Place for Scientific Inquiry*, Missoula, Montana, 23–27 May 1999. *USDA Forest Service Proceedings RMRS-P-15-VOL-3*. US Department of Agriculture, Forest Service, Rocky Mountain Research Station, Ogden, Utah, pp. 23–27.

Williams, K. and Harvey, D. (2001) Transcendent experience in forest environments. *Journal of Environmental Psychology* 21(3), 249–260.

10

Future Actions and Implications: Policy and Research—Take a Park, Not a Pill

A place-sensitive society would do more than minimally tolerate a taste for place in those who can afford to indulge in it: it would nurture relationships to place structurally as the normal case, not the exception of privilege.

Val Plumwood in *Environmental Culture: The Ecological Crisis of Reason* (2002, p. 231)

Throughout this book, we have described the connections between natural environments and human health. As part of this discourse, we have discussed concepts such as what constitutes a natural environment, the early history of the human–environment interface, and worldviews associated with health. We then moved to an exploration of the various theories underlying the human health and natural environment paradigm as well as the variety of ways humans have adapted to this paradigm. We then explored the many outcomes and benefits revealed through research and literature followed by the role that education plays in the health and natural environment construct. We also explored some of the innovative methods that have been developed, in this case, to use adventure-based and outdoor-based programs for enhancing health. In this chapter, we expand our thinking about these issues by asking what implications and policy connections can be anticipated. We begin with policy implications, followed by exploring the concept

of environmental responsibility as it relates to human health. Finally, we analyze the role that research plays and will play in the continued exploration of health and natural environments.

Policy: Action and Implications

As previously mentioned in Chapter 9, the connection between human health and natural settings was formally recognized in 1986 through the President's Commission on Americans Outdoors in its declaration that the outdoors was a 'giant health machine'. Similar declarations have been made in other countries including the UK, Australia, and Canada. With a substantial and growing body of knowledge, much of which supports the contention that natural environments are connected to many health-related issues, what policy implications can be drawn relative to natural environments and human health?

In thinking about policy actions and implications, we must elucidate some of the social determinants that guide our individual and collective health. Some of the factors that influence this connection between natural environments and health would include the following:

- Increasing sedentary lifestyle and changing lifestyles. For example, there are

© CAB International 2014. *Natural Environments and Human Health*
(A.W. Ewert, D.S. Mitten and J.R. Overholt)

aging populations across cultures and societies.

- A growing demand for moving health-care from an institutional setting to a community-based setting. Thus, there is a need to reconsider our social and community structures, and to further awareness of the role that natural environments can play in individual and collective health. This need is compounded by a heightened awareness of the attraction and need for using nature to provide for a variety of physical, reflective, and social engagement experiences.

- The importance of research and evidence-based decision making in the provision of natural environments for individual and community use. This is particularly true in an era of limited resources and tight budgets.

As reflected in this book, natural environments such as forests, parks, gardens, and non-developed open spaces (NDOS) are increasingly recognized as places that provide opportunities to enhance both individual and public health and well-being (Nilsson *et al.*, 2011). While the evidence for this belief is growing, it is not always consistent and involves a research approach that is both transdisciplinary and multidisciplinary, as well as cross-cutting and collaborative, involving the health disciplines, environmental scientists, social scientists, and others.

In addition, while both public health and modern medicine have produced a host of beneficial outcomes, there is an increasing recognition that many health-related issues such as obesity, substance abuse, lack of physical activity, and response to stress are a function of lifestyle and individual decision making. Moreover, health is increasingly thought of as involving more than an absence of disease, and includes a sense of well-being and quality of life. This broader concept of health integrates very clearly with the idea that natural environments can influence human health in positive ways for a variety of reasons.

First, natural environments such as parks and forests are often public resources, implying that they are available for all the public,

not just the financially well-off. Thus, in one sense, the use of natural settings for health purposes cuts across socio-economic status, racial and cultural divides, and differing age cohorts. Second, engaging in natural environments, particularly for experiences related to relaxation, reflection, aesthetic enjoyment, and physical activities such as walking, requires a minimum of skill development or specialized types of equipment. This tends to enhance the accessibility of the activity for a broad range of potential users and visitors. Third, natural settings are often amenable to the integration of a wide variety of social and health determinants that make these resources attractive to both citizens and politicians. Thus, parks come to be viewed as valuable entities that are in need of protection and maintenance because they are used by many different members and strata of a community, often times for health-producing reasons. In the long run, health status and quality of life are mostly influenced by a combination of our genetic makeup, a wide range of social and environmental factors (i.e. social determinants of health), and the day-to-day behaviors we practice, either as individuals or collectively, as a society (Green and Kreuter, 2005).

It should be noted, however, that the positive effects that often emerge from a natural environment–health interaction are often overlooked in lieu of the negative perceptions held toward natural settings (Van den Berg and ter Heijne, 2005). Thus, implications for policy development also need to provide for evidence-based research and best practices in the planning of natural environments specifically focused on the enhancement of human health.

Life stages

Life stage refers to the phases that an individual passes through in her/his life. Typical life stages include young childhood, adolescence, young adulthood, families with children, 'empty nesters' (parents whose children have moved out of the house) or pre-retirement, and retirement. In considering life stage and its relationship to the concept of natural environments and human health, several issues need to be considered. First, while not completely separated, life stage is somewhat autonomous from actual age. This is especially true of young adults, families with children, and empty nesters or pre-retirement. In these cases, there may be wide fluctuations in the actual ages of the individuals. What is important to consider are the changing needs of having an interface with the natural environment as it pertains to health. When in the young childhood and adolescent phases, individuals may be particularly prone to the influence of family and/or friends in the formation of attitudes and subsequent behaviors toward natural environments (Ewert *et al.*, 2005; Berns and Simpson, 2009).

In addition, life stages often can be used to anticipate how a natural environment might be used by the individual. For example, adolescents and young adults typically use natural environments for recreation and personal 'testing' and as such engage in activities that demand a high level of physical activity. Families with children, on the other hand, often look for a place for the family to play or picnic and require a landscape that is not too challenging and offers a sense of safety. Likewise, those in a pre-retirement or retirement phase often seek out natural environments for wildlife observation, relaxation, or a sense of restoration. Thus, the type of natural environment sought out with respect to health and other outcomes such as recreation are often related to the life stage that individuals may find themselves in.

Lifestyle factors

Just as each life stage, particularly later in life, can bring with it certain types of health-related challenges, another variable that plays an important role in personal health are factors related to lifestyles. Examples of lifestyle factors that have particular relevance to health and natural environments include level of exercise and physical activity, type of work, level of education and income, leisure activities, and living environment. In this case, the literature suggests the following realities.

- As previously discussed, the level of exercise and physical activity can play an important role in the health of an individual or of society in general. Natural landscapes come into play with the issue of exercise in two ways: (i) they provide a place for exercise and activity to take place; and (ii) natural settings often serve to attract people to walk or engage in some form of exercise or recreation (de Vries *et al.*, 2011).

- The type of work an individual engages in often influences several health-related issues including level of stress and the amount of physical activity engaged in throughout the work day, and there may be little opportunity for restorative types of activities. Relative to health and natural environments, the concept of self-regulated restorative experiences (SRREs) has appeared. SRREs are activities deliberately selected by the individual for the explicit purpose of providing for recovery or restoration from the demands of everyday life (Degenhardt and Buchecker, 2012). Not surprisingly, Hartig *et al.* (1991) found evidence that natural settings were effective in providing a sense of restoration, both in psychological and physiological terms. While historically natural settings were primarily thought of as being a park, forest, or wilderness area, the concept of nearby outdoor recreation areas (NORAs) has emerged to imply a natural setting or location that is nearby or close to where an individual works or lives. These areas might be relatively small in size (i.e. a municipal park) but represent an offset to a person's home or workplace. Thus, the health-related problems often associated with work can be partially offset, at times, by the presence of NORAs and the practice of SRREs; that is, choosing to engage in restorative activities that can be done in nearby natural settings.

- Although not the same, level of education and income are often viewed synonymously, implying that higher levels of education often result in higher incomes. While not always true, both have implications for health and natural environments. In general terms, people with higher incomes and levels of education are better housed, have better nutrition, more access to healthcare, and live longer. Poudyal *et al.* (2009) found that people living in areas with higher proportions of forests, rangeland, or water exhibited longer life expectancies and higher levels of self-reported quality of life. Likewise, Wells and Evans (2002) found that having nearby nature served as a buffer against the stresses of daily life among children.

- In many cases, the types of leisure activities in which an individual engages are often strongly linked to health-related issues such as level of physical activity, social engagement, and types of natural environments needed. Different types of natural settings offer different opportunities for leisure. For example, the recreational opportunity spectrum (ROS) describes how different landscapes can be linked with specific recreational activities (Clark and Stankey, 1979). From a policy perspective, what this suggests is that specific desired health-related outcomes can be linked to specific leisure and recreational activities, which, in turn, often require specific natural landscapes. For example, health outcomes that are influenced by social support and physical activity can often be achieved by activities such as whitewater rafting which, in turn, requires a specific type of natural environment—namely, a whitewater river. In another example, Kline *et al.* (2011) report that the presence of National Forests and other natural environments significantly contribute to levels of physical activity among the public.

- Living environments can exert a tremendous influence on health. As previously mentioned, higher standards of living often imply greater access to natural areas such as 'greenways' and open space. For example, Maas *et al.* (2006) found that the presence of greenspace (e.g. vegetation, trees, etc.) within a radius of 1 km from a person's residence resulted in enhanced feelings of perceived general health.

Social and community structure

The types of social and community structure that an individual is exposed to can play tremendous roles in human health and this is also true relative to natural environments. Social structure involves family, friends, other participants, and general society. Community structure relates to the infrastructure, both people and facilities, that serves an enabling or disabling function. For example, a local community may not have easy access to a natural or undeveloped setting, or health-related activities such as walking or relaxation in the outdoors are not encouraged or supported. On the other hand, an individual who participates in gardening may experience restoration or physical activity for a short period of time. With support from family or friends, however, the individual might be encouraged to continue participation. This continued participation will provide a greater breadth and depth of positive health outcomes. In a similar fashion, when afforded access to natural environments, social and community structures can serve as adjunct or primary providers of health-related outcomes. Outdoor camps and programs provide numerous examples of how local communities have developed specific opportunities for individuals to experience natural environments in a structured and facilitated way.

Research and Evidence— Considerations for the Future

Despite the growing body of evidence that natural environments can facilitate the enhancement of factors related to improving or maintaining health, there remain a number of research-related concerns. Being cognizant of these issues is important because it leads directly to our confidence in the belief that natural environments are influential in promoting positive and beneficial health outcomes. In turn, this confidence can be related to the types of policies that are adopted in a community- or society-wide effort to promote good health. A sample of these research-related factors includes the following.

1. *People may perceive themselves as having improved health from exposure to natural environments but are they actually healthier?* Distilling the essence of this question potentially has profound impacts for health-related policy. While it is important for people to feel healthy, from a policy development perspective it is crucial to know that they actually are healthy. More specifically, this becomes causal with respect to: are there any noticeable effects traceable to the natural setting or are there other confounding variables that provide some or all of the impact? Once again, from a policy perspective, protecting nature for nature's sake is a worthwhile goal; however, protecting natural settings because they offer avenues for improved human health requires a different set of evidence. This requirement becomes even more critical if natural sites are scarce or limited and could be used for other societal needs (de Vries *et al.*, 2003).

2. *What is the shape of the function between 'naturalness' and health?* That is, are environments that are more natural, such as wilderness areas, more effective in influencing positive health outcomes? Conversely, do more urban environments consistently produce less healthy outcomes? Moreover, is the health effect of natural environments different in the ways that these environments impact specific groups or demographic variables such as socio-economic status? If there is a differential effect, public policy might be developed that focuses on specific groups and/or settings for a greater return on investment.

3. *Relative to the restorative experience as a result of exposure to a natural environment, several questions emerge:*

   a. Is the experience robust enough to be proactive; that is, can these types of experience prepare individuals to better cope and deal with issues such as stress and pressure extant to daily life? For example, several adventure-based programs use natural environments to develop a sense of hardiness in military soldiers in an attempt to help them cope with issues such as post-traumatic stress disorder (PTSD).

   b. Are there differential effects of specific aspects of natural environments, such as the scenery, bodies of water, trees, and vegetation, upon parameters related to health? For example, does moving water provide for a particularly restful and stress- or blood pressure-reducing outcome? If present, these differential effects can be considered in built environments; for example, the presence of fountains, pools, and small natural areas.

4. *Is there a difference in health-related outcomes between 'virtual' and real nature?* For example, does taking a walk in a wooded area result in the same outcomes as viewing a natural scene out of a window or in a mural? Felsten (2009) demonstrated that large murals featuring natural scenes in locations where college students would take a study break were effective in restoration for attentionally fatigued students. If nature-based scenes or views of nature are effective, can built environments be used as a supplement to or made as effective at attention restoration or stress reduction as 'real' natural environments (Mayer *et al.*, 2009)?

5. *Are there confounding variables that serve to explain the effect of natural environments upon human health?* For example, most individuals visit a natural area with others. Thus, does social support serve as a moderating variable that accounts for nature's effect (Wells and Evans, 2003)?

6. *If there is a link between human health and natural environments, what role will the concept of environmental justice play out in society (Evans and Kantrowitz, 2002)?* In what ways will environmental justice be integrated into public policy, particularly in situations of vulnerable populations or underserved communities? It should be noted, however, that issues involving environmental justice usually involve factors such as toxicity levels, crowding levels, and other variables associated with negative health outcomes. Natural environments can be thought of as health-producing settings that lead to positive health, not just combating the effects of poor living conditions.

In sum, because of the growing and relatively robust nature of human health and natural environments research, access to these types of settings should be considered in the development of public policy. Likewise, development initiatives such as land-use planning and urban revitalization need to consider the presence of natural environments, in differing forms and settings, as part of the policy and planning process (see Chapter 6 for a discussion on the socio-ecological approach to human health). There are positive relationships between health and environmental settings and differential effects of these settings for specific groups.

Finally, and related to the above discussion, developing a body of evidence becomes an academic exercise if the findings are not considered and used by both the practitioner and planner (Nilsson *et al.*, 2011). Increasingly, visitation and use of natural environments by the general public must be inculcated into lifestyle choices and sought-out opportunities. Moreover, medical practice and practitioners should build on the presence and potential outcomes from time in natural environments as they strive to increase the health of the general public. Perhaps the best indication of successfully integrating natural environments into the medical lexicon will occur when a common physician-written prescription includes visits to natural environments.

References

Berns, G.N. and Simpson, S. (2009) Outdoor recreation participation and environmental concern: a research summary. *Journal of Experiential Education* 32(1), 79–91.

Clark, R.N. and Stankey, G.H. (1979) *The Recreation Opportunity Spectrum: A Framework for Planning, Management, and Research*. General Technical Report PNW-98. US Department of Agriculture, Forest Service, Pacific Northwest Forest and Range Experiment Station, Seattle, Washington.

Degenhardt, B. and Buchecker, M. (2012) Exploring everyday self-regulation in nearby nature: determinants, patterns, and a framework of nearby outdoor recreation behavior. *Leisure Sciences* 34(5), 450–469.

de Vries, S., Verheij, R.A., Groenewegen, P.P. and Spreeuwenberg, P. (2003) Natural environments – healthy environments? An exploratory analysis of the relationship between greenspace and health. *Environment and Planning A* 35(10), 1717–1732.

de Vries, S., Claßen, T., Eigenheer-Hug, S.-M., Korpela, K., Maas, J., *et al.* (2011) Contributions of natural environments to physical activity: theory and evidence base. In: Nilsson, K., Sangster, M., Gallis, C., Hartig, T., de Vries, S., *et al.* (eds) *Forests, Trees and Human Health*. Springer, New York, pp. 205–243.

Evans, G.W. and Kantrowitz, E. (2002) Socioeconomic status and health: the potential role of environmental risk exposure. *Annual Review of Public Health* 23(1), 202–231.

Ewert, A., Place, G. and Sibthorp, J. (2005) Early-life outdoor experiences and an individual's environmental attitudes. *Leisure Sciences* 27(3), 225–239.

Felsten, G. (2009) Where to take a study break on college campus: an attention restoration theory perspective. *Journal of Environmental Psychology* 29(1), 160–167.

Green, L.W. and Kreuter, M.W. (2005) *Health Program Planning: An Educational and Ecological Approach*. McGraw Hill, Boston, Massachusetts.

Hartig, T., Mang, M. and Evans, G. (1991) Restorative effects of natural environment experiences. *Environment and Behavior* 23(1), 3–26.

Kline, J.T., Rosenberger, R.S. and White, E.M. (2011) A national assessment of physical activity in US National Forests. *Journal of Forestry* 109(6), 343–351.

Maas, J., Verheij, R.A., Groenewegen, P.P., de Vries, S. and Spreeuwenberg, P. (2006) Green space, urbanity, and health: how strong is the relation? *Journal of Epidemiology and Community Health* 60(7), 587–592.

Mayer, F.S., McPherson Frantz, C., Bruehlmen-Senecal, E. and Dolliver, K. (2009) Why is nature beneficial? The role of connectedness to nature. *Environment and Behavior* 41(5), 607–643.

Nilsson, K., Sangster, M., Gallis, C., Hartig, T., de Vries, S., *et al.* (eds) (2011) *Forests, Trees and Human Health*. Springer, New York.

Plumwood, V. (2002) *Environmental Culture: The Ecological Crisis of Reason*. Routledge, New York.

Poudyal, N.C., Hodges, D.G., Bowker, J.M. and Cordell, H.K. (2009) Evaluating natural resource amenities in a human life expectancy production function. *Forest Policy and Economics* 11(4), 253–259.

President's Commission on Americans Outdoors (1986) *Report and Recommendations to the President of the United States*. US Government Printing Office, Washington, DC.

Van den Berg, A.E. and ter Heijne, M. (2005) Fear versus fascination: an exploration of emotional responses to natural threats. *Journal of Environmental Psychology* 25(3), 261–272.

Wells, N.M. and Evans, G.W. (2003) Nearby nature: a buffer of life stress among rural children. *Environment and Behavior* 35(3), 311–330.

11

Resources

All of us share a sense of common purpose.
We represent many, many others; some we
know, and others we have never met.
People throughout the world are increasingly
connected by a resonance and passion, to
create a new common sense for the good
health of children today and generations
to come.

Cheryl Charles, past CEO of Children &
Nature Network

The Researchers

Research connecting human health and natural environments has experienced a recent surge in popularity, due primarily to a number of prominent researchers who have all converged on this topic from a variety of academic disciplines and an array of geographical locations. The following section attempts to provide some biographical information and historical context for some of these individuals. The growing nature of this field makes it impossible to include everyone—a boon for the field itself, but our apologies to anyone who has been left out!

Judy Atkinson

Judy Atkinson is a Jiman-Aboriginal Australian (from central west Queensland)/Bundjalung (northern New South Wales) woman, who also has Anglo-Celtic and German heritage. She is a graduate of the Harvard University, Program for Refugee Trauma, Global Mental Health Trauma and Recovery certificate course. In 2011 she was awarded the Fritz Redlich Memorial Award for Human Rights and Mental Health from the Harvard University Program for Refugee Trauma. Her primary academic and research focus has been in the area of violence, with its relational trauma, and healing or recovery for Indigenous (and indeed all) peoples. She participated in the design and delivery of We Al-li, an educational-healing (educaring) program specifically developed in response to generational trauma by Aboriginal Australians. She co-authored the Aboriginal and Torres Strait Islander Women's Task Force on Violence report for the Queensland government. Her book, *Trauma Trails–Recreating Songlines: The Transgenerational Effects of Trauma in Indigenous Australia*, provides context to the life stories of people who have moved/ been moved from their country in a process that has created trauma trails, and the changes that can occur in the lives of people as they make connections with each other and share their stories of healing. She is a member of the Harvard Global Mental Health Scientific Research Alliance. She presently serves on the Australian Institute of Health and Welfare Scientific Advisory Committee on Closing the Gap Research. She is a Board member of the Aboriginal and Torres Strait Islander Healing Foundation, chairs the Research Advisory Committee, and is a member of the Education and Training Advisory Committee of the Foundation, and is a Patron of the We Al-li Trust.

Louise Chawla

Louise Chawla, an environmental psychologist in the interdisciplinary field of children's environments, is a professor in the Environmental Design Program in the College of Architecture and Planning at the University of Colorado in Boulder, Colorado. With research interests and publications in child- and youth-friendly communities, design for children's access to nature, child and youth participation in urban planning and design, lifespan development of active care for the environment, and effects of contact with nature for human health and well-being, Chawla supports the rights of children and is Co-editor of the *Children, Youth and Environments Journal*, Associate Director of the Children, Youth, and Environments Center for Community Engagement, and a member of the Children & Nature Network Advisory Board. When she served as a Fulbright Scholar at the Norwegian Centre for Child Research, she revived the Growing Up in Cities project of UNESCO, which continues to involve urban children around the world in evaluating and improving their local communities. The project won the 2002 Place Research Award from the Environmental Design Research Association. Chawla is a steering committee member of Growing Up Boulder, an initiative to engage children and youth in the city of Boulder planning process.

Cheryl Charles

Cheryl Charles co-founded the Children & Nature Network with Richard Louv and others in 2006 and served as its first CEO for 7 years. Charles has worked in the area of environmental education for over 40 years, serving close to 20 years as the founding National Director of Project Learning Tree and Project Wild. As an innovator, educator, author, and organizational executive, she served for 8 years as a member of the steering committee for the Commission on Education and Communication of the International Union for the Conservation of Nature (IUCN-CEC). Her latest book, *Coming Home: Community, Creativity and Consciousness*, is co-authored with her late husband Bob Samples.

Susan Clayton

Susan Clayton is the Whitmore-Williams Professor of Psychology at the College of Wooster, Wooster, Ohio. She is a conservation psychologist, interested in human relationships with nature, as well as environmental identity and the psychology of justice. Clayton is involved in understanding and promoting a healthy relationship between humans and nature and developed an environmental identity scale (EID) to assess the degree to which the natural environment plays an important part in the way in which people think about themselves. She has published several books and numerous journal articles in these areas and is a Fellow of the American Psychological Association, the Society for Environmental, Population and Conservation Psychology, and the Society for the Psychological Study of Social Issues.

Marti Erickson

Martha (Marti) Farrell Erickson, Founding Director of the University of Minnesota's Children, Youth & Family Consortium and Director (Emeriti) of the Irving Harris Training Programs in Early Childhood Mental Health, is passionate about the role of nature in children's development. Recipient of honors for her contributions in psychology, she works to translate research for the general public. Erickson researches and writes in the area of parent–child attachment, children's mental health, and strategies for working with children and families in high-risk situations. She is the past director and a co-founder of the Children & Nature Network with Richard Louv and others, and supports the movement to reunite children with nature as well as critical research in how time in nature relates to human development.

Howard Frumkin

In the medical field, Howard Frumkin, PhD, MD, MPH, has been researching and speaking out about the health benefits from contact with nature since the 1990s. As the Director of the National Center for Environmental Health and Agency for Toxic Substances and Disease Registry (NCEH/ATSDR) at the US Centers for Disease Control and Prevention, he specifically works in the area of promoting healthy environments free of toxic substances and other environmental hazards. Frumkin's public health background informs his work. Recently he teamed with Richard Louv, author of the popular and informative book *Last Child in the Woods*. Both were keynote speakers at the Healthy Parks Healthy People Congress in April 2010 in Melbourne.

Terry Hartig

Terry Hartig is Professor of Environmental Psychology at the Institute for Housing and Urban Research and the Department of Psychology at Uppsala University in Sweden. His specific research interests include the public health values of natural environments and the residential context of health. Hartig's work emphasizes the restorative aspects of natural environments and includes a focus on the types of places that can be restorative, as well as the types of interactions necessary for restoration. He uses diverse research methods in studying the social ecology of stress and restoration. As a professor of psychology he integrates the importance of these types of interactions into his teaching and research.

Peter Kahn

Peter Kahn is a psychology professor at the University of Washington, Seattle, Washington and the Director of the Human Interaction with Nature and Technological Systems (HINTS) Lab. His research looks at both the degradation of the natural world and our increasing interaction with technology and technical systems, and he has published in the areas of ecopsychology, human relationship with nature, and children and nature.

Rachel and Stephen Kaplan

Since the early 1970s, and starting with a US Forest Service grant, Rachel and Stephen Kaplan have been researching and writing about attention restoration theory (ART). As environmental psychologists at the University of Michigan, Stephen is in the Department of Psychology while Rachel is in the School of Natural Resources and the Environment. They have hundreds of publications about their ART. The Kaplans' work has been combined with many other disciplines including landscape architecture and urban planning.

Stephen Kellert

Stephen Kellert is the Tweedy/Ordway Professor Emeritus of Social Ecology and senior research scholar at the Yale University School of Forestry and Environmental Studies. He has been involved with many organizations, and has received numerous awards for his work on human–natural environment connections, with a specific focus on sustainable design and development, and environmental conservation. He has authored over 150 books and publications in this area, serves on the board of directors of a sustainable land-use firm, and most recently produced a documentary on biophilic design.

Kalevi Korpela

Kalevi Korpela is Professor of Psychology at the University of Tampere in Finland.

His research in environmental psychology focuses on self-regulation, emotion-regulation, and well-being in a variety of environments. He has been widely published in both English and Finnish, with a variety of articles, books chapters, interviews, and reports. Korpela is a Board member for the International Society of Nature and Forest Medicine (INFOM).

Frances (Ming) Kuo

Frances Kuo, Director of the Landscape and Human Health Laboratory, University of Illinois at Urbana–Champaign, has published extensively documenting the relationship between natural environments and healthy development and healthy communities and the contribution of natural space and horticultural in the socio-ecological approach to health and well-being. Since 1996 Kuo and her colleagues have shown through research the importance of trees and greenspace to stronger, safer communities as well as robust concentration, self-control, and coping in individuals. Primarily an environmental psychologist, Kuo has studied the areas of crime, aggression, and nature; children and nature; inner city and nature; sense of community and nature, and has published extensively in the *Journal of Environmental Psychology* and *Environment & Behavior*. Her numerous publications can be found on the University of Illinois web site.

Cecily Maller

Cecily Maller is a Senior Research Fellow at the Centre for Design, RMIT University in Australia, was the lead author for the *Healthy Parks, Healthy People* research report for Parks Victoria in Australia, and is the founder of the Nature in Community, Health and Environment (NiCHE) Research Group in the Faculty of Health. Maller's PhD, funded by the Victorian Health Promotion Foundation, explored contact with nature and children's mental, emotional, and social health in the context of sustainability education. Currently, she explores the complexities of planning for

health, with a particular focus on tensions between relational and place-based understandings of community. Her interests focus on the social dimensions of health, well-being, and sustainability, examining interactions between people and natural, built, and social environments in the context of everyday life. Recent projects have focused on social practices involving energy and water use in migrant and 'green renovating' households, with the aim of informing policy and moving beyond dominant rationalist and neoliberal paradigms.

David Orr

David W. Orr is the Paul Sears Distinguished Professor of Environmental Studies and Politics and Senior Adviser to the President, Oberlin College, Oberlin, Ohio. He is the executive director of the Oberlin Project, an initiative designed to promote sustainability and community in Oberlin. He has written extensively in the areas of the environment, sustainability, education, politics, ecological design, and climate change, authoring over 200 articles, books, and chapters, and also furthers this mission through speaking, teaching, and entrepreneurship.

Jules Pretty

Jules Pretty is Professor of Environment and Society at the University of Essex, Colchester, UK. He has written and researched widely in the areas of agricultural sustainability, green exercise and green design, biodiversity and ecoliteracy, connections to place, and social capital and natural resources. He set up the Center for Environment and Society at the University of Essex, which is an interdisciplinary program connecting a variety of departments and disciplines.

Carolyn Raffensperger

Carolyn is Executive Director of the Science & Environmental Health Network and co-editor of *Precautionary Tools for Reshaping Environmental Policy* published by MIT Press (2006) and *Protecting Public Health and the Environment: Implementing the Precautionary Principle* published by Island Press (1999). Carolyn coined the term 'ecological medicine' to encompass the broad notions that both health and healing are entwined with the natural world. As an environmental lawyer she specializes in the fundamental changes in law and policy necessary for the protection and restoration of public health and the environment. Carolyn is at the forefront of developing new models for government that depend on these larger ideas of precaution and ecological integrity. The new models include guardianship for future generations, a vision for the courts of the 21st century, and the public trust doctrine.

Kate Rawles

Kate Rawles is a Senior Lecturer in Outdoor Studies at the University of Cumbria, Cumbria, UK who leads the Greening Outdoor Practice postgraduate certificate and the Institute for Leadership and Sustainability. Kate combines her background in ethics and values, environmental and animal welfare issues, sustainability, and the philosophy of nature conservation to argue for using the concept and practice of adventure in sustainability. Her recent book, *The Carbon Cycle: Crossing the Great Divide*, a finalist in the 2013 People's Book Prize for non-fiction, is based on her 4500-mile bicycle trip from Texas to Alaska following the spine of the Rockies, exploring North American attitudes to climate change. Using a scholarship from NESTA she developed a package of short courses that combine critical thinking about humans' relationship with the environment and impact on it, with inspirational experiences of wild places. The aim of this 'outdoor philosophy' is to inspire and support more sustainable ways of living and working. She has been an independent consultant for Nirex UK on ethical issues in radioactive waste management. Kate is a Council member of the Food Ethics Council, an active member of the Adventure and Environmental Awareness Group, and a Fellow of the Royal Geographical Society.

Ted Schettler

Ted Schettler, the Science & Environmental Health Network's Science Director, has worked extensively with community groups and non-governmental organizations throughout the US and internationally, addressing many aspects of human health and the environment. Ted has co-authored a number of books that look at the effects of toxicity and environmental health risks on reproductive health, child development, and healthy aging. He has also published numerous articles in the medical literature, and is frequently quoted in the popular press about environmental health, ecological health, and the precautionary principle. He has served on advisory committees of the US Environmental Protection Agency and National Academy of Sciences and practiced medicine for many years in New England. Ted received his MD from Case-Western Reserve University and a master's degree in public health from the Harvard School of Public Health.

Vandana Shiva

Vandana Shiva is an Indian philosopher, environmental activist, author, and eco-feminist. She was trained as a physicist and received her PhD in philosophy from the University of Western Ontario, Canada. Currently based in Delhi, she has authored more than 20 books and is one of the leaders and board members of the International Forum on Globalization and a figure of the global solidarity movement known as the alter-globalization movement. She has argued for the wisdom of many traditional practices, as is evident from her interview in the book *Vedic Ecology* (by Ranchor Prime) that draws upon India's Vedic heritage. In 2004 Dr Shiva started Bija Vidyapeeth, an international college for sustainable living in Doon Valley, in collaboration with Schumacher College, UK. *Time Magazine* identified Dr Shiva as an environmental 'hero' in 2003, and *Asia Week* has called her one of the five most powerful communicators of Asia. She wrote the introduction to *Women, Ecology, and Health* (1992).

Mardi Townsend

Mardi Townsend is Deputy Head of School for the School of Health & Social Development at Deakin University, Melbourne, Australia. Her research interests include the human health benefits of interaction with nature, park policy implications for health, urban and rural contexts for health and well-being, social and health impact assessment, and housing and homelessness. Currently Townsend is investigating health and well-being outcomes of different levels of access to nature for residents of high-rise housing; a nature-based intervention program for people suffering depression, anxiety, and/or social isolation; health and well-being outcomes of 'bush adventure therapy' programs in Australia; outcomes from nature-based activities in primary schools; the benefits of environmental volunteering; and the value and use of parks in areas of high and low socio-economic status. She has supervised many graduate students researching human health and the natural environment, including Cecily Maller's work mentioned above, Anita Pryor's 'Wild adventures in wellbeing: foundations, features and wellbeing impacts of Australian outdoor adventure interventions', and Cathryn Carpenter's 'Changing spaces: contextualising outdoor experiential programs for health and wellbeing'.

Roger Ulrich

Roger Ulrich is Professor of Architecture at the Center for Healthcare Building Research at Chalmers University of Technology in Sweden, and is Adjunct Professor of Architecture at Aalborg University in Denmark. His research focuses on the various aspects of healthcare environments as they impact patient well-being, health, and recovery. His involvement in this field crosses national boundaries, including the co-founding of the Center for Health Systems and Design at Texas A&M University, and service as a senior adviser on patient care environments at the invitation of Britain's National Health Service from 2005 to 2006.

Nancy Wells

Nancy M. Wells, a professor in Design and Environmental Analysis, College of Human Ecology Cornell University in Ithaca, New York, is an environmental psychologist who studies people's relationship to the built and natural environment through the life course. Wells studies school and residential environments and health-related outcomes, including the effects of school gardens on children's diet and physical activity, the impact of nearby nature on cognitive functioning, the influence of neighborhood design on physical activity, and the effects of housing quality on psychological well-being. Her work has included the study of children, adults, and elders.

E.O. Wilson

E.O. Wilson is a biologist and Research Professor Emeritus at Harvard University, Cambridge, Massachusetts. Wilson's career began with the study of ants, but in the 1970s he began publishing on biodiversity and conservation issues. His work, *Biophilia*, was published in 1984, and in 1993 he joined forces with Stephen Kellert to further refine his work. He has won numerous awards in the areas of science, literature, and conservation. Wilson continues to publish books and articles to this day, uniting his vast knowledge of ants and insects with the broader mission of protecting the biodiversity on Earth.

Kathleen Wolf

Kathleen Wolf is the Director of the Human Dimensions of Urban Forestry and Urban Greening in the College of Forest Resources at the University of Washington, Seattle, Washington. Her work focuses primarily on the human dimensions of open space, urban forestry, and natural systems. This includes a focus on the importance of urban trees and plants and the resulting costs, benefits, and perceptions of urban nature.

Books of Interest

Clayton, S. and Opotow, S. (2003) *Identity and the Natural Environment: The Psychological Significance of Nature*. MIT Press, Cambridge, Massachusetts.

Goodenough, E.N. (2003) *Secret Spaces of Childhood*. University of Michigan Press, Ann Arbor, Michigan.

Hamrell, S. and Nordberg, O. (1992) *Women, Ecology, and Health: Rebuilding Connections*. Dag Hammarskjöld Foundation, Uppsala, Sweden. Introduction by Vandana Shiva.

Henderson, B. and Vikander, N. (2007) *Nature First: Outdoor Life the Friluftsliv Way*. Natural Heritage Books, Toronto, Canada.

Kahn, P.H. Jr (1999) *The Human Relationship with Nature: Development and Culture*. The MIT Press, Cambridge, Massachusetts.

Kahn, P.H. Jr and Hasbach, P.H. (2012) *Ecopsychology: Science, Totems, and the Technological Species*. MIT Press, Cambridge, Massachusetts.

Kahn, P.H. Jr and Kellert, S.R. (2002) *Children and Nature: Psychological, Sociocultural, and Evolutionary Investigations*. MIT Press, Cambridge, Massachusetts.

Kellert, S.R. (2005) *Building for Life: Designing and Understanding the Human–Nature Connection*. Island Press, Washington, DC.

Kellert, S.R. (2012) *Birthright: People and Nature in the Modern World*. Yale University Press, New Haven, Connecticut.

Kellert, S.R. and Wilson, E.O. (eds) (1993) *The Biophilia Hypothesis*. Island Press, Washington, DC.

Linden, S. and Grut, J. (2002) *The Healing Fields: Working with Psychotherapy and Nature To Rebuild Shattered Lives*. Frances Lincoln, London.

Louv, R. (2005) *Last Child in the Woods: Saving our Children from Nature-Deficit Disorder*. Algonquin Books of Chapel Hill, Chapel Hill, North Carolina.

Louv, R. (2011) *The Nature Principle: Human Restoration and the End of Nature-Deficit Disorder*. Algonquin Books of Chapel Hill, Chapel Hill, North Carolina.

Marten, G.G. (2001) *Human Ecology: Basic Concepts for Sustainable Development*. Earthscan/James & James, London.

Moran, E.F. (2006) *People and Nature: An Introduction to Human Ecological Relations*. Blackwell Primers in Anthropology, Volume 1. Wiley-Blackwell, Hoboken, New Jersey.

Nilsson, K., Sangster, M., Gallis, C., Hartig, T., de Vries, S., *et al.* (eds) (2011) *Forests, Trees and Human Health*. Springer, New York.

Orr, D.W. (1992) *Ecological Literacy: Education and the Transition to a Postmodern World*. State University of New York Press, Albany, New York.

Orr, D.W. (2004) *Earth in Mind: On Education, Environment, and the Human Prospect*. Island Press, Washington, DC.

Schettler, T., Solomon, G., Valenti, M. and Huddle, A. (2000) *Generations at Risk: Reproductive Health and the Environment*. MIT Press, Cambridge, Massachusetts.

Selhub, E.M. and Logan, A.C. (2012) *Your Brain on Nature: The Science of Nature's Influence on Your Health, Happiness, and Vitality*. John Wiley & Sons Canada, Ltd, Mississauga, Canada.

Shiva, V. (1994) *Close to Home: Women Reconnect Ecology, Health and Development Worldwide*. New Society, Philadelphia, Pennsylvania.

Sobel, D. (1993) *Children's Special Places: Exploring the Role of Forts, Dens, and Bush Houses in Middle Childhood*. Wayne State University Press, Detroit, Michigan.

Sobel, D. (2004) *Place-based Education: Connecting Classroom and Community*. The Orion Society and the Myrin Institute, Great Barrington, Massachusetts.

Spencer, C. and Blades, M. (2006) *Children and Their Environments: Learning, Using and Designing Spaces*. Cambridge University Press, Cambridge, UK.

Sternberg, E.M. (2009) *Healing Spaces: The Science of Place and Well-being.* Harvard University Press, Cambridge, Massachusetts.

Suzuki, D.T., McConnell, A. and Mason, A. (2007) *The Sacred Balance: Rediscovering our Place in Nature.* Greystone/David Suzuki Foundation, Vancouver, Canada.

Thomashow, M. (1996) *Ecological Identity: Becoming a Reflective Environmentalist.* MIT Press, Cambridge, Massachusetts.

Trimble, S. and Nabhan, G. (1995) *The Geography of Childhood: Why Children Need Wild Places.* Beacon Press, Boston, Massachusetts.

Weigert, A. (1997) *Self, Interaction, and Natural Environment: Refocusing Our Eyesight.* State University of New York Press, Albany, New York.

Wilson, E.O. (1984) *Biophilia: The Human Bond with Other Species.* Harvard University Press, Cambridge, Massachusetts.

Wilson, R. (2012) *Nature and Young Children: Encouraging Creative Play and Learning in Natural Environments.* Routledge, Abingdon, UK.

Films and Documentaries

Biophilic Design: The Architecture of Life. Tamarack Media and Stephen Kellert, 2011. Available at: http://www.biophilicdesign.net; http://www.bullfrogfilms.com

Call of Life: Facing the Mass Extinction. Species Alliance, 2010. Available at http://calloflife.org/

Chasing Ice. Chasing Ice, LLC, 2012. Available at http://www.chasingice.com/

Green Fire: Aldo Leopold and a Land Ethic for our Time. The Aldo Leopold Foundation, 2011. Available at: http://www.greenfiremovie.com/

Mother Nature's Child: Growing Outdoors in the Media Age. Fuzzy Slippers Productions, 2011. Available at: http://www.mothernaturesmovie.com

Play Again: What are the consequences of a childhood removed from nature? Ground Productions, 2010. Available at: http://www.playagainfilm.com

Young Voices for the Planet. Available at: http://www.youngvoicesonclimatechange.com/

Web Resources

The Aldo Leopold Foundation: http://www.aldoleopold.org/

American Horticultural Therapy Association: http://ahta.org/ and http://ahta.org/horticultural-therapy/suggested-readings

The Bioneers: http://www.bioneers.org/

The Center for Ecoliteracy: http://www.ecoliteracy.org/

Children & Nature Network: http://www.childrenandnature.org/

Children's Environmental Health Network: http://www.cehn.org

E.O. Wilson Biodiversity Foundation: http://eowilsonfoundation.org/mission-statement

Healthy Parks, Healthy People Australia: http://www.hphpcentral.com/

Human Dimensions of Urban Forestry and Urban Greening: http://www.naturewithin.info/

Human Interaction with Nature and Technological Systems (HINTS) Lab: http://depts.washington.edu/hints/

International Society of Nature and Forest Medicine (INFOM): http://infom.org/

Landscape and Human Health Laboratory: http://lhhl.illinois.edu/

Lothlorien Therapeutic Community: http://www.lothlorien.tc/index.html

National Environmental Education Foundation—Health and Environment, Children and Nature Initiative: http://www.neefusa.org/health/index.htm

National Park Service—Healthy Parks, Healthy People US: http://www.nps.gov/public_health/hp/hphp/resources.htm

No Child Left Inside Coalition: http://www.cbf.org/ncli/landing

The North American Chapter of the International Society of Nature and Forest Medicine (INFOM): http://hikingresearch.wordpress.com/infom-2/

Plants for People: http://www.plants-for-people.org/eng/

The Rocky Mountain Institute: http://www.rmi.org/

Science & Environmental Health Network: http://sehn.org/

The Society of Forest Medicine within the Japanese Society for Hygiene: http://forest-medicine.com/epage04.html

The Wild Foundation: http://www.wild.org/

UK Forestry Commission—People, Trees and Woodlands: http://www.forestresearch.gov.uk/peopleandtrees

University of Minnesota, Center for Spirituality and Healing: http://www.csh.umn.edu/program-areas-section/nature-based-therapeutics/index.htm

US Department of Agriculture Forest Service—Natural Environments for Urban Populations: http://www.ncrs.fs.fed.us/4902/

Bibliography

Abraham, A., Sommerhalder, K. and Abel, T. (2010) Landscape and well-being: a scoping study on the health-promoting impact of outdoor environment. *International Journal of Public Health* 55(5), 59–69.

Ackoff, R. and Greenberg, D. (2008) *Turning Learning Right Side Up: Putting Education Back on Track.* Prentice Hall, Upper Saddle River, New Jersey.

Adhikari, S. (2009) Commonly used ethno-medicinal plants by the indigenous people of Nepalese Himalaya. Presented at *Ecological Society of America 94th Annual Meeting*, Albuquerque, New Mexico, 2–7 August 2009.

Aerts, D., Apostel, L., De Moor, B., Hellemans, S., Maex, E., *et al.* (1994) *Worldviews: From Fragmentation to Integration.* VUBPRESS, Brussels.

Ainsworth, M.D.S., Blehar, M.C., Waters, E. and Wall, S. (1978) *Patterns of Attachment: A Psychological Study of the Strange Situation.* Erlbaum, Hillsdale, New Jersey.

Altman, I. and Low, S.M. (eds) (1992) *Place Attachment.* Plenum, New York.

Altman, I. and Rogoff, B. (1992) World views in psychology: trait, interactional, organismic, and transactional perspective. In: Stokols, D. and Altman, I. (eds) *Handbook of Environmental Psychology.* Krieger Publishing, Melbourne, Australia, pp. 7–40.

Altman, N. (1994) *Sacred Trees.* Sierra Club Books, San Francisco, California.

American Academy of Pediatrics (2000) Clinical practice guideline: diagnosis and evaluation of children with attention deficit/hyperactivity disorder. *Pediatrics* 105(5), 1158–1170.

American Academy of Pediatrics, Committee on Public Education (2001) Children, adolescents, and television. *Pediatrics* 107(2), 423–426.

Anderson, A. (2004) *Real Medicine Real Health.* Holographic Health Press, Waynesville, North Carolina.

Anderson, J. (2008) *The Second Journey: The Road Back to Yourself.* Hyperion, New York.

Andrews, M. and Gatersleben, B. (2010) Variations in perceptions of danger, fear and preference in a simulated natural environment. *Journal of Environmental Psychology* 30(4), 481.

Anon. (1964) *Wilderness Act of 1964. 16 U.S.C. 1131–1136, 78 Stat. 890.* United States of America.

Anon. (2009) Video: Sanatorium Scenes. *Harvard Magazine*, March–April 2009. Available at: http://harvardmagazine.com/2009/03/video-sanatorium-scenes (accessed 8 September 2013).

Anon. (2013) An Interview with Republic of Korea Secretary of Forestry Won Sop Shin posted July 21, 2013. Available at: http://hikingresearch.wordpress.com/2013/07/21/an-interview-with-republic-of-korea-secretary-of-forestry-won-sop-shin/ (accessed 23 July 2013).

Apelian, N.M. (2013) Restorative ecotourism as a solution to intergenerational knowledge retention: an exploratory study with two communities of San Bushmen in Botswana. PhD thesis, Prescott College, Prescott, Arizona.

Appalachian Trail Conservancy (2013) 2000 milers. Available at: http://www.appalachiantrail.org/about-the-trail/2000-milers (accessed 8 September 2013).

Appleton, J. (1975) *The Experience of Landscape*. Wiley, London.

Arbor Day Foundation and Dimensions of Educational Research Foundation (2008) Nature Explore. Available at: http://www.natureexplore.org/ (accessed 22 July 2013).

Armstrong, D.A. (2000) Survey of community gardens in upstate New York: implications for health promotion and community development. *Health & Place* 6(4), 319–327.

Arnould, E.J. and Price, L.L. (1993) River magic: extraordinary experience and the extended service encounter. *Journal of Consumer Research* 20, 24–45.

Asah, S.T., Bengston, D.N. and Westphal, L.M. (2011) The influence of childhood: operational pathways to adulthood participation in nature-based activities. *Environment and Behavior* 44(4), 545–569.

Atchley, R.A., Strayer, D.L. and Atchley, P. (2012) Creativity in the wild: improving creative reasoning through immersion in natural settings. *PLoS ONE* 7(12), e51474.

Atlantis, E., Chow, C.M., Kirby, A. and Fiatarone Singh, M. (2004) An effective exercise-based intervention for improving mental health and quality of life measures: a randomized controlled trial. *Preventive Medicine* 39(2), 424–434.

Atran, S. (1998) Folk biology and the anthropology of science: cognitive universals and cultural particulars. *Behavioral and Brain Sciences* 21(4), 547–569.

Atran, S. (1999) Itzaj Maya folkbiological taxonomy: cognitive universals and cultural particulars. In: Medin, D.L. and Atran, S. (eds) *Folkbiology*. MIT Press, Cambridge, Massachusetts, pp. 119–213.

Aung, S.K.H. and Lee, M.H.M. (2004) Music, sounds, medicine and meditation. *Alternative & Complementary Therapies* 10(5), 266–270.

Autry, C.E. (2001) Adventure therapy with girls at-risk: responses to outdoor experiential activities. *Therapeutic Recreation Journal* 35(4), 289–306.

Backster, C. (2003) *Primary Perception: Biocommunication with Plants, Living Foods, and Human Cells*. White Rose Millennium Press, Anza, California.

Baker, M. (2008) Landfullness in adventure-based programming: promoting reconnection to the land. In: Warren, K., Mitten, D. and Loeffler, T.A. (eds) *Theory and Practice of Experiential Education*. Association of Experiential Education, Boulder, Colorado, pp. 359–367.

Baldwin, C., Persing, J. and Magnuson, D. (2004) The role of theory, research, and evaluation in adventure education. *Journal of Experiential Education* 26(3), 167–183.

Bandura, A. (1986) *Social Functions of Thought and Action*. Prentice-Hall, Englewood Cliffs, New Jersey.

Bandura, A. (1994) *Self-efficacy*. Academic Press, New York.

Bardwell, L. (1992) A bigger piece of the puzzle: the restorative experience and outdoor education. In: Henderson, K. (ed.) *Coalition for Education in the Outdoors Research Symposium Proceedings, Bradford Woods, Indiana, 17–19 January 1992*. Coalition for Education in the Outdoors, Cortland, New York, pp. 15–20.

Barker, E.T. (1987) The critical importance of mothering. *Mothering Magazine* vol. 47. Available at: http://www.naturalchild.org/elliott_barker/mothering.html (accessed 14 October 2013).

Barlow, Z. and Crabtree, M. (2000) *Ecoliteracy. Mapping the Terrain*. Center for Ecoliteracy, Berkeley, California.

Barnes, P.M., Bloom, B. and Nahin, R.L. (2007) *Complementary and Alternative Medicine Use Among Adults and Children: United States, 2007. National Health Statistics Report Number 12*. National Center for Health Statistics, Hyattsville, Maryland.

Barnett, R. (2009) Serpent of pleasure: emergence and difference in the medieval garden of love. *Landscape Journal* 28(2), 137–150.

Barton, J. and Pretty, J. (2010) What is the best dose of nature and green exercise for improving mental health? A multi-study analysis. *Environmental Science & Technology* 44(10), 3947–3955.

Bauer, M.W., Petkova, K. and Boyadjieva, P. (2000) public knowledge of and attitudes to science: alternative measures that may end the 'science war'. *Science, Technology & Human Values* 25, 30–51.

Bauman, A.E. (2004) Updating the evidence that physical activity is good for health: an epidemiological review 2000–2003. *Journal of Science and Medicine in Sport* 7(1), 6–19.

BBC News Africa (2011) Somalia famine: UN warns of 750,000 deaths. *BBC News*, 5 September 2011. Available at: http://www.bbc.co.uk/news/world-africa-14785304 (accessed 15 April 2013).

Beatley, T. (2013) Biophilic Cities. Available at: http://biophiliccities.org/ (accessed 8 September 2013).

Beaton, A.A. and Funk, D.C. (2008) An evaluation of theoretical frameworks for studying physically active leisure. *Leisure Sciences* 30(1), 53–70.

Beightol, J., Jevertson, J., Carter, S., Gray, S. and Gass, M. (2012) Adventure education and resilience enhancement. *Journal of Experiential Education* 35(2), 307–325.

Bell, A.C. and Dyment, J.E. (2006) *Grounds for Action: Promoting Physical Activity through School Ground Greening in Canada*. Evergreen, Toronto, Ontario.

Bell, J.F., Wilson, J.S. and Liu, G.C. (2008) Neighborhood greenness and 2-year changes in body mass index of children and youth. *American Journal of Preventive Medicine* 35(6), 547–553.

Berlin, A.A., Kop, W.J. and Deuster, P.A. (2006) Depressive mood symptoms and fatigue after exercise withdrawal: the potential role of decreased fitness. *Psychosomatic Medicine* 68(2), 224–230.

Berlyne, D. (1960) *Conflict, Arousal, and Curiosity*. McGraw-Hill, New York.

Berman, M.G., Jonides, J. and Kaplan, S. (2008) The cognitive benefits of interacting with nature. *Psychological Science* 19(12), 1207–1212.

Berman, M.G., Kross, E., Krpan, K.M., Askren, M.K., Burson, A., *et al.* (2012) Interacting with nature improves cognition and affect for individuals with depression. *Journal of Affective Disorders* 140(3), 300–305.

Berns, G.N. and Simpson, S. (2009) Outdoor recreation participation and environmental concern: a research summary. *Journal of Experiential Education* 32(1), 79–91.

Bernstein, H.M., Russo, M.A. and Laquidara-Carr, D. (2013) *New and Retrofit Green Schools: The Cost Benefits and Influence of a Green School on its Occupants*. McGraw-Hill Construction, Bedford, Massachusetts.

Berry, W. (1998) The peace of wild things. *The Selected Poems of Wendell Berry*. Counterpoint Press, Berkeley, California.

Bixler, R.D. and Floyd, M.F. (1997) Nature is scary, disgusting, and uncomfortable. *Environment and Behavior* 29(4), 17–22.

Bixler, R.D., Floyd, M.F. and Hammitt, W.E. (2002) Environmental socialization: quantitative tests of the childhood play hypothesis. *Environment and Behavior* 34(6), 795–818.

Blair, D. (2009) The child in the garden: an evaluative review of the benefits of school gardening. *Journal of Environmental Education* 40(2), 15–38.

Blanchard, L.T., Gurka, M. and Blackman, J.A. (2006) Emotional, developmental, and behavioral health of American children and their families: a report from the 2003 National Survey of Children's Health. *Pediatrics* 117(6), 1202–1212.

Blonna, R. (2011) *Coping with Stress in a Changing World*, 5th edn. McGraw-Hill, Boston, Massachusetts.

Blue, F. (1979) Aerobic running as a treatment for moderate depression. *Perceptual and Motor Skills* 48(1), 228.

Blumenthal, J.A., Babyak, M.A., Doraiswamy, P.M., Watkins, L., Hoffman, B.M., *et al.* (2007) Exercise and pharmacotherapy in the treatment of major depressive disorder. *Psychosomatic Medicine* 69(7), 587–596.

Blumer, H. (1969) *Symbolic Interactionism: Perspective and Method*. Prentice-Hall, Englewood Cliffs, New Jersey.

Bobilya, A.J., Akey, L. and Mitchell, D. Jr (2009) Outcomes of a spiritually focused wilderness orientation program. *Journal of Experiential Education* 31(3), 440–443.

Bonney, R. and LaBranche, M. (2004) Citizen science: involving the public in research. *ASTC Dimensions* May/June issue, 13.

Bonney, R., Ballard, H., Jordan, R., McCallie, E., Phillips, T., *et al.* (2009) *Public Participation in Scientific Research: Defining the Field and Assessing its Potential for Informal Science Education. A CAISE Inquiry Group Report*. Center for Advancement of Informal Science Education (CAISE), Washington, DC.

Booth, M.L., Owen, N., Bauman, A., Clavisi, O. and Leslie, E. (2000) Social-cognitive and perceived environment influences associated with physical activity in older Australians. *Preventive Medicine* 31(1), 15–22.

Bossen, A. (2010) The importance of getting back to nature for people with dementia. *Journal of Gerontological Nursing* 36(2), 17–22.

Bowers, C. (2001) *Educating for Eco-justice and Community*. University of Georgia Press, Athens, Georgia.

Bowers, C.A. (2006) *Transforming Environmental Education: Making Renewal of the Cultural and Environmental Commons the Focus of Educational Reform*. Ecojustice Press, Eugene, Oregon.

Bowlby, J. (1951) *Maternal Care and Mental Health*. World Health Organization, Geneva, Switzerland.

Bowlby, J. (1969) *Attachment and Loss*, 2nd edn. Basic Books, New York.

Bowler, D.E., Buyung-Ali, L.M., Knight, T.M. and Pullin, A.S. (2010) A systematic review of evidence for the added benefits to health of exposure to natural environments. *BMC Public Health* 10, 456.

Bratton, S.C., Ray, D., Rhine, T. and Jones, L. (2005) The efficacy of play therapy with children: a meta-analytic review of treatment outcomes. *Professional Psychology, Research and Practice* 36(4), 376–390.

Breitenbach, E., Stumpf, E., Ferson, L.V. and Harald, E. (2009) Dolphin-assisted therapy: changes in interaction and communication between children with severe disabilities and their caregivers. *Anthrozoös* 22(3), 277–289.

Breunig, M., O'Connell, T., Todd, S., Anderson, L. and Young, A. (2010) The impact of outdoor pursuits on college students' perceived sense of community. *Journal of Leisure Research* 42(4), 551–572.

Bronson, M.B. (2000) *Self-regulation in Early Childhood: Nature and Nurture*. Guilford Press, New York.

Brossard, D., Lewenstein, B. and Bonney, R. (2005) Scientific knowledge and attitude change: the impact of a citizen science project. *International Journal of Science Education* 27(9), 1099–1121.

Bruni, C.M., Chance, R., Schultz, P.W. and Nolan, J. (2012) Natural connections: bees sting, snakes bite, but they still are nature. *Environment and Behavior* 44(2), 197–215.

Bukatko, D. and Daehler, M. (2011) *Child Development: A Thematic Approach*. Cengage Learning, Belmont, California.

Bullock, C. and Mahon, M. (2001) *Introduction to Recreation Services for Persons with Disabilities: A Person-centered Approach*, 2nd edn. Sagamore, Champaign, Illinois.

Bunting, C., Tolson, H., Kuhn, C., Suarez, E. and Williams, R.B. (2000) Physiological stress response of the neuroendocrine system during outdoor adventure tasks. *Journal of Leisure Research* 32(2), 191–207.

Bunting, T.E. and Cousins, L.R. (1985) Environmental dispositions among school-age children. *Environment and Behavior* 17(6), 725–768.

Burch, W.R. Jr (1969) The social circles of leisure: competing explanations. *Journal of Leisure Research* 1(2), 125–147.

Burdette, H. and Whitaker, R. (2005) Resurrecting free play in young children: looking beyond fitness and fatness to attention, affiliation, and affect. *Archives of Pediatrics & Adolescent Medicine* 159(1), 46–50.

Burton-Christie, D. (1999) Into the body of another: Eros, embodiment and intimacy with the natural world. *Anglican Theological Review* 81(1), 13–37.

Bushnell, R. (2003) *Green Desire: Imagining Early Modern English Gardens*. Cornell University Press, Ithaca, New York.

Caduto, M. (1998) Ecological education: a system rooted in diversity. *Journal of Environmental Education* 29(4), 11–16.

Canin, L.H. (1991) Psychological restoration among AIDS caregivers: maintaining self care. Doctoral dissertation, University of Michigan, Ann Arbor, Michigan.

Carson, R. (1962) *Silent Spring*. Houghton-Mifflin, Boston, Massachusetts.

Carson, R. (1998) *The Sense of Wonder*. HarperCollins Publishers, New York.

Caulkins, M., White, D. and Russell, K. (2006) The role of physical exercise in wilderness therapy for troubled adolescent women. *Journal of Experiential Education* 29(1), 18–37.

Cavill, N., Kahlmeier, S. and Racioppi, F. (eds) (2006) *Physical Activity and Health in Europe: Evidence for Action*. World Health Organization, Geneva, Switzerland.

CDC (2008) 2008 Physical Activity Guidelines for Americans. Available at: http://www.cdc.gov/physicalactivity/everyone/guidelines/index.html (accessed 15 October 2013).

Celizic, M. (2007) Dolphins save surfer from becoming shark bait. *NBC News*, 8 November 2007. Available at: http://www.today.com/id/21689083#.UWf48b8jlFt (accessed 12 April 2013).

Center for Ecoliteracy (2000) *Ecoliteracy. Mapping the Terrain*. Center for Ecoliteracy, Berkeley, California.

Centers for Disease Control and Prevention (2009) Physical activity and beyond. Available at: http://www.cdc.gov/Features/ParksAndTrails/ (accessed 11 November 2009).

Cerulo, K. (2009) Nonhumans in social interaction. *Annual Review of Sociology* 35, 531–552.

Cervinka, R., Roderer, K. and Hefler, E. (2012) Are nature lovers happy? On various indicators of well-being and connectedness to nature. *Journal of Health Psychology* 17(3), 379–388.

Chalquist, C. (2007) *Terrapsychology: Re-engaging the Soul of Place*. Spring Journal Books, New Orleans, Louisiana.

Chalquist, C. (2009) A look at the ecotherapy research evidence. *Ecopsychology* 1(2), 64–74.

Chalquist, C. (2013) Mind and environment: a psychological survey of perspectives literal, wide, and deep. Keynote speaker, *Fifth Annual Sustainability Education Symposium*, Prescott College, Prescott, Arizona, 16–18 May 2013.

Charles, C. (2013) Cheryl Charles welcomes C&NN's new Executive Director. Web log comment, 27 June 2013. Available at: http://blog.childrenandnature.org/2013/06/27/cheryl-charles-welcomes-cnns-new-executive-director/ (accessed 9 September 2013).

Chawla, L. (1988) Children's concern for the natural environment. *Children's Environments Quarterly* 5(3), 13–20.

Chawla, L. (1990) Ecstatic places. *Children's Environments Quarterly* 7(4), 18–23.

Chawla, L. (1994) Editors' note. *Children's Environments Quarterly* 11(3), 175–176.

Chawla, L. (1999) Life paths into effective environmental action. *Journal of Environmental Education* 31(1), 15–26.

Chawla, L. (2002) Spots of time: manifold ways of being in nature in childhood. In: Kahn, P.H. Jr and Kellert, S.R. (eds) *Children and Nature: Psychological, Sociocultural, and Evolutionary Investigations*. MIT Press, Cambridge, Massachusetts, pp. 199–225.

Chawla, L. (2006) Learning to love the natural world enough to protect it. *Barn* 2, 57–78.

Chawla, L. and Hart, R. (1988) Roots of environmental concern. In: Lawrence, D., Habe, R., Hacker, A. and Sherrod, D. (eds) *People's Needs/Planet Management: Paths to Coexistence*. Environmental Design Research Association, Pomona, California, pp. 15–18.

Chen, Y.-L. (1996) Conformity with nature: a theory of Chinese American elders' health promotion and illness prevention processes. *Advances in Nursing Science* 19(2), 17–26.

Cheng, J. and Monroe, M.C. (2012) Connection to nature: children's affective attitude toward nature. *Environment and Behavior* 44(1), 31–49.

Chenoweth, R.E. and Gobster, P.H. (1990) The nature and ecology of aesthetic experiences in the landscape. *Landscape Journal* 9(1), 1–8.

Chipeniuk, R.C. (1994) Naturalness in landscape: an inquiry from a planning perspective. PhD thesis, University of Waterloo, Waterloo, Canada.

Chopra, D. (2001) *How to Know God: The Soul's Journey into the Mystery of Mysteries*. Random House, New York.

Cicchetti, D. and Beeghly, M. (1990) Developmental perspectives on the self: a historical review. In: Cicchetti, D. and Beeghly, M. (eds) *The Self in Transition: Infancy to Childhood*. University of Chicago Press, Chicago, Illinois, pp. 1–15.

Cimprich, B.E. (1990) Attentional fatigue and restoration in individuals with cancer. Doctoral dissertation, University of Michigan, Ann Arbor, Michigan. *Dissertation Abstracts International* 51(4): 1740B.

Clark, R.N. and Stankey, G.H. (1979) *The Recreation Opportunity Spectrum: A Framework for Planning, Management, and Research. General Technical Report PNW-98*. US Department of Agriculture, Forest Service, Pacific Northwest Forest and Range Experiment Station, Seattle, Washington.

Clarke, J.I. (1978) *Self-Esteem: A Family Affair*. Winston Press, Minneapolis, Minnesota.

Clarke, J.I. (1998) *Self-Esteem: A Family Affair*. Hazelden Foundation, Center City, Minnesota.

Clarke, J.I. and Dawson, J. (1989/1998) *Growing Up Again: Parenting Ourselves, Parenting Our Children*, 2nd edn. Hazelden Foundation, Center City, Minnesota.

Clay, R. (2001) Green is good for you. *Monitor on Psychology* 32(4), 40–42.

Clayton, S. (2003) Environmental identity. In: Clayton, S. and Opotow, S. (eds) *Identity and the Natural Environment: The Psychological Significance of Nature*. MIT Press, Cambridge, Massachusetts, pp. 45–66.

Clayton, S. and Opotow, S. (2003) *Identity and the Natural Environment: The Psychological Significance of Nature*. MIT Press, Cambridge, Massachusetts.

Clements, R. (2004) An investigation of the status of outdoor play. *Contemporary Issues in Early Childhood* 5(1), 68–80.

Cobb, E. (1977) *The Ecology of Imagination in Childhood*. Routledge & Kegan Paul Publishers, London.

Cobb, E. (1998) *The Ecology of Imagination in Childhood*. Spring, Dallas, Texas.

Cobb, E. (1977/1998) *The Ecology of Imagination in Childhood*. John Wiley, New York.

Cobb, E. and Mead, M. (1977) *The Ecology of Imagination in Childhood*. Columbia University Press, New York.

Cole, D.N. and Hall, T.E. (2010) Experiencing the restorative components of wilderness environments: does congestion interfere and does length of exposure matter? *Environment and Behavior* 42(6), 806–823.

Cole, E., Erdman, E. and Rothblum, E.D. (eds) (1994) *Wilderness Therapy for Women: The Power of Adventure*. Harrington Press, New York.

Coleman, D. and Iso-Ahola, S.E. (1993) Leisure and health: the role of social support and self-determination. *Journal of Leisure Research* 25(2), 111–128.

Coley, J., Solomon, G. and Shafto, P. (2002) The development of folkbiology: a cognitive science perspective on children's understanding of the biological world. In: Kahn P.H. Jr and Kellert, S.R. (eds) *Children and Nature: Psychological, Sociocultural, and Evolutionary Investigations*. MIT Press, Cambridge, Massachusetts, pp. 65–89.

Colin, V.L. (1996) *Human Attachment*. Temple University Press, Philadelphia, Pennsylvania.

Collins, S.D. (1974) *A Different Heaven and Earth*. Judson Press, Valley Forge, Pennsylvania.

Committee on Environmental Health and Tester, J.M. (2009) The built environment: designing communities to promote physical activity in children. *Pediatrics* 123(6), 1591–1598.

Connor, S. (2010) Mankind leaves mark on the planet with the end of the 12,000-year Holocene age. Landmark in the Earth's 4.7bn-year history as geologists hail dawn of the 'human epoch'.

The Independent, 6 April 2010. Available at: http://www.independent.co.uk/news/science/mankind-leaves-mark-on-the-planet-with-the-end-of-the-12000year-holocene-age-1936725.html (accessed 17 September 2013).

Cooper, A.R., Page, A.S., Wheeler, B.W., Hillsdon, M., Griew, P., *et al.* (2010) Patterns of GPS measured time outdoors after school and objective physical activity in English children: the PEACH project. *International Journal of Behavioral Nutrition and Physical Activity* 7, 31.

Cosgriff, M., Little, D. and Wilson, E. (2010) The nature of nature: how New Zealand women in middle to later life experience nature-based leisure. *Leisure Sciences* 32(1), 15–32.

Crain, W. (2001) How nature helps children develop. *Montessori Life* 9(2), 41–43.

Crain, W. (2011) *Theories of Development: Concepts and Applications*, 6th edn. Pearson Education, Inc., Upper Saddle River, New Jersey.

Cramer, J.R. (2008) Reviving the connection between children and nature through service-learning restoration partnerships. *Native Plants Journal* 9(3), 278–286.

Crews, D.J., Lochbaum, M.R. and Landers, D.M. (2004) Aerobic physical activity effects on psychological well-being in low-income Hispanic children. *Perceptual and Motor Skills* 98(1), 319–324.

Crompton, T. and Kasser, T. (2009) *Meeting Environmental Challenges: The Role of Human Identity*. WWF-UK, Godalming, UK.

Crutzen, P.J. (2002) Geology of mankind. *Nature* 415(6867), 23.

Crutzen, P.J. and Stoermer, E.F. (2000) The 'Anthropocene'. *Global Change Newsletter* 41(May), 17–18.

Csikszentmihalyi, M. (1975) *Beyond Boredom and Anxiety: The Experience of Play in Work and Games.* Jossey-Bass, San Francisco, California.

Csikszentmihalyi, M. (1992) *Flow: The Psychology of Happiness*. Rider, London.

Csikszentmihalyi, M. (1997) *Finding Flow: The Psychology of Engagement with Everyday Life*. Basic Books, New York.

Csikszentmihalyi, M. (2003) *Good Business: Leadership, Flow and the Making of Meaning*. Penguin Putnam Inc., New York.

Csikszentmihalyi, M. and Csikszentmihalyi, M. (1990) *Flow: The Psychology of Optimal Experience*. Harper & Row, New York.

Cumes, D. (1998) Nature as medicine: the healing power of the wilderness. *Alternative Therapies in Health and Medicine* 4(2), 79–86.

Davidson, L. (2012) The calculable and the incalculable: narratives of safety and danger in the mountains. *Leisure Sciences* 34(4), 298–313.

Davis, J. (2008) Psychological benefits of nature experiences: an outline of research and theory with special reference to transpersonal psychology. Available at: http://www.soulcraft.co/essays/PSYCHOLOGICAL_BENEFITS_OF_NATURE_EXPERIENCES.pdf (accessed 6 April 2013).

Davis, J., Lockwood, L. and Wright, C. (1991) Reasons for not reporting peak experiences. *Journal of Humanistic Psychology* 31(1), 86–94.

d'Eaubonne, F. (1974) Feminism or death. In: Marks, E. and de Courtivron, I. (eds) *New French Feminisms: An Anthology*. University of Massachusetts Press, Amherst, Massachusetts, pp. 64–67.

d'Eaubonne, F. (1974) *Le féminisme ou la mort*, vol. 2. P. Horay, Paris.

d'Eaubonne, F. (1978) *Écologie-Féminisme: Révolution ou Mutation?* (*Ecology-Feminism: Revolution or Mutation?*). Éditions ATP, Paris.

Degenhardt, B. and Buchecker, M. (2012) Exploring everyday self-regulation in nearby nature: determinants, patterns, and a framework of nearby outdoor recreation behavior. *Leisure Sciences* 34(5), 450–469.

Degenhardt, B., Frick, J., Buchecker, M. and Gutscher, H. (2011) Influences of personal, social, and environmental factors on workday use frequency of the nearby outdoor recreation areas by working people. *Leisure Sciences* 33(5), 420–440.

Delaware State News (2013) Dog sounds alarm in Magnolia house blaze. *Delaware State News*, Friday 4 January 2013.

DeLoache, J.S. (2010) Early development of the understanding and use of symbolic artifacts. In: Goswami, U. (ed.) *The Wiley-Blackwell Handbook of Childhood Cognitive Development*, 2nd edn. John Wiley & Sons, Inc., Hoboken, New Jersey, pp. 312–336.

DeLoache, J.S. and LoBue, V. (2009) The narrow fellow in the grass: human infants associate snakes and fear. *Developmental Science* 12(1), 201–207.

Devine-Wright, P. and Clayton, S. (2010) Introduction to the special issue: Place, identity and environmental behaviour. *Journal of Environmental Psychology* 30(3), 267–270.

de Vries, S. (2010) Nearby nature and human health: looking at the mechanisms and their implications. In: Ward Thompson, C., Aspinall, P. and Bell, S. (eds) *Innovative Approaches to Researching Landscape and Health. Open Space: People Space 2.* Routledge, Abingdon, UK, pp. 77–96.

de Vries, S., Verheij, R.A., Groenewegen, P.P. and Spreeuwenberg, P. (2003) Natural environments – healthy environments? An exploratory analysis of the relationship between greenspace and health. *Environment and Planning A* 35(10), 1717–1732.

de Vries, S., Claßen, T., Eigenheer-Hug, S.-M., Korpela, K., Maas, J., *et al.* (2011) Contributions of natural environments to physical activity: theory and evidence base. In: Nilsson, K., Sangster, M., Gallis, C., Hartig, T., de Vries, S., *et al.* (eds) *Forests, Trees and Human Health.* Springer, New York, pp. 205–243.

Dewey, J. (1938/1997) *Experience and Education.* Touchstone, New York.

Dickinson, J.L. and Bonney, R. (2012) Introduction: why citizen science? In: Dickinson, J.L. and Bonney, R. (eds) *Citizen Science: Public Participation in Environmental Research.* Cornell University, Ithaca, New York, pp. 1–14.

Dickinson, J.L., Shirk, J., Bonter, D., Bonney, R., Crain, R.L., *et al.* (2012) The current state of citizen science as a tool for ecological research and public engagement. *Frontiers in Ecology & the Environment* 10(6), 291–297.

Diette, G.B., Lechtzin, N., Haponik, E., Devrotes, A. and Rubin, H.R. (2003) Distraction therapy with nature sights and sounds reduces pain during flexible bronchoscopy: a complementary approach to routine analgesia. *Chest Journal* 123(3), 941–948.

Docksai, R. (2011) The sounds of wellness. *Futurist* 45(5), 13–16.

Donovan, G.H., Butry, D.T., Michael, Y.L., Prestemon, J.P., Liebhold, A.M., *et al.* (2013) The relationship between trees and human health: evidence from the spread of the emerald ash borer. *American Journal of Preventive Medicine* 44(2), 139–145.

Douillard, J. (2004) *Perfect Health for Kids: Ten Ayurvedic Health Secrets Every Parent Must Know.* North Atlantic Books, Berkeley, California.

Doxey, J.S., Waliczek, T.M. and Zajicek, J.M. (2009) The impact of interior plants in university classrooms on student course performance and on student perceptions of the course and instructor. *HortScience* 44(2), 384–391.

Doyne, E.J., Chambless, D.L. and Beutler, L.E. (1983) Aerobic exercise as a treatment for depression in women. *Behavior Therapy* 14(3), 434–440.

Driskell, J.E., Johnston, J.H. and Salas, E. (2001) Does stress training generalize to novel settings? *Human Factors: Journal of the Human Factors and Ergonomics Society* 43(1), 99–110.

Driver, B., Tinsley, E. and Manfredo, M. (1991) Leisure and recreation experience preference scales: two inventories designed to assess the breadth of perceived psychological benefits of leisure. In: Driver, B., Brown, P. and Peterson, G. (eds) *Benefits of Leisure.* Venture Publishing Inc., State College, Pennsylvania, pp. 263–286.

Dunn, A.L., Trivedi, M.H. and O'Neal, H.A. (2001) Physical activity dose–response effects on outcomes of depression and anxiety. *Medicine and Science in Sports and Exercise* 33(6 Suppl.), S587–S597.

Dunn, A.L., Trivedi, M.H., Kampert, J.B., Clark, C.G. and Chambliss, H.O. (2005) Exercise treatment for depression: efficacy and dose response. *American Journal of Preventive Medicine* 28(1), 1–8.

Dunn, H.L. (1961) *High-level Wellness: A Collection of Twenty-nine Short Talks on Different Aspects of the Theme.* Mt. Vernon, Washington, DC.

Dustin, D., Bricker, K. and Scwab, K. (2010) People and nature: toward an ecological model of health promotion. *Leisure Sciences* 32(1), 3–14.

Dustman, R.E., Ruhling, R.O., Russell, E.M., Shearer, D.E., Bonekat, H.W., *et al.* (1984) Aerobic exercise training and improved neuropsychological function of older individuals. *Neurobiology of Aging* 5(1), 35–42.

Dyment, J.E. and Bell, A.C. (2008) Grounds for movement: green school grounds as sites for promoting physical activity. *Health Education Research* 23(6), 952–962.

Earhart, B.H. (ed.) (2001) *Religious Traditions of the World: A Journey through Africa, Mesoamerica, North America, Judaism, Christianity, Islam, Hinduism, Buddhism, China, and Japan.* HarperCollins, New York.

Earp, S.E. and Maney, D.L. (2012) Birdsong: is it music to their ears? *Frontiers in Evolutionary Neuroscience* 4(14), 1–10.

Earth Ministry (2013) Creation Care Sermons. Available at: http://earthministry.org/resources/worship-aids/sermons/creation-care/Creation%20Care/?searchterm=creation care sermon (accessed 21 September 2013).

Eckersley, R. (2004) A new worldview struggles to emerge. *The Futurist* 38(5), 20–24.

Ecklund, E.H. and Long, E. (2011) Scientists and spirituality. *Sociology of Religion* 72(3), 253–274.

Edwards, T. (2008) Cancer's Achilles' heel. *Ode Magazine*, June 2008 issue. Available at: http://odewire. com/61867/cancers-achilles-heel.html (accessed 8 September 2013).

Ehrenreich, B. and English, D. (1973) *Witches, Midwives, and Nurses: A History of Women Healers.* The Feminist Press at the City University of New York, New York.

Eisenhauer, B., Krannich, R. and Blahna, D. (2000) Attachments to special places on public lands: an analysis of activities, reason for attachments, and community connections. *Society & Natural Resources* 13(5), 421–441.

Elgin, D. (2009) *The Living Universe.* Berrett-Koehler Publishers, San Francisco, California.

Elmer, P. (ed.) (2004) *The Healing Arts: Health, Disease and Society in Europe 1500–1800.* Manchester University Press, Manchester, UK.

Engeser, S. and Rheinberg, F. (2008) Flow, performance and moderators of challenge–skill balance. *Motivation and Emotion* 32(3), 158–172.

Erikson, E. (1998) *The Life Cycle Completed.* W.W. Norton & Company, New York.

Erikson, E.H. (1950) *Childhood and Society.* Norton, New York.

Erzen, J. (2011) Reading mosques: meaning and architecture in Islam. *Journal of Aesthetics & Art Criticism* 69(1), 125–131.

Esterl, M. (2008) German tots learn to answer call of nature. *The Wall Street Journal*, 14 April 2008. Available at: http://online.wsj.com/article/SB120813155330311577.html (accessed 16 September 2013).

Evans, G.W. and Kantrowitz, E. (2002) Socioeconomic status and health: the potential role of environmental risk exposure. *Annual Review of Public Health* 23(1), 202–231.

Ewert, A. (1988) Reduction of trait anxiety through participation in Outward Bound. *Leisure Sciences* 10(2), 107–117.

Ewert, A. (1989) *Outdoor Adventure Pursuits: Foundations, Models, and Theories.* Publishing Horizons, Scottsdale, Arizona.

Ewert, A. (ed.) (1996) *Natural Resource Management: The Human Dimension. Social Behavior and Natural Resources Series.* Westview Press, Boulder, Colorado.

Ewert, A. and Galloway, G. (2009) Socially desirable responding in an environmental context: development of a domain specific scale. *Environmental Education Research* 15(1), 55–70.

Ewert, A. and Galloway, G. (2012) Take a park, not a pill: promoting health and wellness through adventure programming. In: Martin, B. and Wagstaff, M. (eds) *Controversial Issues in Adventure Programming.* Human Kinetics, Champaign, Illinois, pp. 130–137.

Ewert, A. and Garvey, D. (2007) Philosophy and theory of adventure education. In: Prouty, D., Panicucci, J. and Collinson, R. (eds) *Adventure Education: Theory and Applications.* Human Kinetics, Champaign, Illinois, pp. 19–32.

Ewert, A. and McAvoy, L. (2000) The effects of wilderness settings on organized groups: a state-of-knowledge paper. In: McCool, S.F., Cole, D.N., Borrie, W.T. and O'Loughlin, J. (comps) *Wilderness Science in a Time of Change Conference—Volume 3: Wilderness as a Place for Scientific Inquiry*, Missoula, Montana, 23–27 May 1999. *USDA Forest Service Proceedings RMRS-P-15-VOL-3.* US Department of Agriculture, Forest Service, Rocky Mountain Research Station, Ogden, Utah, pp. 13–26.

Ewert, A. and Yoshino, A. (2011) The influence of short-term adventure-based experiences on levels of resilience. *Journal of Adventure Education and Outdoor Learning* 11(1), 35–50.

Ewert, A., Place, G. and Sibthorp, J. (2005) Early-life outdoor experiences and an individual's environmental attitudes. *Leisure Sciences* 27(3), 225–239.

Ewert, A., Attarian, A., Hollenhorst, S., Russell, K. and Voight, A. (2006) Evolving adventure pursuits on public lands: emerging challenges for management and public policy. *Journal of Park and Recreation Administration* 24(2), 125–140.

Ewert, A., Overholt, J., Voight, A. and Wang, C.C. (2010) Understanding the transformative aspects of the wilderness and protected lands experience upon human health. In: Watson, A., Murrieta-Saldivar, J. and McBride, B. (comps) *Science and Stewardship to Protect and Sustain Wilderness Values. Ninth World Wilderness Congress Symposium*, Merida, Yucatan, Mexico, 6–13 November 2009. *Proceedings RMRS-P-64.* US Department of Agriculture, Forest Service, Rocky Mountain Research Station, Fort Collins, Colorado, pp. 140–146.

Ewert, A., Davidson, C., Overholt, J., Luo, Y., Billingham, R. & Bishop, C. (2011) Retying the knot: enhancing father/son relationships through adventure education. Presented at *13th Canadian Conference on Leisure Research*, St. Catharines, Ontario, Canada, 18–21 May 2011.

Faarlund, N. (2010) Challenging modernity by way of the Norwegian friluftsliv tradition. *Norwegian Journal of Friluftsliv.*

Farmer, J., Knapp, D., Meretsky, V.J., Chancellor, C. and Fischer, B.C. (2011) Motivations influencing the adoption of conservation easements. *Conservation Biology* 25(4), 827–834.

Feinstein, B.C. (2004) Learning and transformation in the context of Hawaiian traditional ecological knowledge. *Adult Education Quarterly* 54(2), 105–120.

Felsten, G. (2009) Where to take a study break on the college campus: an attention restoration theory perspective. *Journal of Environmental Psychology* 29(1), 160–167.

Fenton, L. (2008) Evolutionary tails of the dis-stressing response to nature and future research directions. Presented at *Canadian Parks for Tomorrow: 40th Anniversary Conference*, University of Calgary, Calgary, Alberta, Canada, 8–11 May 2008.

Fishwick, L. and Vining, J. (1992) Toward a phenomenology of recreation place. *Journal of Environmental Psychology* 12(1), 57–63.

Fisman, L. (2001) *Child's Play: An Empirical Study of the Relationship between the Physical Form of Schoolyards and Children's Behavior.* Yale School of Forestry & Environmental Studies, Hixon Center for Urban Ecology, New Haven, Connecticut.

Fjørtoft, I. (2001) The natural environment as a playground for children: the impact of outdoor play activities in pre-primary school children. *Early Childhood Education Journal* 29(2), 111–117.

Fjørtoft, I. (2004) Landscape as playscape: the effects of natural environments on children's play and motor development. *Children, Youth and Environments* 14(2), 21–44.

Flach, F.F. (1997) *Resilience: How to Bounce Back when the Going Gets Tough!* Hatherleigh Press, New York.

Flanagan, S. (1989) *Hildegard of Bingen, a Visionary Life.* Routledge, London.

Flinders, C. (2003) *Rebalancing the World: Why Women Belong and Men Compete and How to Restore the Ancient Equilibrium.* HarperCollins, San Francisco, California. Originally published as: Flinders, C. (2002) *The Values of Belonging: Rediscovering Balance, Mutuality, Intuition, and Wholeness in a Competitive World.* HarperOne, New York.

Foote, K.E. and Azaryahu, M. (2009) Sense of place. In: Rob, K. and Nigel, T. (eds) *International Encyclopedia of Human Geography.* Elsevier, Oxford, UK, pp. 96–100.

Fosar, G. and Bludorf, F. (2011) DNA can be influenced and reprogrammed by words and frequencies (taken from the book "Vernetzte Intelligenz" von Grazyna Fosar und Franz Bludorf and summarised by Baerbel), 13 June 2011. Available at: http://www.alchemyoflight.org/2011/06/dna-can-be-influenced-and-reprogrammed-by-words-and-frequencies-taken-from-the-book-%E2%80%9Cvernetzte-intelligenz%E2%80%9D-von-grazyna-fosar-und-franz-bludorf-and-summarised-by-baerbel/ (accessed 25 September 2013).

Fosnot, C.T. (2005) *Constructivism: Theory, Perspectives and Practice,* 2nd edn. Teacher's College, New York.

Foundation for Critical Thinking (2009) Our concept of critical thinking. Available at http://www.criticalthinking.org/aboutCT/ourConceptCT.cfm (accessed 26 September 2013).

Fox, R. (1999) Enhancing spiritual experience in adventure programs. In: Miles, J.C. and Priest, S. (eds) *Adventure Programming.* Venture Publishing, State College, Pennsylvania, pp. 455–461.

Fox, R.J. (1997) Women, nature and spirituality: a qualitative study exploring women's wilderness experience. In: Rowe, D. and Brown, P. (eds) *Proceedings of ANZALS Conference 1997.* Australian and New Zealand Association for Leisure Studies and Department of Leisure and Tourism Studies, University of Newcastle, Newcastle, Australia, pp. 59–64.

Fox, S. (1981) *John Muir and his Legacy: The American Conservation Movement.* Little, Brown, Boston, Massachusetts.

Francis, C. and Cooper-Marcus, C. (1991) Places people take their problems. In: *Proceedings of 22nd Annual Conference of the Environmental Design Research Association.* Environmental Design Research Association, Oklahoma City, Oklahoma, pp. 178–184.

Francis, M. and Devereaux, K. (1991) Children of nature. *UC Davis Magazine* 9(2). University of California–Davis, Davis, California.

Fredrickson, B.L. (1998) What good are positive emotions? *Review of General Psychology* 2(3), 300–319.

Fredrickson, B.L. (2001) The role of positive emotions in positive psychology: the broaden-and-build theory of positive emotions. *American Psychologist* 56(3), 218–226.

Fredrickson, B.L. and Joiner, T. (2002) Positive emotions trigger upward spirals toward emotional well-being. *Psychological Science* 13(2), 172–175.

Fredrickson, L.M. (1996) Exploring spiritual benefits of person–nature interactions through an ecosystem management approach. Doctoral dissertation, University of Minnesota, Minneapolis, Minnesota.

Fredrickson, L.M. and Anderson, D.H. (1999) A qualitative exploration of the wilderness experience as a source of spiritual inspiration. *Journal of Environmental Psychology* 19(1), 21–39.

Freeman, P.A. and Zabriskie, R.B. (2002) The role of outdoor recreation in family enrichment. *Journal of Adventure Education & Outdoor Learning* 2(2), 131–146.

Friedmann, E. and Thomas, S.A. (1995) Pet ownership, social support, and one-year survival after acute myocardial infarction in the Cardiac Arrhythmia Suppression Trial (CAST). *American Journal of Cardiology* 76(17), 1213–1217.

Frohoff, T. and Dudzinski, K. (2011) *Dolphin Mysteries: Unlocking the Secrets of Communication.* Yale University Press, New Haven, Connecticut.

Fromm, E. (1964) *The Heart of Man.* Harper and Row, New York.

Frumkin, H. (2001) Beyond toxicity: human health and the natural environment. *American Journal of Preventive Medicine* 20(3), 234–240.

Frumkin, H. and Fox, J. (2011) Healthy schools. In: Dannenberg, A., Frumkin, H. and Jackson, R. (eds) *Making Healthy Places: Designing and Building for Health, Well-being, and Sustainability.* Island Press, Washington, DC, pp. 216–228.

Frumkin, H. and Louv, R. (2007) *The Powerful Link between Conserving Land and Preserving Health.* Land Trust Alliance, Washington, DC.

Furnass, B. (1979) Health values. In: Messer, J. and Mossley, J.G. (eds) *The Value of National Parks to the Community: Values and Ways of Improving the Contribution of Australian National Parks to the Community.* Australian Conservation Foundation, University of Sydney, Sydney, Australia, pp. 60–69.

Gaarder, J. (1995) *Sophie's World: A Novel about the History of Philosophy.* Phoenix/Orion Books Ltd, London.

Gallagher, W. (1993) *The Power of Place.* Poseiden Press, New York.

Gardner, H.E. (2000) *Intelligence Reframed: Multiple Intelligences for the 21st Century.* Basic Books, New York.

Garst, B.A., Williams, D.R. and Roggenbuck, J.W. (2010) Exploring early twenty-first century developed forest camping experiences and meaning. *Leisure Sciences* 32(1), 90–107.

Gass, M.A. (2007) Bringing adventure into marriage and family therapy: an innovative experiential approach. *Journal of Marital and Family Therapy* 19(3), 273–286.

Gass, M.A., Gillis, H.L. and Russell, K. (2012) *Adventure Therapy: Theory, Research, and Practice.* Routledge, New York.

Gasser, K. and Kaufmann-Hayoz, R. (2004) *Woods, Trees and Human Health & Well-being.* Interfakultäre Koordinationsstelle für Allgemeine Ökologie (IKAO), Bern.

Gelter, H. (2000) Friluftsliv: the Scandinavian philosophy of outdoor life. *Canadian Journal of Environmental Education* 5(1), 77–92.

Gierlach-Spriggs, N., Kaufman, R.E. and Warner, S.B. Jr (1998) *Restorative Garden: The Healing Landscape.* Yale University Press, New Haven, Connecticut.

Ginsburg, K.R. and American Academy of Pediatrics Committee on Communications and Committee on Psychosocial Aspects of Child and Family Health (2007) The importance of play in promoting healthy child development and maintaining strong parent–child bonds. *Pediatrics* 119(1), 182–191.

Gladwell, M. (2002) *The Tipping Point: How Little Things Can Make a Big Difference.* Back Bay Books, Boston, Massachusetts.

Gladwell, M. (2008) *Outliers: The Story of Success.* Little, Brown and Company, New York.

Glendinning, C. (1994) *Hello My Name is Chellis and I'm Recovering from Western Civilization.* Shambhala Publications, Boston, Massachusetts.

Glendinning, C. (1995) Technology, trauma, and the wild. In: Roszak, T., Gomes, M.E. and Kanner, A.D. (eds) *Ecopsychology: Restoring the Earth, Healing the Mind.* Sierra Club Books, San Francisco, California, pp. 41–54.

Goldenberg, M. and Martin, B. (eds) (2008) *Outdoor Adventures: Hiking and Backpacking.* Wilderness Education Association and Human Kinetics, Champaign, Illinois.

Goleman, D. (1995/2006) *Emotional Intelligence: Why It Can Matter More Than IQ,* 10th anniversary edn. Bantam, New York.

Goodenough, E.N. (2003) *Secret Spaces of Childhood.* University of Michigan Press, Ann Arbor, Michigan.

Gould, S.J. (1991) Enchanted evening. *Natural History* 100(September), 4–14.

Grafanaki, S., Pearson, D., Cini, F., Godula, B., McKenzie, B., et al. (2005) Sources of renewal: a qualitative study of the experience and role of leisure in the life of counselors and psychologists. *Counselling Psychology Quarterly* 18(1), 31–40.

Green, L.W. and Kreuter, M.W. (2005) *Health Program Planning: An Educational and Ecological Approach.* McGraw Hill, Boston, Massachusetts.

Greene, T. (1996) Cognition and the management of place. In: Driver, B.L., Dustin, D., Baltic, T., Elsner, G. and Peterson, G. (eds) *Nature and the Human Spirit: Toward an Expanded Land Management Ethic*. Venture Publishing, State College, Pennsylvania, pp. 301–310.

Greenway, R. (1996) Wilderness experience and ecopsychology. *International Journal of Wilderness* 2(1), 26–30.

Greenway, R. (1999) Ecopsychology: a personal history. *Gatherings: Journal of the International Community for Ecopsychology* 1(1). Available at: http://www.ecopsychology.org/journal/gatherings/personal.htm (accessed 15 September 2013).

Griffin, S. (1978) *Woman and Nature: The Roaring Inside Her*. Harper & Row, New York.

Griffin, S. (1980) *Woman and Nature: The Roaring Inside Her*. Harper Colophon, New York.

Grinde, B. and Patil, G.G. (2009) Biophilia: does visual contact with nature impact on well-being? *International Journal of Environmental Research and Public Health* 6(9), 2332–2343.

Grocott, A. and Hunter, J. (2009) Increases in global and domain specific self-esteem following a 10 day developmental voyage. *Social Psychology of Education* 12(4), 443–459.

Groff, D. and Dattilo, J. (2000) Adventure therapy. In: Dattilo, J. (ed.) *Facilitation Techniques in Therapeutic Recreation*. Venture Publishing, State College, Pennsylvania, pp. 13–37.

Gruenewald, D. (2008) The best of both worlds: a critical pedagogy of place. *Environmental Education Research* 14(3), 308–324.

Haller, R. and Kramer, C. (eds) (2006) *Horticultural Therapy Methods: Making Connections in Health Care, Human Service, and Community Programs*. The Haworth Press, Binghamton, New York.

Halstead, M. and Roscoe, S. (2002) Music as an intervention for oncology nurses. *Clinical Journal of Oncology Nursing* 6(6), 332–336.

Haluza-Delay, R. (1999) Navigating the terrain: helping care for the Earth. In: Miles, J.C. and Priest, S. (eds) *Adventure Education*. Venture Publishing, Inc., State College, Pennsylvania, pp. 445–454.

Hamrell, S. and Nordberg, O. (1992) *Women, Ecology, and Health: Rebuilding Connections*. Dag Hammarskjöld Foundation, Uppsala, Sweden.

Han, K. (2001) A review: theories of restorative environments. *Journal of Therapeutic Horticulture* 12, 30–43.

Hand, E. (2010) Citizen science: people power. *Nature* 466(7307), 685–687.

Hans, T. (2000) A meta-analysis of the effects of adventure programming on locus of control. *Journal of Contemporary Psychotherapy* 30(1), 33–60.

Harshaw, H.W. and Tindall, D.B. (2005) Social structure, identities, and values: a network approach to understanding people's relationships to forests. *Journal of Leisure Research* 37(4), 426–449.

Hart, R. (1997) *Children's Participation: The Theory and Practice of Involving Young Citizens in Community Development and Environmental Care*. Earthscan Publications Limited, Oxford, UK.

Hartig, T. (2004) Restorative environments. In: Spielberger, C. (ed.) *Encyclopedia of Applied Psychology*. Academic Press, San Diego, California, pp. 273–279.

Hartig, T. (2008) Green space, psychological restoration, and health inequality. *Lancet* 372(9650), 1614–1615.

Hartig, T., Mang, M. and Evans, G. (1991) Restorative effects of natural environment experiences. *Environment and Behavior* 23(1), 3–26.

Hartig, T., Korpela, K., Evans, G.W. and Gärling, T. (1997) A measure of restorative quality in environments. *Scandinavian Housing and Planning Research* 14(4), 175–194.

Hartig, T., Nyberg, L., Nilsson, L.-G. and Gärling, T. (1999) Testing for mood congruent recall with environmentally induced mood. *Journal of Environmental Psychology* 19(4), 353–367.

Hartig, T., van den Berg, A.E., Hagerhall, C.M., Tomalak, M., Bauer, N., et al. (2011) Health benefits and nature experiences: psychological, social and cultural processes. In: Nilsson, K., Sangster, M., Gallis, C., Hartig, T., de Vries, S., et al. (eds) *Forests, Trees and Human Health*. Springer, New York, pp. 127–168.

Harwood Group (1995) Yearning for Balance: Views of Americans on Consumption, Materialism, and the Environment. Prepared for the Merck Family Fund by The Harwood Group. Available at: http://www.iisd.ca/consume/harwood.html (accessed 7 September 2013).

Hattie, J.A., Myers, J.E. and Sweeney, T.J. (2004) A factor structure of wellness: theory, assessment, analysis, and practice. *Journal of Counseling & Development* 82(3), 354–364.

Havitz, M.E., Kaczynski, A.T. and Mannell, R.C. (2013) Exploring relationships between physical activity, leisure involvement, self-efficacy, and motivation via participant segmentation. *Leisure Sciences* 35(1), 45–62.

Hawken, P. (2007) *Blessed Unrest: How the Largest Social Movement in History is Restoring Grace, Justice and Beauty to the World*. Penguin Books, London.

Hay, R. (1998) Sense of place in developmental context. *Journal of Environmental Psychology* 18(1), 5–29.

Heft, H. (2010) Affordances and the perception of landscape: an inquiry into environmental perception and aesthetics. In: Ward Thompson, C., Aspinall, P. and Bell, S. (eds) *Innovative Approaches to Researching Landscape and Health. Open Space: People Space 2*. Routledge, Abingdon, UK, pp. 9–32.

Heintzman, P. (2000) Leisure and spiritual well-being relationships: a qualitative study. *Society and Leisure* 23(1), 41–69.

Heintzman, P. (2007) Men's wilderness experience and spirituality: a qualitative study. In: Burns, R. and Robinson, K. (comps) *Proceedings of the 2006 Northeastern Recreation Research Symposium. General Technical Report NRS-P-14*. US Department of Agriculture, Forest Services, Northern Research Station, Newton Square, Pennsylvania, pp. 432–439.

Heintzman, P. (2007) Rowing, sailing, reading, discussing, praying: the spiritual and lifestyle impact of an experientially based, graduate, environmental education course. Paper presented at the *Trails to Sustainability Conference*, Kananaskis, Canada, 24–27 May 2007.

Heintzman, P. (2008) Leisure-spiritual coping: a model for therapeutic recreation and leisure services. *Therapeutic Recreation Journal* 42(1), 56–73.

Heintzman, P. (2008) Men's wilderness experience and spirituality: further explorations. In: LeBlanc, C. and Vogt, C. (eds) *Proceedings of the 2007 Northeastern Recreation Research Symposium. General Technical Report NRS-P-23*. US Department of Agriculture, Forest Services, Northern Research Station, Newton Square, Pennsylvania, pp. 55–59.

Heintzman, P. (2009) Nature-based recreation and spirituality: a complex relationship. *Leisure Sciences* 32(1), 72–89.

Heise, U. (2008) *Sense of Place and Sense of Planet: The Environmental Imagination of the Global*. Oxford University Press, New York.

Henderson, B. and Vikander, N. (2007) *Nature First: Outdoor Life the Friluftsliv Way*. Natural Heritage Books, Toronto, Canada.

Hendrix, S.E. (2011) Natural philosophy or science in premodern epistemic regimes? The case of the astrology of Albert the Great and Galileo Galilei. *Teorie vědy/Theory of Science* 33(1), 111–132.

Herron, S. and Byers, J. (2011) TEK combined with laboratory research on wild rice is driving the restoration of the wild rice culture and ecosystems. Presented at *96th Annual Meeting: The Ecological Society of America*, Austin, Texas, 7–12 August 2011.

Herron, S., Byers, J. and Mitten, L. (2010) Reflections on a multi-year study of stratification needs and germination rates in wild rice. Presented at *95th Annual Meeting: The Ecological Society of America*, Pittsburgh, Pennsylvania, 1–6 August 2010.

Herzog, T.R. and Strevey, S.J. (2008) Contact with nature, sense of humor, and psychological well-being. *Environment and Behavior* 40(6), 747–776.

Herzog, T.R., Black, A.M., Fountaine, K.A. and Knotts, D.J. (1997) Reflection and attentional recovery as distinctive benefits of restorative environments. *Journal of Environmental Psychology* 17(2), 165–170.

Hilgenkamp, K. (2006) *Environmental Health: Ecological Perspectives*. Jones and Bartlett Learning, Boston, Massachusetts.

Hill, B.J., Freeman, P.A. and Huff, C. (2001) The influence of challenging family recreation on family functioning. In: Havitz, M.E. and Floyd, M.F. (eds) *Book of Abstracts: 2001 Leisure Research Symposium*. National Recreation and Park Association, Ashburn, Virginia, abstr. 65.

Hill, T. (1577) *The Gardeners Labyrinth; or, A New Art of Gardening*. Reprint: 1652, Jane Bell, London.

Hines, J.M., Hungerford, H. and Tomera, A. (1987) Analysis and synthesis of research on responsible environmental behavior: a meta-analysis. *Journal of Environmental Education* 18(2), 1–8.

His Holiness the 14th Dalai Lama of Tibet (1992) A Buddhist Concept of Nature. Transcript of an address on February 4, 1992, at New Delhi, India. Available at: http://www.dalailama.com/messages/environment/buddhist-concept-of-nature (accessed 17 February 2013).

Hockett, K.S., McClafferty, J.A. and McMullin, S.L. (2005) *The Making of a Resource Steward: Defining the Relationship between Aquatic Recreation and Aquatic Stewardship*. Virginia Tech, Blacksburg, Virginia.

Holman, T. and McAvoy, L. (2005) Transferring benefits of participation in an integrated wilderness adventure program to daily life. *Journal of Experiential Education* 27(3), 322–325.

Horsburgh, C.R. (1995) Healing by design. *New England Journal of Medicine* 333(11), 735–740.

Howell, A.J., Dopko, R.L., Passmore, H. and Buro, K. (2011) Nature connectedness: associations with well-being and mindfulness. *Personality and Individual Differences* 51(2), 166–171.

HRH The Prince of Wales (2010) *Harmony: A New Way of Looking at our World*. Blue Door, London.

Hu, Z., Liebens, J. and Rao, K.R. (2008) Linking stroke mortality with air pollution, income, and green-ness in northwest Florida: an ecological geographical study. *International Journal of Health Geographics* 7, 20.

Hug, S., Hansmann, R., Monn, C., Krütli, P. and Seeland K. (2008) Restorative effects of physical activity in forests and indoor settings. *International Journal of Fitness* 4(2), 25–38.

Hulmes, D.F. (2009) Sacred trees of Norway and Sweden: a friluftsliv voyage. Presented at *Henrik Ibsen: The Birth of 'Friluftsliv'. A 150 Year International Dialogue Conference Jubilee Celebration*. North Troendelag University College, Levanger, Norway, Mountains of Norwegian/Swedish Border, 14–19 September 2009.

Hume, D. (1739) *A Treatise of Human Nature*. John Noon, London.

Hunter, M.C. and Hunter, M.D. (2008) Designing for conservation of insects in the built environment. *Insect Conservation and Diversity* 1(4), 189–196.

Inglehart, R., Basáñez, M., Díez-Medrano, J., Halman, L. and Luijkx, R. (eds) (2004) *Human Beliefs and Values: A Cross-cultural Sourcebook based on the 1999–2002 Values Surveys*. Siglo XXI Editores, Coyoacan, Mexico.

Ingram, J. (2009) The role of ecological systems and natural resource management in reducing vulnerability to hazards. Presented at *Ecological Society of America 94th Annual Meeting*, Albuquerque, New Mexico, 2–7 August 2009.

Ip, J.M., Rose, K.A., Morgan, I.G., Burlutsky, G. and Mitchell, P. (2008) Myopia and the urban environment: findings in a sample of 12-year-old Australian school children. *Investigative Ophthalmology & Visual Science* 49(9), 3858–3863.

Irvine, K.N. and Warber, S.L. (2002) Greening healthcare: practicing as if the natural environment really mattered. *Alternative Therapies in Health and Medicine* 8(5), 76–83.

Iso-Ahola, S.E. (1999) Motivational foundations of leisure. In: Jackson, E.L. and Burton, T.L. (eds) *Leisure Studies: Prospects for the Twenty-first Century*. Venture Publishing, Inc., State College, Pennsylvania, pp. 35–51.

James, J.J., Bixler, R.D. and Vadala, C.E. (2010) From play in nature, to recreation then vocation: a develop-mental model for natural history-oriented environmental professionals. *Children, Youth and Environments* 20(1), 231–256.

Jefferies, K. and Lepp, A. (2012) An investigation of extraordinary experiences. *Journal of Park and Recreation Administration* 30(3), 37–51.

Jelalian, E., Mehlenbeck, R., Lloyd-Richardson, E.E., Birmaher, V. and Wing, R.R. (2006) 'Adventure therapy' combined with cognitive-behavioral treatment for overweight adolescents. *International Journal of Obesity* 30(1), 31–39.

Johansson, M., Hartig, T. and Staats, H. (2011) Psychological benefits of walking: moderation by company and outdoor environment. *Applied Psychology: Health and Well-Being* 3(3), 261–280.

Kaczynski, A.T., Potwarka, L.R. and Saelens, B.E. (2008) Association of park size, distance and features with physical activity in neighborhood park. *American Journal of Public Health* 98(8), 1451–1456.

Kahn, P.H. Jr (1997) Developmental psychology and the biophilia hypothesis. *Developmental Review* 17(1), 1–61.

Kahn, P.H. Jr (1999) *The Human Relationship with Nature: Development and Culture*. The MIT Press, Cambridge, Massachusetts.

Kahn, P.H. Jr (2002) Children's affiliations with nature: structure, development, and the problem of environ-mental generational amnesia. In: Kahn, P.H. Jr and Kellert, S.R. (eds) *Children and Nature: Psychological, Sociocultural, and Evolutionary Investigations*. MIT Press, Cambridge, Massachusetts, pp. 93–116.

Kahn, P.H. Jr and Hasbach, P.H. (2012) *Ecopsychology: Science, Totems, and the Technological Species*. MIT Press, Cambridge, Massachusetts.

Kahn, P.H. Jr and Kellert, S.R. (2002) *Children and Nature: Psychological, Sociocultural, and Evolutionary Investigations*. MIT Press, Cambridge, Massachusetts.

Kail, R.V. and Cavanaugh, J.C. (2010) *Human Development: A Life-span View*, 5th edn. Wadsworth, Belmont, California.

Kals, E. and Ittner, H. (2003) Children's environmental identity, indicators and behavioral impacts. In: Clayton, S. and Opotow, S. (eds) *Identity and the Natural Environment, The Psychological Significance of Nature*. MIT Press, Cambridge, Massachusetts, pp. 135–157.

Kals, E., Schumacher, D. and Montada, L. (1999) Emotional affinity towards nature as a motivational basis to protect nature. *Environment and Behavior* 31(2), 178–202.

Kaltenborn, B. (1997) Nature of place attachment: a study among recreation homeowners in southern Norway. *Leisure Sciences* 19(3), 175–189.

Kaplan, R. (2001) The nature of the view from home: psychological benefits. *Environment and Behavior* 33(4), 507–542.

Kaplan, R. and Kaplan, S. (1989) *The Experience of Nature: A Psychological Perspective*. Cambridge University Press, Cambridge, UK.

Kaplan, R. and Kaplan, S. (2005) Preference, restoration, and meaningful action in the context of nearby nature. In: Barlett, P.F. (ed.) *Urban Place: Reconnecting with the Natural World*. MIT Press, Cambridge, Massachusetts, pp. 271–298.

Kaplan, R., Kaplan, S. and Ryan, R.L. (1998) *With People in Mind: Design and Management of Everyday Nature*. Island Press, Washington, DC.

Kaplan, S. (1995) The restorative benefits of nature: toward an integrative framework. *Journal of Environmental Psychology* 15(3), 169–182.

Kaplan, S. and Berman, M.G. (2010) Directed attention as a common resource for executive functioning and self-regulation. *Perspectives on Psychological Science* 5(1), 43–57.

Karen, R. (1998) *Becoming Attached: First Relationships and How They Shape Our Capacity to Love*. Oxford University Press, New York.

Katcher, A. and Beck, A. (1987) Health and caring for living things. *Anthrozoös* 1(3), 175–183.

Kats, G. (2006) *Greening America's Schools*. Capital E, Washington, DC.

Kearns, R.A. and Gesler, W.M. (1998) *Putting Health into Place: Landscape, Identity, and Well-being*. Syracuse University Press, Syracuse, New York.

Keilhofner, G. (1997) *Conceptual Foundations of Occupational Therapy*. F.A. Davis Company, Philadelphia, Pennsylvania.

Kellert, S.R. (1993) Introduction. In: Kellert, S.R. and Wilson, E.O. (eds) *The Biophilia Hypothesis*. Island Press, Washington, DC, pp. 20–31.

Kellert, S.R. (1993) The biological basis for human values of nature. In: Kellert, S.R. and Wilson, E.O. (eds) *The Biophilia Hypothesis*. Island Press, Washington, DC, pp. 42–69.

Kellert, S.R. (2002) Experiencing nature: affective, cognitive, and evaluative development in children. In: Kahn, P.H. Jr and Kellert, S.R. (eds) *Children and Nature: Psychological, Sociocultural, and Evolutionary Investigations*. MIT Press, Cambridge, Massachusetts, pp. 117–151.

Kellert, S.R. (2003) *Kinship to Mastery: Biophilia in Human Evolution and Development*. Island Press, Washington, DC.

Kellert, S.R. (2005) *Building for Life: Designing and understanding the Human–Nature Connection*. Island Press, Washington, DC.

Kellert, S.R. (2012) *Birthright: People and Nature in the Modern World*. Yale University Press, New Haven, Connecticut.

Kellert, S.R. and Wilson, E.O. (eds) (1993) *The Biophilia Hypothesis*. Island Press, Washington, DC.

Kellert, S.R., Heerwagen, J.H. and Mador, M.L. (eds) (2008) *Biophilic Design: The Theory, Science, and Practice of Bringing Buildings to Life*. John Wiley & Sons, Hoboken, New Jersey.

Kelley, M.P., Coursey, R.D. and Selby, P.M. (1997) Therapeutic adventures outdoors: a demonstration of benefits for people with mental illness. *Psychiatric Rehabilitation Journal* 20(4), 61–73.

Kelly, P. (1988) Linking arms, dear sisters, brings hope! In: Plant, J. (ed.) *Healing the Wounds*. New Society Publishers, Philadelphia, Pennsylvania, pp. ix–xi.

Keysers, C. (2011) *The Empathic Brain*. CreateSpace Independent Publishing Platform.

Khatri, P., Blumenthal, J.A., Babyak, M.A., Craighead, W., Herman, S., *et al.* (2001) Effects of exercise training on cognitive functioning among depressed older men and women. *Journal of Aging and Physical Activity* 9(1), 43–57.

Kimbro, R.T. and Schachter, A. (2011) Neighborhood poverty and maternal fears of children's outdoor play. *Family Relations* 60(4), 461–475.

Ki-moon, B. (2012) Secretary-General stresses negative impact of climate change in drylands to International Conference on Food Security in Drylands, Qatar, 14 November 2012. Available at: http://www.un.org/News/Press/docs/2012/sgsm14631.doc.htm (accessed 7 September 2013).

King, B.K. (2008) What Binti Jua knew. *The Washington Post*, 15 August 2008. Available at: http://www.washingtonpost.com/wp-dyn/content/article/2008/08/14/AR2008081403049.html (accessed 12 April 2013).

Kline, J.T., Rosenberger, R.S. and White, E.M. (2011) A national assessment of physical activity in US National Forests. *Journal of Forestry* 109(6), 343–351.

Klokov, K. (2007) Reindeer husbandry in Russia. *International Journal of Entrepreneurship and Small Business* 4(6), 726–784.

Knapp, C.E. (1999) *In Accord with Nature: Helping Students Form an Environmental Ethic Using Outdoor Experience and Reflection*. ERIC/CRESS, Appalachia Educational Laboratory, Charleston, West Virginia.

Koelsch, S. (2010) Toward a neural basis of music-evoked emotions. *Trends in Cognitive Sciences* 14(3), 131–137.

Kolb, D.A. (1983) *Experiential Learning: Experience as the Source of Learning and Development*. Prentice-Hall, Upper Saddle River, New Jersey.

Korpela, K. (1989) Place-identity as a product of environmental self-regulation. *Journal of Environmental Psychology* 9(3), 241–256.

Korpela, K. (1995) *Developing the Environmental Self-Regulation Hypothesis*. University of Tampere, Tampere, Finland.

Korpela, K., Hartig, T., Kaiser, F. and Fuhrer, U. (2001) Restorative experience and self-regulation in favorite places. *Environment and Behavior* 33(4), 572–589.

Korpela, K., Kyttä, M. and Hartig, T. (2002) Restorative experience, self-regulation, and children's place preferences. *Journal of Environmental Psychology* 22(4), 387–398.

Korpela, K.M., Klemettilä, T. and Hietanen, J.K. (2002) Evidence for rapid affective evaluation of environmental scenes. *Environment and Behavior* 34(5), 634–650.

Korpela, K.M., Ylén, M., Tyrväinen, L. and Silvennoinen, H. (2010) Favorite green, waterside and urban environments, restorative experiences and perceived health in Finland. *Health Promotion International* 25(2), 200–209.

Korten, D. (2006) *The Great Turning: From Empire to Earth Community*. Kumarian Press, Bloomfield, Connecticut.

Kortenkamp, K.V. and Moore, C.F. (2001) Ecocentrism and anthropocentrism: moral reasoning about ecological commons dilemmas. *Journal of Environmental Psychology* 21(3), 261–272.

Kreitzer, M.J., Mitten, D., Harris, I. and Shandeling, J. (2002) Attitudes toward CAM among medical, nursing, and pharmacy faculty and students: a comparative analysis. *Alternative Therapies in Health and Medicine* 8(6), 44–47.

Kruger, J., Nelson, K., Klein, P., McCurdy, L.E., Pride, P., *et al.* (2010) Building on partnerships: reconnecting kids with nature for health benefits. *Health Promotion Practice* 11(3), 340–346.

Kubesch, S., Bretschneider, V., Freudenmann, R., Weidenhammer, N., Lehmann, M., *et al.* (2003) Aerobic endurance exercise improves executive functions in depressed patients. *Journal of Clinical Psychiatry* 64(9), 1005–1012.

Kuo, F.E. (2001) Coping with poverty: impacts of environment and attention in the inner city. *Environment and Behavior* 33(1), 5–34.

Kuo, F.E. (2003) Book review: Children and nature: psychological, sociocultural, and evolutionary investigations. *Children, Youth and Environments* 13(1).

Kuo, F.E. (2004) Horticulture, well-being, and mental health: from intuitions to evidence. In: Relf, D. (ed.) Proceedings of the XXVI International Horticulture Congress: Expanding Roles for Horticulture in Improving Human Well-being and Life Quality. *Acta Horticulturae* 639, 27–36.

Kuo, F.E. and Sullivan, W.C. (2001) Environment and crime in the inner city: does vegetation reduce crime? *Environment and Behavior* 33(3), 343–367.

Kuo, F.E. and Taylor, A.F. (2004) A potential natural treatment for attention-deficit/hyperactivity disorder: evidence from a national study. *American Journal of Public Health* 94(9), 1580–1586.

Kuo, F.E., Bacaicoa, M. and Sullivan, W.C. (1998) Transforming inner-city neighborhoods: trees, sense of safety, and preference. *Environment and Behavior* 30(1), 28–59.

Kweon, B.S., Sullivan, W.C. and Wiley, A.R. (1998) Green common spaces and the social integration of inner-city older adults. *Environment and Behavior* 30(6), 832–858.

Kyttä, M. (2004) The extent of children's independent mobility and the number of actualized affordances as criteria for child-friendly environments. *Journal of Environmental Psychology* 24(2), 179–198.

Lane-Zucker, L. (2005) Forward. In: Sobel, D. (ed.) *Place-Based Education: Connecting Classrooms and Communities*. The Orion Society, Great Barrington, Massachusetts.

Larivière, M., Couture, R., Ritchie, S.D., Côté, D., Oddson, B., *et al.* (2012) Behavioural assessment of wilderness therapy participants: exploring the consistency of observational data. *Journal of Experiential Education* 35(1), 290–302.

Laski, M. (1961) *Ecstasy: A Study of Some Secular and Religious Experiences*. The Cressett Press, London.

Lear, L. (ed.) (1998) *Lost Woods: The Discovered Writing of Rachel Carson*. Beacon Press, Boston, Massachusetts.

Learning through Landscapes (2013) Homepage. Available at: http://www.ltl.org.uk/index.php (accessed 2 September 2013).

Lee, J. (2009) The secret place of Thunder. *The Nag Hammadi Library*. Available at: http://www.youtube.com/watch?v=39U_zBOVbF0 (accessed 31 July 2009).

Leopold, A. (1949) *A Sand County Almanac*. Oxford University Press, London.

Leopold, A. (1966) *A Sand County Almanac*. Oxford University Press, New York.

Levine, S.B. (2004) *Inventing the Rest of Our Lives: Women in Second Adulthood*. Penguin, New York.

Lewicki, J., Goyette, A. and Marr, K. (1996) Family camp: a multimodal treatment strategy for linking process and content. *Journal of Child and Youth Care* 10(4), 51–66.

Lewis, C.A. (1976) The evolution of horticulture therapy in the US. Presented at *Fourth Annual Meeting of the National Council for Therapy and Rehabilitation through Horticulture*, Philadelphia, Pennsylvania, 6 September 1976.

Lewis, J. (1985) The birth of EPA. *EPA Journal* 11, 6.

Leyden, K.M. (2003) Social capital and the built environment: the importance of walkable neighborhoods. *American Journal of Public Health* 93(9), 1546–1551.

Li, Q., Morimoto, K., Nakadai, A., Inagaki, H., Katsumata, M., *et al.* (2007) Forest bathing enhances human natural killer activity and expression of anti-cancer proteins. *International Journal of Immunopathology and Pharmacology* 20(2 Suppl. 2), 3–8.

Li, Q., Kobayashi, M. and Kawada, T. (2008) Relationships between percentage of forest coverage and standardized mortality ratios (SMR) of cancers in all prefectures in Japan. *The Open Public Health Journal* 1, 1–7.

Li, Q., Morimoto, K., Kobayashi, M., Inagaki, H., Katsumata, M., *et al.* (2008) A forest bathing trip increases human natural killer activity and expression of anti-cancer proteins in female subject. *Journal of Biological Regulators and Homeostatic Agents* 22(1), 45–55.

Linden, S. and Grut, J. (2002) *The Healing Fields: Working with Psychotherapy and Nature To Rebuild Shattered Lives*. Frances Lincoln, London.

Linneweber, V., Hartmuth, G. and Fritsche, I. (2003) Representations of the local environment as threatened by global climate change: toward a contextualized analysis of environmental identity in a coastal area. In: Clayton, S. and Opotow, S. (eds) *The Psychological Significance of Nature*. MIT Press, Cambridge, Massachusetts, pp. 227–246.

Litz, K. (2010) Inspiring environmental stewardship: developing a sense of place, critical thinking skills, and ecoliteracy to establish an environmental ethic of care. MA dissertation, Prescott College, Prescott, Arizona.

Litz, K. and Mitten, D. (2013) Inspiring environmental stewardship: developing a sense of place, critical thinking skills, and ecoliteracy to establish an environmental ethic of care. *Pathways, The Ontario Journal of Outdoor Education* 25(2), 4–8.

Livengood, J. (2009) The role of leisure in the spirituality of New Paradigm Christians. *Leisure/Loisir* 33(1), 389–417.

LoBue, V. and DeLoache, J.S. (2008) Detecting the snake in the grass attention to fear-relevant stimuli by adults and young children. *Psychological Science* 19(3), 284–289.

Lodewyk, K., Chunlei, L. and Kentel, J. (2009) Enacting the spiritual dimension in physical education. *Physical Educator* 66(4), 170–179.

Loeffler, T.A. (2004) A photo elicitation study of the meanings of outdoor adventure experiences. *Journal of Leisure Research* 36(4), 536–556.

Louv, R. (1998) *The Web of Life: Weaving the Values That Sustain Us*. Red Wheel Weiser & Conari Press, Newburyport, Massachusetts.

Louv, R. (2005) *Last Child in the Woods: Saving Our Children from Nature-Deficit Disorder*. Algonquin Books of Chapel Hill, Chapel Hill, North Carolina.

Louv, R. (2010) Grow Outside! Keynote Address to the American Academy of Pediatrics National Conference, 4 October 2010. Available at: http://blog.childrenandnature.org/2010/10/04/grow-outside-keynote-address-to-the-american-academy-of-pediatrics-national-conference/ (accessed 8 September 2013).

Louv, R. (2011) *The Nature Principle: Human Restoration and the End of Nature-deficit Disorder*. Algonquin Books of Chapel Hill, Chapel Hill, North Carolina.

Lovasi, G.S., Quinn, J.W., Neckerman, K.M., Perzanowski, M.S. and Rundle, A. (2008) Children living in areas with more street trees have lower prevalence of asthma. *Journal of Epidemiology and Community Health* 62(7), 647–649.

Lynch, P. and Moore, K. (2004) Adventures in paradox. *Australian Journal of Outdoor Education* 8(2), 3–12.

Lyytimaki, J. (2012) Indoor ecosystem services: bringing ecology and people together. *Human Ecology Review* 19(1), 70–76.

Maas, J., Verheij, R.A., Groenewegen, P.P., de Vries, S. and Spreeuwenberg, P. (2006) Green space, urbanity, and health: how strong is the relation? *Journal of Epidemiology and Community Health* 60(7), 587–592.

Mackay, G.J. and Neill, J.T. (2009) The effect of 'green exercise' on state anxiety and the role of exercise duration, intensity, and greenness: a quasi-experimental study. *Psychology of Sport and Exercise* 11(3), 238–245.

MacKaye, B. (1921) An Appalachian Trail: a project in regional planning. *Journal of the American Institute of Architects* 9, 325–330.

Mactavish, J.B. and Schleien, S.J. (1998) Playing together growing together: parents' perspectives on the benefits of family recreation in families that include children with a developmental disability. *Therapeutic Recreation Journal* 32(3), 207–230.

Macy, J. (2006) The great turning as compass and lens. *Yes! Magazine*, 10 May 2006. Available at: http://www.yesmagazine.org/issues/5000-years-of-empire/the-great-turning-as-compass-and-lens (accessed 7 September 2013).

Maddux, J.E. and Gosselin, J.T. (2003) Self-efficacy. In: Leary, M.R. and Tangney, J.P. (eds) *Handbook of Self and Identity*. Guilford Press, New York, pp. 218–238.

Malkin, J. (1992) *Hospital Interior Architecture*. Van Nostrand Reinhold, New York.

Maller, C. (2005) Hands-on contact with nature in primary schools as a catalyst for developing a sense of community and cultivating mental health and wellbeing. *Eingana* 28(3), 16–21.

Maller, C. and Townsend, M. (2006) Children's mental health and wellbeing and hands-on contact with nature. *International Journal of Learning* 12(4), 359–372.

Maller, C., Townsend, M., St, Leger, L., Henderson-Wilson, C., Pryor, A., *et al.* (2002) *Healthy Parks, Healthy People: The Health Benefits of Contact with Nature in a Park Context. A Review of the Current Literature*. Deakin University, Melbourne, Australia.

Maller, C., Townsend, M., Pryor, A., Brown, P. and St. Leger, L. (2006) Healthy nature healthy people: 'contact with nature' as an upstream health promotion intervention for populations. *Health Promotion International* 21(1), 45–54.

Maller, C., Henderson-Wilson, C. and Townsend, M. (2009) Rediscovering nature in everyday settings: or how to create healthy environments and healthy people. *EcoHealth* 6(4), 553–556.

Malone, K. and Tranter, P. (2003) Children's environmental learning and the use, design and management of school grounds. *Children, Youth and Environments* 13(2), 1–30.

Mandondo, A. (1997) Trees and spaces as emotion and norm laden components of local ecosystems in Nyamaropa communal land, Nyanga District, Zimbabwe. *Agriculture and Human Values* 14(4), 353–372.

Manger, T.A. and Motta, R.W. (2005) The impact of an exercise program on posttraumatic stress disorder, anxiety, and depression. *International Journal of Emergency Mental Health* 7(1), 49–57.

Mannell, R. (1980) Social psychological techniques and strategies for studying leisure experience. In: Iso-Ahola, S.E. (ed.) *Social Psychological Perspectives on Leisure and Recreation*. Charles C. Thomas, Springfield, Illinois, pp. 62–88.

Mannell, R. (1996) Approaches in the social and behavioural sciences to the systematic study of hard-to-define human values and experience. In: Driver, B.L., Dustin, D., Baltic, T., Elsner, G. and Peterson, G. (eds) *Nature and the Human Spirit: Toward and Expanded Land Management Ethic*. Venture Publishing, Inc., State College, Pennsylvania, pp. 405–416.

Marapana, R.A.U.J., Hewamanage, D.S., Seresinhe, R.T. and Senaratne, R. (2012) Study on behavioral changes of animals prior to a tsunami natural disaster. In: Senaratne, R., Filson, G.C. and Janakiram, J. (eds) *Rebuilding of Tsunami Affected Areas in the Southern and Eastern Provinces of Sri Lanka: Workshop Proceedings*. Tharanjee Prints, Maharagama, Sri Lanka, pp. 65–72.

Marcus, B. and Forsyth, L. (2008) *Motivating People to be Physically Active*. Human Kinetics, Champaign, Illinois.

Marcus, C.C. and Barnes, M. (eds) (1999) *Healing Gardens: Therapeutic Benefits and Design Recommendations*. John Wiley and Sons, New York.

Margulis, L. (1998) *Symbiotic Planet: A New Look at Evolution*. Basic Books, New York.

Margulis, L. and Sagan, D. (eds) (2007) *Dazzle Gradually: Reflections on the Nature of Nature*. Chelsea Green, White River Junction, Vermont.

Marques, R. (2011) Individuation within the context of ecological identity: outdoor education and the passage rite into adulthood. MSc dissertation, University of Edinburgh, Edinburgh, UK.

Marsh, P. (1988) *Tribes*. Peregrine Smith Books, Salt Lake City, Utah.

Marten, G.G. (2001) *Human Ecology: Basic Concepts for Sustainable Development*. Earthscan/James & James, London.

Marten, G.G. (2012) *Human Ecology: Basic Concepts for Sustainable Development*. Earthscan, Abingdon, UK.

Marti, B., *et al.* (2002) Bekanntheit, Nutzung und Bewertung des Vitaparcours: Vergleich zwischen 1997 und 2001 (Attitude towards use of the fitness trail Vitaparcours: comparison between 1997 and 2001). *Schweizerische Zeitschrift für Sportmedizin und Sporttraumatologie* 50(4), 161–163.

Martinez, D.E. (2008) Indigenous consciousness and the production of knowingness. Presented at the *American Sociological Association Annual Meeting*, Boston, Massachusetts, 31 July 2008.

Martusewicz, R. (2005) Eros in the commons: educating for eco-ethical consciousness in a poetics of place. *Ethics, Place & Environment: A Journal of Philosophy & Geography* 8(3), 331–348.

Maslow, A. (1962) *Toward a Psychology of Being*. Van Nostrand, New York.

Maslow, A.H. (1964) *Religion, Values, and Peak-Experiences*. The Ohio State University Press, Columbus, Ohio.

Massey, D. (1991) A global sense of place. *Marxism Today* 35(6), 24–29.

Mathur, V.A., Harada, T., Lipke, T. and Chiao, J.Y. (2010) Neural basis of extraordinary empathy and altruistic motivation. *Neuroimage* 51(4), 1468–1475.

Matsuoka, R.H. (2010) Student performance and high school landscapes: examining the links. *Landscape and Urban Planning* 97(4), 273–282.

Mayer, F.S., McPherson Frantz, C., Bruehlman-Senecal, E. and Dolliver, K. (2009) Why is nature beneficial? The role of connectedness to nature. *Environment and Behavior* 41(5), 607–643.

McAuley, E., Pena, M.M. and Jerome, G.J. (2001) Self-efficacy as a determinant and an outcome of exercise. In: Roberts, G.C. (ed.) *Advances in Motivation in Sport and Exercise*. Human Kinetics, Champaign, Illinois, pp. 235–261.

McAvoy, L., Holman, T., Goldenberg, M. and Klenosky, D. (2006) Wilderness and persons with disabilities. *International Journal of Wilderness* 12(2), 23–31.

McAvoy, L.D., Mitten, J., Steckart, L. and Stringer, L. (1992) Research in outdoor education: group development and group dynamics. In: Henderson, K. (ed.) *Coalition for Education in the Outdoors Research Symposium Proceedings, Bradford Woods, Indiana, 17–19 January 1992*. Coalition for Education in the Outdoors, Cortland, New York, pp. 23–34.

McBrien, N.A., Morgan, I.G. and Mutti, D.O. (2009) What's hot in myopia research – The 12th International Myopia Conference, Australia, July 2008. *Optometry & Vision Science* 86(1), 2–3.

McCormick, R. and Gerlitz, J. (2009) Nature as healer: Aboriginal ways of healing through nature. *Counseling and Spirituality/Counseling et Spiritualité* 28(1), 55–72.

McDermott, L. (2004) Exploring intersections of physicality and female-only canoeing experiences. *Leisure Sciences* 23(3), 283–301.

McDonald, H. (1993) India: the Parsi dilemma. *Far Eastern Economic Review* 156(40), 36.

McDonald, J. (1995) A comparative study of horticulture therapy profession in the United Kingdom and the United States of America. MSc thesis, University of Reading, Reading, UK.

McDonald, M.G., Wearing, S. and Ponting, J. (2009) The nature of the peak experience in wilderness. *The Humanistic Psychologist* 37(4), 370–385.

McGregor, C. (2010) *Partnering with Nature: The Wild Path to Reconnecting with the Earth*. Atria Books/Beyond Words, Hillsboro, Oregon.

McGregor, D. (2004) Indigenous knowledge, environment, and our future. *The American Indian Quarterly* 28(3&4) 385–410.

McKenzie, M. (2000) How are adventure education program outcomes achieved? A review of the literature. *Australian Journal of Outdoor Education* 5(1), 19–28.

McKibben, B. (2007) *Deep Economy: The Wealth of Communities and the Durable Future*. Macmillan, London.

McKibben, B. (2009) Human nature, community, and 'deep economy'. In: Buzzell, L. and Chalquist, C. (eds) *Ecotherapy: Healing with Nature in Mind*. Sierra Club Books, San Francisco, California, pp. 186–191.

McLeroy, K.R., Bibeau, D., Steckler, A. and Glanz, K. (1988) An ecological perspective on health promotion programs. *Health Education Quarterly* 15(4), 351–377.

MdYusof, M.S.B. and Chia, N.K.H. (2012) Dolphin encounter for special children (DESC) program: effectiveness of dolphin-assisted therapy for children with autism. *International Journal of Special Education* 27(3), 1–14.

Mead, G.H. (1934) *Mind, Self, and Society*. University of Chicago Press, Chicago, Illinois.

Medin, D.L. and Atran, S. (eds) (1999) *Folkbiology*. MIT Press, Cambridge, Massachusetts.

Medina, J. (2008) *Brain Rules*. Pear Press, Seattle, Washington.

Memishevikj, H. and Hodzhikj, S. (2010) The effects of equine-assisted therapy in improving the psychosocial functioning of children with autism. *Journal of Special Education & Rehabilitation* 11(3–4), 67–67.

Merchant, C. (1980) *The Death of Nature: Women, Ecology, and the Scientific Revolution*. Harper & Row, San Francisco, California.

Miles, J. (1987) Wilderness as healing place. *Journal of Experiential Education* 10(3), 4–10.

Miller, J. (2012) Smarter than the average bear: bears use nightfall to avoid hunters. *Yale Environment Review*, 10 December 2012. Available at: http://environment.yale.edu/yer/article/smarter-than-the-average-bear (accessed 6 April 2013).

Miller-Rushing, A., Primack, R. and Bonney, R. (2012) The history of public participation in ecological research. *Frontiers in Ecology & the Environment* 10(6), 285–290.

Milligan, C., Gatrell, A. and Bingley, A. (2004) 'Cultivating health': therapeutic landscapes and older people in northern England. *Social Science & Medicine* 58(9), 1781–1793.

Minnesota Department of Natural Resources (2008) Natural Wild Rice in Minnesota. A Wild Rice Study document submitted to the Minnesota Legislature by the Minnesota Department of Natural Resources, February 15, 2008. Available at: http://files.dnr.state.mn.us/fish_wildlife/wildlife/shallowlakes/natural-wild-rice-in-minnesota.pdf (accessed 15 September 2013).

Miranda, W. (1987) The genteel radicals. *Camping Magazine* 59(4), 12–15, 31.

Miranda, W. and Yerkes, R. (1987) Women's outdoor adventure programming. In: Meier, J., Morash, T. and Welton, G. (eds) *High Adventure Outdoor Pursuits*. Publishing Horizons, Inc., Columbus, Ohio, pp. 259–267.

Miranda, W. and Yerkes, R. (1996) The history of camping women in the professionalization of experiential education. In: Warren, K. (ed.) *Women's Voices in Experiential Education*. Kendall/Hunt Publishing Company, Dubuque, Iowa, pp. 24–32.

Mitchell, R. and Popham, F. (2008) Effect of exposure to natural environment on health inequalities: an observational population study. *Lancet* 372(9650), 1655–1660.

Mitchell, R.G. (1983) *Mountain Experience: The Psychology and Sociology of Adventure*. University of Chicago Press, Chicago, Illinois.

Mitten, D. (1985) A philosophical basis for a women's outdoor adventure program. *Journal of Experiential Education* 8(2), 20–24.

Mitten, D. (1992) Empowering girls and women in the outdoors. *Journal of Physical Education, Recreation & Dance* 63(2), 56–60.

Mitten, D. (1994) Ethical considerations in adventure therapy: a feminist critique. In: Cole, E., Erdman, E. and Rothblum, E.D. (eds) *Wilderness Therapy for Women: The Power of Adventure*. The Haworth Press, Binghamton, New York, pp. 55–84.

Mitten, D. (1996) The value of feminist ethics in experiential education teaching and leadership. In: Warren, K. (ed.) *Women's Voices in Experiential Education*. Kendall/Hunt, Dubuque, Iowa, pp. 159–171.

Mitten, D. (1998) 'You ain't gonna get me on that rock'. *ZipLines: The Voice for Adventure Education* (34), 46–50.

Mitten, D. (2004) Adventure therapy as a complementary and alternative therapy. In: Bandoroff, S. and Newes, S. (eds) *Coming of Age: The Evolving Field of Adventure Therapy*. Association of Experiential Education, Boulder, Colorado, pp. 240–257.

Mitten, D. (2008) Getting fit for hiking and backpacking. In: Goldenberg, M. and Martin, B. (eds) *Hiking and Backpacking*. Wilderness Education Association and Human Kinetics, Champaign, Illinois, pp. 21–50.

Mitten, D. (2009) Under our noses: the healing power of nature. *Taproot Journal* 19(1), 20–26.

Mitten, D. (2010) Friluftsliv and the healing power of nature: the need for nature for human health, development, and wellbeing. In: *Henrik Ibsen: The Birth of 'Friluftsliv'. A 150 Year International Dialogue Conference Jubilee Celebration*. North Troendelag University College, Levanger, Norway, Mountains of Norwegian/Swedish Border, 14–19 September 2009. Available at: http://norwegianjournaloffriluftsliv.com/doc/122010.pdf (accessed 8 September 2013).

Mitten, D. (2012) Systems theory and thinking [course handout]. Master of Arts Program, Prescott College, Prescott, Arizona.

Mitten, D. and Woodruff, S.L. (2009) Women's adventure history and education programming in the United States favors friluftsliv. In: *Henrik Ibsen: The Birth of 'Friluftsliv'. A 150 Year International Dialogue Conference Jubilee Celebration*. North Troendelag University College, Levanger, Norway, Mountains of

Norwegian/Swedish Border, 14–19 September 2009. Available at: http://norwegianjournaloffriluftsliv. com/doc/212010.pdf (accessed 8 September 2013).

Mitten, D., Ady, J.C. and D'Amore, C. (2013) The healing power of nature: the need for human health, development, and well being. Poster presented at the *42nd Annual North American Association of Environmental Education Conference*, Baltimore, Maryland, 9–12 October 2013.

Mitten, L. and Herron, S.M. (2010) Elucidating the life cycle of *Claviceps zizaniae*, a fungal smut pathogen of wild rice. Presented at *95th Annual Meeting: The Ecological Society of America*, Pittsburgh, Pennsylvania, 1–6 August 2010.

Montes, S. (1996) Uses of natural settings to promote, maintain and restore human health. In: Driver, B.L., Dustin, D., Baltic, T., Elsner, G. and Peterson, G. (eds) *Nature and the Human Spirit: Toward and Expanded Land Management Ethic*. Venture Publishing Inc., State College, Pennsylvania, pp. 105–115.

Moore, E.O. (1981) A prison environment's effect on health care service demands. *Journal of Environmental Systems* 11(1), 17–34.

Moore, R.C. (1986) The power of nature orientations of girls and boys toward biotic and abiotic play settings on a reconstructed schoolyard. *Children's Environments Quarterly* 3(3), 52–69.

Moore, R.C. and Cosco, N.G. (2000) Developing an earth-bound culture through design of childhood habitats. Presented at the *International Conference on People, Land and Sustainability*, University of Nottingham, Nottingham, UK, 13–16 September 2000.

Moore, R.C. and Wong, H.H. (1997) *Natural Learning: The Life History of an Environmental Schoolyard. Creating Environments for Rediscovering Nature's Way of Teaching*. MIG Communications, Berkeley, California.

Moran, E.F. (2006) *People and Nature: An Introduction to Human Ecological Relations*. Blackwell Primers in Anthropology, Volume 1. Wiley-Blackwell, Hoboken, New Jersey.

Mortlock, C. (2002) *Beyond Adventure: An Inner Journey*. Cicerone Press, Milnthorpe, UK.

Muñoz, S.A. (2009) Children in the outdoors: a literature review. Available at: http://www.educationscotland. gov.uk/images/Children%20in%20the%20outdoors%20literature%20review_tcm4-597028.pdf (accessed 15 October 2013).

Murad, M.M. (2012) Inner and outer nature: an Islamic perspective on the environmental crisis. *Islam & Science* 10(2), 117–137.

Murray, A.L. (1967) Frederick Law Olmsted and the design of Mount Royal Park, Montreal. *Journal of the Society of Architectural Historians* 26(3), 163–171.

Murray, M.A. (1921) *The Witch-cult in Western Europe: A Study in Anthropology*. Clarendon Press, Oxford, UK.

Myers, J.E., Sweeney, T.J. and Witmer, J.M. (2000) The wheel of wellness counseling for wellness: a holistic model for treatment planning. *Journal of Counseling & Development* 78(3), 251–266.

Myers, O.E. Jr and Saunders, C.D. (2002) Animals as links toward developing caring relationships with the natural world. In: Kahn, P.H. Jr and Kellert, S.R. (eds) *Children and Nature: Psychological, Sociocultural, and Evolutionary Investigations*. MIT Press, Cambridge, Massachusetts, pp. 153–178.

Mygind, E. (2007) A comparison between children's physical activity levels at school and learning in an outdoor environment. *Journal of Adventure Education & Outdoor Learning* 7(2), 161–176.

Nakamura, J. and Csikszentmihalyi, M. (2009) Flow: theory and research. In: Snyder, C.R. and Lopez, S.J. (eds) *Oxford Handbook of Positive Psychology*. Oxford University Press, New York, pp. 195–206.

Naropa University (2009) What is ecopsychology? Available at: http://www.naropa.edu/academics/gsp/grad/ ecopsychology-ma/what-is-ecopsychology.php (accessed 6 April 2013).

Nash, R. (2001) *Wilderness and the American Mind*, 4th edn. Yale University Press, New Haven, Connecticut.

Nathanson, D.E. (1998) Long-term effectiveness of dolphin-assisted therapy for children with severe disabilities. *Anthrozoös* 11(1), 22–32.

National Park Service (2013) *The National Parks & Public Health: A NPS Healthy Parks, Healthy People Science Plan*. US Department of the Interior, Washington, DC.

National Research Council (1981) *Indoor Pollutants*. National Academy Press, Washington, DC, p. 537.

Natural England (2009) *Childhood and Nature: A Survey on Changing Relationships with Nature across Generations*. England Marketing, Warboys, UK.

Naumov, G. (2007) Housing the dead: burials inside houses and vessels in the Neolithic Balkans. In: Barrowclough, D.A. and Malone, C. (eds) *Cult in Context: Reconsidering Ritual in Archaeology*. Oxbow Books Limited, Oxford, UK, pp. 257–268.

Neill, J. (2009) Green exercise: the psychological effects of exercising in nature. Presented at *2009 Outdoor Recreation Industry Council Annual Conference*, Sydney, New South Wales, Australia, 15–16 August 2009.

Nemeroff, C. and Rozin, P. (2000) 'You are what you eat': applying the demand-free 'impressions' technique to an unacknowledged belief. In: Rosengren, K.S., Johnson, C.N. and Harris, P.L. (eds) *Imagining the Impossible: Magical, Scientific and Religious Thinking in Children.* Cambridge University Press, New York, pp. 1–33.

Newman, F. and Holzman, L. (2013) *Lev Vygotsky: A Revolutionary Scientist.* Psychology Press, London.

Newman, G., Wiggins, A., Crall, A., Graham, E., Newman, S., *et al.* (2012) The future of citizen science: emerging technologies and shifting paradigms. *Frontiers in Ecology & the Environment* 10(6), 298–304.

Nightingale, F. (1996) *Notes on Nursing (Revised with additions).* Ballière Tindall, London.

Nilsson, K., Sangster, M., Gallis, C., Hartig, T., de Vries, S., *et al.* (eds) (2011) *Forests, Trees, and Human Health.* Springer, New York.

Nilsson, K., Sangster, M. and Konijnendijk, C.C. (2011) Forests, trees, and human health and well-being: Introduction. In: Nilsson, K., Sangster, M., Gallis, C., Hartig, T., de Vries, S., *et al.* (eds). *Forests, Trees, and Human Health.* Springer, New York, pp. 1–19.

Nisbet, E.K. and Zelenski, J.M. (2011) Underestimating nearby nature: affective forecasting errors obscure the happy path to sustainability. *Psychological Science* 22(9), 1101–1106.

Nisbet, E.K., Zelenski, J.M. and Murphy, S.A. (2011) Happiness is in our nature: exploring nature relatedness as a contributor to subjective well-being. *Journal of Happiness Studies* 12(2), 303–322.

Noble, V. (ed.) (1993) *Uncoiling the Snake: Ancient Patterns in Contemporary Women's Lives.* HarperCollins, San Francisco, California.

Noddings, N. (1984) *Caring: A Feminine Approach to Ethics and Moral Education.* University of California Press, Berkeley, California.

Noddings, N. (2005) Place-based education to preserve the Earth and its people. In: Noddings, N. (ed.) *Educating Citizens for Global Awareness.* Teachers College Press, New York, pp. 57–68.

Noddings, N. (2005) *The Challenge to Care in Schools: An Alternative Approach to Education.* Teachers College Press, New York.

Norman, J., Annerstedt, M., Boman, M. and Mattson, L. (2010) Influence of outdoor recreation on self-rated health: comparing three categories of Swedish recreationists. *Scandinavian Journal of Forest Research* 25(3), 234–244.

Norton, B. (1992) Sustainability, human welfare, and ecosystem health. *Environmental Values* 1(2), 97–111.

Nosich, G. (2012) *Learning to Think Things Through: A Guide to Critical Thinking in the Curriculum.* Prentice Hall, New York.

O'Brien, L., Burls, A., Bentsen, P., Hilmo, I., Holter, K., *et al.* (2011) Outdoor education, lifelong learning and skills development in woodlands and green spaces: the potential links to health and well-being. In: Nilsson, K., Sangster, M., Gallis, C., Hartig, T., de Vries, S., *et al.* (eds) *Forests, Trees and Human Health.* Springer, New York, pp. 343–372.

Oelschlaeger, M. (1991) *The Idea of Wilderness.* Yale University Press, New Haven, Connecticut.

Ogden, C.L., Carroll, M.D., Kit, B.K. and Flegal, K.M. (2012) *Prevalence of Obesity in the United States 2009–2010. NCHS Data Brief No. 82.* National Center for Health Statistics, Hyattsville, Maryland.

Olmsted, F.L. (1967) Frederick Law Olmsted and the design of Mount Royal Park, Montreal. *Journal of the Society of Architectural Historians* 26(3), 163–171.

Olpin, M. and Hesson, M. (2007) *Stress Management for Life: A Research-based Experiential Approach.* Wadsworth Publishers, Belmont, California.

Olson, S. (1946) We need wilderness. *National Parks Magazine* January–March issue. Available at: http://www4.uwm.edu/letsci/research/sigurd_olson/articles/1940s/1946-01-00–We%20Need%20Wilderness–National%20Parks%20Mag.htm (accessed 15 September 2013).

Olson, S. (2012) *From Neurons to Neighborhoods: An Update. Workshop Summary.* Institute of Medicine and National Research Council of the National Academies, Washington, DC.

OpenScientist (2011) Finalizing a definition of 'citizen science' and 'citizen scientists'. Available at: http://www.openscientist.org/2011/09/finalizing-definition-of-citizen.html (accessed 20 December 2012).

Ordiz, A., Støen, O.-G., Sæbø, S., Kindberg, J., Delibes, M., *et al.* (2012) Do bears know they are being hunted? *Biological Conservation* 152, 21–28.

Orians, G.H. (1980) Habitat selection: general theory and applications to human behaviour. In: Lockard, J.S. and Lowenthal, D. (eds) *Landscape Meanings and Values.* Allen and Unwin, London, pp. 86–94.

Orr, D.W. (1992) *Ecological Literacy: Education and the Transition to a Postmodern World.* State University of New York Press, Albany, New York.

Orr, D.W. (2004) *Earth in Mind: On Education, Environment, and the Human Prospect.* Island Press, Washington, DC.

Orsega-Smith, E., Mowen, A.J., Payne, L.L. and Godbey, G. (2004) The interaction of stress and park use on psycho-physiological health in older adults. *Journal of Leisure Research* 36(2), 232–257.

Outdoor Foundation (2011) *Outdoor Recreation Participation Report 2011.* Outdoor Foundation, Boulder, Colorado.

Outward Bound (n.d.) Outward Bound philosophy. Available at: http://www.outwardbound.org/about-outward-bound/philosophy/ (accessed 17 September 2013).

Park, B.J., Furuya, K., Kasetani, T., Takayama, N., Kagawa, T., *et al.* (2011) Relationship between psychological responses and physical environments in forest settings. *Landscape and Urban Planning* 102(1), 24–32.

Parsons, R. (1991) The potential influences of environmental perception on human health. *Journal of Environmental Psychology* 11(1), 1–23.

Partridge, E. (1993) The philosophical foundations of Aldo Leopold's 'Land Ethic'. Available at: http://gadfly.igc.org/papers/leopold.htm (accessed 3 December 2012).

Passarelli, A., Hall, E. and Anderson, M. (2010) A strength-based approach to outdoor and adventure education: possibilities for personal growth. *Journal of Experiential Education* 33(2), 120–135.

Pavlov, I.P. (1960/2003) *Conditioned Reflexes.* Dover, New York.

Paxton, T. and McAvoy, L. (2000) Social psychological benefits of a wilderness adventure program. In: McCool, S.F., Cole, D.N., Borrie, W.T. and O'Loughlin, J. (comps) *Wilderness Science in a Time of Change Conference—Volume 3: Wilderness as a Place for Scientific Inquiry,* Missoula, Montana, 23–27 May 1999. *USDA Forest Service Proceedings RMRS-P-15-VOL-3.* US Department of Agriculture, Forest Service, Rocky Mountain Research Station, Ogden, Utah, pp. 23–27.

Pergams, O.R.W. and Zaradic, P.A. (2008) Evidence for a fundamental and pervasive shift away from nature-based recreation. *Proceedings of the National Academy of Sciences USA* 105(7), 2295–2300.

Perlman, M. (1994) *The Power of Trees.* Spring Publications, Dallas, Texas.

Pescosolido, B.A. and Georgianna, S. (1989) Durkheim, suicide, and religion: toward a network theory of suicide. *American Sociological Review* 54(1), 33–48.

Phenice, L. and Griffore, R. (2003) Young children and the natural world. *Contemporary Issues in Early Childhood* 4(2), 167–178.

Phillips, W.T., Kiernan, M. and King, A.C. (2003) Physical activity as a nonpharmacological treatment for depression: a review. *Complementary Health Practice Review* 8(2), 139–152.

Philo, C. (2009) Medical geography. In: Gregory, D., Johnston, R., Pratt, G., Watts, M. and Whatmore, S. (eds) *The Dictionary of Human Geography,* 5th edn. Blackwell, Oxford, UK, pp. 451–453.

Pirages, D. and Ehrlich, P. (1974) *Ark II: Social Response to Environmental Imperatives.* Viking Press, New York.

Platt, S.H. (1608) *Floraes Paradise.* London.

Plumwood, V. (1993) *Feminism and the Mastery of Nature.* Routledge, New York.

Plumwood, V. (2002) *Environmental Culture: The Ecological Crisis of Reason.* Routledge, New York.

Plumwood, V. (2009) Nature in the active voice. *Australian Humanities Review* 46(May). Available at: http://www.australianhumanitiesreview.org/archive/Issue-May-2009/plumwood.html (accessed 16 September 2013).

Pollan, M. (2002) *The Botany of Desire: A Plant's-Eye View of the World.* Random House Trade Paperbacks, New York.

Poole, R. (2008) *Earthrise: How Man First Saw the Earth.* Yale University Press, New Haven, Connecticut.

Pooley, J.A. and O'Connor, M. (2000) Environmental education and attitudes: emotions and beliefs are what is needed. *Environment and Behavior* 32(5), 711–723.

Poudyal, N.C., Hodges, D.G., Bowker, J.M. and Cordell, H.K. (2009) Evaluating natural resource amenities in a human life expectancy production function. *Forest Policy and Economics* 11(4), 253–259.

Prescott, J. (1975) Body pleasure and the origins of violence. *The Bulletin of the Atomic Scientists* (November), 10–20.

President's Commission on Americans Outdoors (1986) *Report and Recommendations to the President of the United States.* US Government Printing Office, Washington, DC.

Pretty, J., Griffin, M., Sellens, M. and Pretty, C. (2003) *Green Exercise: Complementary Roles of Nature, Exercise and Diet in Physical and Emotional Well-being and Implications for Public Health Policy. CES Occasional Paper 2003-1.* University of Essex, Colchester, UK.

Pretty, J., Griffin, M., Peacock, J., Hine, R., Sellens, M., *et al.* (2005) *Countryside for Health and Wellbeing: The Physical and Mental Health Benefits of Green Exercise.* Countryside Recreation Network, Sheffield, UK.

Pretty, J., Peacock, J., Sellens, M. and Griffin, M. (2005) The mental and physical health outcomes of green exercise. *International Journal of Environmental Health Research* 15(5), 319–337.

Pretty, J., Hine, R. and Peacock, J. (2006) Green exercise: the benefits of activities in green places. *Biologist* 53(3), 143–148.

Priest, S. (1986) Outdoor leadership preparation in five nations. PhD thesis, University of Oregon, Eugene, Oregon.

Priest, S. and Gass, M. (1997) *Effective Leadership in Adventure Programming*. Human Kinetics, Champaign, Illinois.

Prior, V. and Glaser, D. (2006) *Understanding Attachment and Attachment Disorders: Theory, Evidence and Practice*. Jessica Kingsley Publishers, Philadelphia, Pennsylvania.

Propst, D.B. and Koesler, R.A. (1998) Bandura goes outdoors: role of self-efficacy in the outdoor leadership development process. *Leisure Sciences* 20(4), 319–344.

Proshansky, H., Fabian, A. and Kaminoff, R. (1983) Place-identity: physical world socialization of the self. *Journal of Environmental Psychology* 3(1), 57–83.

Prothmann, A., Ettrich, C. and Prottman, S. (2009) Preference for, and responsiveness to people, dogs, and objects in children with autism. *Anthrozoös* 22(2), 161–171.

Pryor, A. (2009) Does adventure therapy have wings? In: Mitten, D. and Itin, C.M. (eds) *Connecting with the Essence of Adventure Therapy: Proceedings from the 4th International Adventure Therapy Conference (2006)*. Association for Experiential Education, Boulder, Colorado, pp. 46–71.

Pryor, A., Carpenter, C. and Townsend, M. (2005) Outdoor education and bush adventure therapy: a social-ecological approach to health and wellbeing. *Australian Journal of Outdoor Education* 9(1), 3–13.

Pryor, A., Townsend, M., Maller, C. and Field, K. (2006) Health and well-being naturally: 'contact with nature' in health promotion for targeted individuals, communities and populations. *Health Promotion Journal of Australia* 17(2), 114–123.

Public Health Agency of Canada (2008) Ottawa charter for health promotion. Available at: http://www.phac-aspc.gc.ca/ph-sp/docs/charter-chartre/index-eng.php (accessed 2 March 2009).

Pumariega, A.J. and Cross, T.L. (1997) Cultural competence in child psychiatry. In: Noshpitz, J. and Alessi, N. (eds) *Basic Handbook of Child and Adolescent Psychiatry*, vol. 4. John Wiley & Sons, New York, pp. 473–484.

Putney, A. (2003) Introduction. In: Putney, A. and Harmon, D. (eds) *The Full Value of Parks: From Economics to the Intangible*. Rowman & Littlefield Publishers, Lanham, Maryland, pp. 3–12.

Pyle, R. (1993) *The Thunder Trees: Lessons from an Urban Wildland*. Houghton Mifflin, Boston, Massachusetts.

Pyle, R. (2002) Eden in the vacant lot: special places, species and kids in the neighborhood of life. In: Kahn, P.H. Jr and Kellert, S.R. (eds) *Children and Nature: Psychological, Sociocultural, and Evolutionary Investigations*. MIT Press, Cambridge, Massachusetts, pp. 305–327.

Raffan, J. (2000) *Nature Nurtures: Investigating the Potential of School Grounds*. Evergreen, Toronto, Canada.

Raffensperger, C. and Tickner, J.A. (eds) (1999) *Protecting Public Health and the Environment: Implementing the Precautionary Principle*. Island Press, Washington, DC.

Ramsay, C.E. and Rickson, R.E. (1976) Environmental knowledge and attitudes. *Journal of Environmental Education* 8(1), 10–18.

Raskin, P., Banuri, T., Gallopín, G., Gutman, P., Hammond, A., *et al.* (2002) *The Great Transition: The Promise and the Lure of the Times Ahead*. Tellus Institute, Boston, Massachusetts.

Ray, P. (2008) *The Potential for a New, Emerging Culture in the US: Report on the 2008 American Values Survey*. Wisdom University, Mill Valley, California.

Reese, R.F. and Myers, J.E. (2012) EcoWellness: the missing factor in holistic wellness models. *Journal of Counseling & Development* 90(4), 400–406.

Reiter, P. (2001) Climate change and mosquito-borne disease. *Environmental Health Perspectives* 109(Suppl. 1), 141–161.

Relph, E. (1976) *Place and Placelessness*. Pion, London.

Relph, E. (2001) Place in geography. In: Neil, J.S. and Paul, B.B. (eds) *International Encyclopedia of the Social & Behavioral Sciences*. Pergamon, Oxford, UK, pp. 11448–11451.

Rest, J., Narvaez, D., Bebeau, M. and Thoma, S. (1999) *Postconventional Moral Thinking: A Neo-Kohlbergian Approach*. Lawrence Erlbaum Associates, Mahwah, New Jersey.

Rhode, C.L.E. and Kendle, A.D. (1997) Nature for people. In: Kendle, A.D. and Forbes, S. (eds) *Urban Nature Conservation – Landscape Management in the Urban Countryside*. E & FN Spon, London, pp. 319–335.

Rishbeth, C. and Finney, N. (2006) Novelty and nostalgia in urban greenspace: refugee perspectives. *Tijdschrift voor Economische en Sociale Geografie* 97(3), 281–295.

Rivkin, M. (1990) *The Great Outdoors: Restoring Children's Rights to Play Outside*. National Association for the Education of Young Children, Washington, DC.

Roberts, E. (1996) Place and spirit in public land management. In: Driver, B.L., Dustin, D., Baltic, T., Elsner, G. and Peterson, G. (eds) *Nature and the Human Spirit: Toward an Expanded Land Management Ethic*. Venture Publishing, State College, Pennsylvania, pp. 61–80.

Robinson, P., Herron, S., Power, R. and Zak, D. (2009) A regional multicultural approach to sustaining wild rice. *Journal of Extension* 47(6), 1–5.

Roggenbuck, J. and Driver, B.L. (2000) Benefits of nonfacilitated uses of wilderness. In: McCool, S.F., Cole, D.N., Borrie, W.T. and O'Loughlin, J. (comps) *Wilderness Science in a Time of Change Conference—Volume 3: Wilderness as a Place for Scientific Inquiry*, Missoula, Montana, 23–27 May 1999. *USDA Forest Service Proceedings RMRS-P-15-VOL-3*. US Department of Agriculture, Forest Service, Rocky Mountain Research Station, Ogden, Utah, pp. 33–49.

Rolston, H. (2000) The land ethic at the turn of the millennium. *Biodiversity and Conservation* 9(8), 1045–1058.

Rose, K.A., Morgan, I.G., Ip, J., Kifley, A., Huynh, S., *et al.* (2008) Outdoor activity reduces the prevalence of myopia in children. *Ophthamology* 115(8), 1279–1285.

Rose, S. and Massey, P. (1993) Adventurous outdoor activities: an investigation into the benefits of adventure for seven people with severe learning difficulties. *Mental Handicap Research* 6(4), 287–302.

Roszak, T. (1995) Where psyche meets Gaia. In: Roszak, T., Gomes, M. and Kanner, A. (eds) *Ecopsychology: Healing the Earth, Restoring the Mind*. Sierra Club Books, San Francisco California, pp. 1–20.

Roszak, T., Gomes, M.E. and Kanner, A.D. (eds) (1995) *Ecopsychology: Restoring the Earth, Healing the Mind*. Sierra Club Books, San Francisco, California.

Russell, C.L. and Bell, A.C. (1996) A politicized ethic of care: environmental education from an ecofeminist perspective. In: Warren, K. (ed.) *Women's Voices in Experiential Education*. Kendall/Hunt Publishing Co., Dubuque, Iowa, pp. 172–181.

Russell, K. and Phillips-Miller, D. (2002) Perspectives on the wilderness therapy process and its relation to outcome. *Child and Youth Care Forum* 31(6), 415–437.

Russell, K.C. (2003) An assessment of outcomes in outdoor behavioral healthcare treatment. *Child and Youth Care Forum* 32(6), 355–381.

Ryan, R.M., Weinstein, N., Bernstein, J., Brown, K.W., Mistretta, L. and Gagne, M. (2010) Vitalizing effects of being outdoors and in nature. *Journal of Environmental Psychology* 30(2), 159–168.

Salimpoor, V.N., Benovoy, M., Larcher, K., Dagher, A. and Zatorre, R.J. (2011) Anatomically distinct dopamine release during anticipation and experience of peak emotion to music. *Nature Neuroscience* 14(2), 257–262.

Sameroff, A.J. and Emde, R.N. (eds) (1989) *Relationship Disturbances in Early Childhood: A Developmental Approach*. Basic Books, New York.

Santmire, H. (1985) *The Travails of Nature: The Ambiguous Ecological Promise of Christian Theology*. Fortress Press, Minneapolis, Minnesota.

Schaefer, C. and DiGeronimo, T. (2000) *Ages and Stages: Tips and Techniques for Building Your Child's Social, Emotional, Interpersonal and Cognitive Skills*. John Wiley & Sons, Inc., New York.

Scherl, L. (1989) Self in wilderness: understanding the psychological benefits of individual–wilderness interaction through self-control. *Leisure Sciences* 11(2), 123–135.

Schettler, T., Solomon, G., Valenti, M. and Huddle, A. (2000) *Generations at Risk: Reproductive Health and the Environment*. MIT Press, Cambridge, Massachusetts.

Schettler, T., Stein, J., Reich, F., Valenti, M. and Wallinga, D. (2000) *In Harm's Way: Toxic Threats to Child Development*. Physicians for Social Responsibility, Greater Boston Chapter, Somerville, Massachusetts.

Schickendanz, J. (2001) Theories of child development and methods of studying children. In: Schickendanz, J., Schickendanz, D.I., Forsyth, P.D. and Forsyth, G.A. (eds) *Understanding Children and Adolescents*, 4th edn. Allyn and Bacon, Boston, Massachusetts, pp. 12–13.

Schmidt, C. and Little, D.E. (2007) Qualitative insights into leisure as a spiritual experience. *Journal of Leisure Research* 39(2), 222–247.

Schoenfeld, T.J., Rada, P., Pieruzzini, P.R., Hsueh, B. and Gould, E. (2013) Physical exercise prevents stress-induced activation of granule neurons and enhances local inhibitory mechanisms in the dentate gyrus. *The Journal of Neuroscience* 33(18), 7770–7777.

Schroeder, H.W. (1996) Psyche, nature, and mystery: some psychological perspectives on the values of natural environment. In: Driver, B.L., Dustin, D., Baltic, T., Elsner, G. and Peterson, G. (eds) *Nature and the Human Spirit: Toward and Expanded Land Management Ethic*. Venture Publishing, Inc., State College, Pennsylvania, pp. 81–96.

Schultz, P.W., Shriver, C., Tabanico, J.J. and Khazian, A.M. (2004) Implicit connections with nature. *Journal of Environmental Psychology* 24(1), 31–42.

Schumacher, E.F. (2009) Small is beautiful. *The Top 50 Sustainability Books* 1(116), 38–41.

Science & Environmental Health Network (1998) Wingspread Statement on the Precautionary Principle. Available at: http://www.sehn.org/wing.html (accessed 15 April 2013).

Scott, N.R. (1974) Toward a psychology of wilderness experience. *Natural Resources Journal* 14, 231–237.

Scriven, M. and Paul, R. (1987) Critical thinking as defined by the National Council for Excellence in Critical Thinking. Presented at 8th Annual International Conference on Critical Thinking and Education Reform, Summer 1987. Available at: http://www.criticalthinking.org/pages/defining-critical-thinking/766 (accessed 16 October 2013).

Seamon, D. and Sowers, J. (2008) Place and placelessness, Edward Relph. In: Hubbard, P., Kitchen, R. and Vallentine, G. (eds) *Key Texts in Human Geography*. Sage, London, pp. 43–51.

Searles, H. (1960) *The Nonhuman Environment in Normal Development and in Schizophrenia*. International Universities Press, Madison, Connecticut.

Seeland, K. and Ballesteros, N. (2004) *Cultural Comparative Studies of the Potential for Social Inclusion within Green Spaces in the Agglomerations of Geneva, Lugano and Zurich. Forstwissenschaftliche Beiträge No. 31.* ETH Zürich, Zürich, Switzerland.

Selhub, E.M. and Logan, A.C. (2012) *Your Brain on Nature: The Science of Nature's Influence on Your Health, Happiness, and Vitality*. John Wiley & Sons Canada, Ltd, Mississauga, Canada.

Shadyac, T. (2010) *I Am* [film]. Los Angeles, California.

Sharp, L.B. (1943) Outside the classroom. *The Educational Forum* 7(4), 361–368.

Sheard, M. and Golby, J. (2006) The efficacy of an outdoor adventure education curriculum on selected aspects of positive psychological development. *Journal of Experiential Education* 29(2), 187–209.

Sheldrake, R. (2005) Listen to the animals: why did so many animals escape December's tsunami? *The Ecologist* 35(2), 18–20.

Shepard, P. (1995) Nature and madness. In: Roszak, T., Gomes, M. and Kanner, A. (eds) *Ecopsychology: Healing the Earth, Restoring the Mind*. Sierra Club Books, San Francisco California, pp. 21–40.

Shepard, P. (1996) *The Others: How Animals Made Us Human*. Island Press, Washington, DC.

Shiva, V. (1994) *Close to Home: Women Reconnect Ecology, Health, and Development Worldwide*. New Society, Philadelphia, Pennsylvania.

Shonkoff, J. and Phillips, D. (2000) *From Neurons to Neighborhoods: The Science of Early Childhood Development*. National Academy Press, Washington, DC.

Shorb, T.L. and Schnoeker-Shorb, Y.A. (2010) *What's Nature Got to Do with Me: The Kellert–Shorb Biophilic Values Indicator—a Continuing History of its Creative Development and Design*. Prescott College, Prescott, Arizona.

Shore, A. (1977) *Outward Bound: A Reference Volume*. Topp Litho, New York.

Sibley, B.A. and Etnier, J.L. (2003) The relationship between physical activity and cognition in children: a meta-analysis. *Pediatric Exercise Science* 15(3), 243–256.

Sibthorp, J. (2003) An empirical look at Walsh and Golins' adventure education process model: relationships between antecedent factors, perceptions of characteristics of an adventure education experience, and changes in self-efficacy. *Journal of Leisure Research* 35(1), 80–106.

Sibthorp, J., Paisley, K. and Gookin, J. (2007) Exploring participant development through adventure-based programming: a model from the National Outdoor Leadership School. *Leisure Sciences* 29(1), 1–18.

Siegel, B.S. (2011) *Love, Medicine and Miracles: Lessons Learned about Self-Healing from a Surgeon's Experience with Exceptional Patients*. HarperCollins, New York.

Siegel, D. (1999) *The Developing Mind: How Relationships and the Brain Interact to Shape Who We Are*. The Guilford Press, New York.

Siegel, R. (producer) (2012) Vanishing vultures: a grave matter for India's Parsis. *National Public Radio*, 12 September 2012. Available at: http://keranews.org/post/vanishing-vultures-grave-matter-indias-parsis (accessed 24 September 2013).

Silvertown, J. (2009) A new dawn for citizen science. *Trends in Ecology & Evolution* 24(9), 467–471.

Simkins, R. (1994) *Creator & Creation: Nature in the Worldview of Ancient Israel*. Hendrickson Publishers, Peabody, Massachusetts.

Simpson, L.R. (2004) Anticolonial strategies for the recovery and maintenance of Indigenous knowledge. *The American Indian Quarterly* 28(3), 373–384.

Skinner, J.B. (1974/1976) *About Behaviorism*. Random House Vintage Editions, New York.

Skutnabb-Kangas, T., Maffi, L. and Harmon, D. (2003) *Sharing a World of Difference: The Earth's Linguistic, Cultural and Biological Diversity*. UNESCO–Tarralingua–WorldWide Fund for Nature, Paris.

Smith, B.J., Tang, K.C. and Nutbeam, D. (2006) WHO health promotion glossary: new terms. *Health Promotion International* 21(4), 340–345.

Snegroff, S. (2012) Psychoneuroimmunology and performance anxiety: mind and body health and adaptation. Presented at *Second International Conference on Health, Wellness, & Society*, University Center, Chicago, Illinois, 10–11 March 2012.

Sobel, D. (1993) *Children's Special Places: Exploring the Role of Forts, Dens, and Bush Houses in Middle Childhood*. Wayne State University Press, Detroit, Michigan.

Sobel, D. (1996) *Beyond Ecophobia: Reclaiming the Heart in Nature Education*. The Orion Society and the Myrin Institute, Great Barrington, Massachusetts.

Sobel, D. (1998) Beyond ecophobia. *Yes! Magazine*, 2 November 1998. Available at: http://www.yesmagazine.org/issues/education-for-life/803 (accessed 12 April 2013).

Sobel, D. (2002) *Children's Special Places: Exploring the Role of Forts, Dens, and Bush Houses in Middle Childhood*. Wayne State University Press, Detroit, Michigan.

Sobel, D. (2004) *Place-based Education: Connecting Classrooms and Communities*. The Orion Society and the Myrin Institute, Great Barrington, Massachusetts.

Sobel, D. (2005) *Place-based Education: Connecting Classrooms and Communities*, 2nd edn. The Orion Society, Great Barrington, Massachusetts.

Sobel, D. (2008) *Childhood and Nature: Design Principles for Educators*. Stenhouse Publishers, Portland, Maine.

Sommer, R. (2003) Trees and human identity. In: Clayton, S. and Opotow, S. (eds) *Identity and the Natural Environment: The Psychological Significance of Nature*. MIT Press, Cambridge, Massachusetts, pp. 179–204.

Spencer, C. and Blades, M. (2006) *Children and Their Environments: Learning, Using and Designing Spaces*. Cambridge University Press, Cambridge, UK.

Speth, G. (2009) *The Bridge at the Edge of the World: Capitalism, the Environment, and Crossing from Crisis to Sustainability*. Yale University Press New Haven, Connecticut.

St. Antoine, S. (2012) *Together in Nature: Pathways to a Stronger, Closer Family*. Children and Nature Network, Santa Fe, New Mexico.

Stedman, R.C. (2002) Toward a social psychology of place: predicting behavior from place-based cognitions, attitude, and identity. *Environment and Behavior* 34(5), 561–581.

Stedman, R.C. (2003) Is it really just a social construction? The contribution of the physical environment to sense of place. *Society & Natural Resources* 16(8), 671–685.

Stein, M.B. (2005) Sweating away the blues: can exercise treat depression? *American Journal of Preventive Medicine* 28(1), 140–141.

Steinbeck, J. (1937/1994) *Of Mice and Men*. Penguin, New York.

Stella, S.G., Vilar, A.P., Lacroix, C.C., Fisberg, M.M., Santos, R.F., *et al.* (2005). Effects of type of physical exercise and leisure activities on the depression scores of obese Brazilian adolescent girls. *Brazilian Journal of Medical and Biological Research* 38(11), 1683–1689.

Sterling, S.R. (2001) *Sustainable Education*. Green Books for the Schumacher Society, Great Barrington, Massachusetts.

Stern, P., Dietz, T., Abel, T., Guagnano, G.A. and Kalof, L. (1999) A value-belief-norm theory of support for social movements: the case of environmentalism. *Human Ecology Review* 6(2), 81–97.

Sternberg, E.M. (2009) *Healing Spaces: The Science of Place and Well-being*. Harvard University Press, Cambridge, Massachusetts.

Stets, J. and Biga, C. (2003) Bringing identity theory into environmental sociology. *Sociological Theory* 21(4), 398–423.

Stevens, B., Kagan, S., Yamada, J., Epstein, I., Beamer, M., *et al.* (2004) Adventure therapy for adolescents with cancer. *Pediatric Blood & Cancer* 43(3), 278–284.

Stigsdotter, U.K., Palsdottir, A.M., Burls, A., Chermaz, A., Ferrini, F., *et al.* (2011) Nature-based therapeutic interventions. In: Nilsson, K., Sangster, M., Gallis, C., Hartig, T., de Vries, S., *et al.* (eds) *Forests, Trees and Human Health*. Springer, New York, pp. 309–342.

Stokowski, P.A. (1990) Extending the social groups model: social network analysis in recreation research. *Leisure Sciences* 12(3), 251–263.

Street, G., James, R. and Cutt, H. (2007) The relationship between organized physical recreation and mental health. *Health Promotion Journal of Australia* 18(3), 236–239.

Stringer, L.A, and McAvoy, L.H. (1992) The need for something different: spirituality and wilderness adventure. *Journal of Experiential Education*, 15(1), 13–20.

Stryker, S. and Serpe, R.T. (1982) Commitment, identity salience, and role behavior: theory and research example. In: Ickes, W. and Knowles, E.S. (eds) *Personality, Roles, and Social Behavior*. Springer, New York, pp. 192–216.

Suedfeld, P. (1992) Extreme and unusual environments. In: Stokols, D. and Altman, I. (eds) *Handbook of Environmental Psychology*. Krieger Publishing, Melbourne, Australia, pp. 863–887.

Suzuki, D.T., McConnell, A. and Mason, A. (2007) *The Sacred Balance: Rediscovering our Place in Nature*. Greystone/David Suzuki Foundation, Vancouver, Canada.

Swan, J. (1992) *Nature as Teacher and Healer: How to Awaken Your Connection with Nature*. Random House, New York.

Swiderski, M.J. (2011) Maria Montessori. In: Smith, T. and Knapp, C.E. (eds) *Sourcebook of Experiential Education: Key Thinkers and Their Contributions*. Routledge, New York, pp. 197–207.

Takano, T., Fu, J., Nakamura, K., Uji, K., Fukuda, Y., *et al.* (2002) Age-adjusted mortality and its association to variations in urban conditions in Shanghai. *Health Policy* 61(3), 239–253.

Taylor, A.F. and Kuo, F.E. (2009) Children with attention deficits concentrate better after walk in the park. *Journal of Attention Disorders* 12(5), 402–409.

Taylor, A.F., Kuo, F.E. and Sullivan, W.C. (2001) Coping with ADD: the surprising connection to green play settings. *Environment and Behavior* 33(1), 54–77.

Taylor, A.F., Kou, F.E. and Sullivan, W.C. (2002) Views of nature and self-discipline: evidence from inner city children. *Journal of Environmental Psychology* 22(1–2), 49–63.

Taylor, S. (2006) Family camps: strengthening family relationships at camp and at home. *Camping Magazine* 79(4), 54–59.

Taylor, S.E. (2006) Tend and befriend biobehavioral bases of affiliation under stress. *Current Directions in Psychological Science* 15(6), 273–277.

Taylor, S.M. (2009) *Green Sisters: A Spiritual Ecology*. Harvard University Press, Cambridge, Massachusetts.

Tennessen, C.M. and Cimprich, B. (1995) Views to nature: effects on attention. *Journal of Environmental Psychology* 15(1), 77–85.

Therapeutic Landscapes Resource Center, Inc. (2000) Therapeutic Landscapes Database. Available at: http://www.healinglandscapes.org/sites.html (accessed 1 May 2009).

Thich Nhat Hanh (2009) *Happiness: Essential Mindfulness Practices*. Parallax Press, Berkley, California.

Thomashow, M. (1996) *Ecological Identity: Becoming a Reflective Environmentalist*. MIT Press, Cambridge, Massachusetts.

Thompson, C.W. and Aspinall, P.A. (2011) Natural environments and their impact on activity, health, and quality of life. *Applied Psychology: Health and Well-Being* 3(3), 230–260.

Thompson Coon, J., Boddy, K., Stein, K., Whear, R., Barton, J., *et al.* (2011) Does participating in physical activity in outdoor natural environments have a greater effect on physical and mental wellbeing than physical activity indoors? A systematic review. *Environmental Science & Technology* 45(5), 1761–1772.

Tolle, E. (2005) *A New Earth: Awakening to Your Life's Purpose*. Penguin Books, New York.

Tompkins, P. and Bird, C. (1974) *The Secret Life of Plants*. Harper & Row, New York.

Torem, M.S., Gilbertson, A. and Light, V. (1990) Indications of physical, sexual, and verbal victimization in projective tree drawings. *Journal of Clinical Psychology* 46(6), 900–906.

Townsend, M. (2006) Feel blue? Touch green! Participation in forest/woodland management as a treatment for depression. *Urban Forestry & Urban Greening* 5(3), 111–120.

Trimble, S. and Nabhan, G. (1995) *The Geography of Childhood: Why Children Need Wild Places*. Beacon Press, Boston, Massachusetts.

Trotter, W. (1914) *Instincts of the Herd in Peace and War*. University of Michigan Press, Ann Arbor, Michigan.

Tsunetsugu, Y., Park, B. and Miyazaki, Y. (2010) Trends in research related to 'Shinrin-yoku' (taking in the forest atmosphere or forest bathing) in Japan. *Environmental Health and Preventative Medicine* 15(1), 27–37.

Tuan, Y. (2001) *Space and Place: The Perspective of Experience*. University of Minnesota Press, Minneapolis, Minnesota.

Turner, N.J., Ignace, M.B. and Ignace, R. (2000) Traditional ecological knowledge and wisdom of aboriginal peoples in British Columbia. *Ecological Applications* 10(5), 1275–1287.

Ulrich, R.S. (1981) Natural versus urban scenes: some psychophysiological effects. *Environment and Behavior* 13(5), 523–556.

Ulrich, R.S. (1983) Aesthetic and affective response to natural environment. In: Altman, J. and Wohlwill, F. (eds) *Behavior and Natural Environment: Advances in Theory and Research*, vol. 6. Plenum, New York, pp. 85–125.

Ulrich, R.S. (1984) View through a window may influence recovery from surgery. *Science* 224(4647), 420–421.

Ulrich, R.S. (1992) Effects of interior design on wellness: theory and recent scientific research. *Journal of Healthcare Design* 3, 97–109.

Ulrich, R.S. (1999) Effects of gardens on health outcomes: theory and research. In: Marcus, C.C. and Barnes, M. (eds) *Healing Gardens: Therapeutic Benefits and Design Recommendations*. John Wiley and Sons, New York, pp. 27–86.

Ulrich, R.S. (2001) Effects of healthcare environmental design on medical outcomes. In: Dilani, A. (ed.) *Design and Health*. Svensk Byggtjanst, Stockholm, pp. 49–59.

Ulrich, R.S. (2002) Health benefits of gardens in hospitals. Presented at *Plants for People International Exhibition, Floriade 2002*, The Netherlands.

Ulrich, R.S. and Parsons, R. (1992) Influences of passive experiences with plants on individual well-being and health. In: Relf, D. (ed.) *The Role of Horticulture in Human Well-being and Social Development*. Timber Press, Portland, Oregon, pp. 93–105.

Ulrich, R.S., Dimberg, U. and Driver, B.L. (1991) Psychophysiological indicators of leisure benefits. In: Driver, B.L., Brown, P.J. and Peterson, G.L. (eds) *Benefits of Leisure*. Venture Publishing, State College, Pennsylvania, pp. 73–89.

Ulrich, R.S., Simons, R.F., Losito, B.D., Fiorito, E., Miles, M.A., *et al*. (1991) Stress recovery during exposure to natural and urban environments. *Journal of Environmental Psychology* 11(3), 201–230.

UNESCO (n.d.) Local and Indigenous Knowledge System. Available at: http://www.unesco.org/new/en/natural-sciences/priority-areas/sids/enabling-environments/local-and-indigenous-knowledge-systems/ (accessed 29 January 2013).

United Church of Christ Commission for Racial Justice (1987) Toxic Wastes and Race in The United States: A National Report on the Racial and Socio-economic Characteristics of Communities with Hazardous Waste Sites. Available at: http://www.ucc.org/about-us/archives/pdfs/toxwrace87.pdf (accessed 7 September 2013).

United Nations Environment Programme (n.d.) Convention on Biological Diversity, Traditional knowledge, innovations and practices. Available at: http://www.cbd.int/traditional/ (accessed 6 April 2013).

University Tübingen (2012) Skilled hunters 300,000 years ago. *Science Daily*, 17 September 2012. Available at: http://www.sciencedaily.com/releases/2012/09/120917085535.htm (accessed 5 July 2013).

US Department of Education (2008) Public School Data Files, 2003–04 and 2007–08. National Center for Education Statistics, Schools and Staffing Survey (SASS). Available at: http://nces.ed.gov/surveys/sass/tables_list.asp (accessed 16 September 2013).

US Department of Health and Human Services (2000) *Healthy People 2010*. Government Printing Office, Washington, DC.

Vaagboe, O. (1993) De forskjelige naturbrukeres verdipreferanser (The values and preferences of Norwegians who spend their time off in nature). In: FRIFO (ed.) *Frisk i Friluft* (*Fresh in the Open Air*). FRIFO, Oslo, pp. 29–38.

Van den Berg, A.E. and ter Heijne, M. (2005) Fear versus fascination: an exploration of emotional responses to natural threats. *Journal of Environmental Psychology* 25(3), 261–272.

Van den Berg, A.E., Hartig, T. and Staats, H. (2007) Preference for nature in urbanized societies: stress, restoration, and the pursuit of sustainability. *Journal of Social Issues* 63(1), 79–96.

Van Puymbroeck, M., Ewert, A.W., Luo, Y. and Frankel, J. (2012) The influence of the Outward Bound Veterans Program on sense of coherence. *American Journal of Recreation Therapy* 11(3), 31–38.

Van Sluijs, E.M.F., Van Poppel, M.N., Twisk, J.W., Brug, J. and Van Mechelen, W. (2005) The positive effect on determinants of physical activity of a tailored, general practice-based physical activity intervention. *Health Education Research* 20(3), 345–356.

Velarde, Ma.D., Fry, G. and Tveit, M. (2007) Health effects of viewing landscapes-landscape types in environmental psychology. *Urban Forestry & Urban Greening* 6(4), 199–212.

Vest, J.H.C. (2008) The philosophical significance of wilderness solitude. *Environmental Ethics* 9(4), 303–330.

Virden, R.J. and Walker, G.J. (1999) Ethnic/racial and gender variations among meanings given to, and preferences for, the natural environment. *Leisure Sciences* 21(3), 219–239.

Vitebsky, P. (2005) *Reindeer People: Living with Animals and Spirits in Siberia*. HarperCollins UK, London.

Walant, K. (1999) *Creating the Capacity for Attachment: Treating Addictions and the Alienated Self*. Jason Aronson, Inc., Northvale, New Jersey.

Walsh, V. and Golins, G. (1976) *The Exploration of the Outward Bound Process*. Outward Bound Publications, Denver, Colorado.

Warburg, O. (1966) The prime cause and prevention of cancer with two prefaces on prevention. Revised lecture at the Meeting of the Nobel-Laureates on June 30, 1966, Lindau, Lake Constance, Germany. Available at: http://www.whale.to/a/warburg.html (accessed 8 September 2013).

Ward Thompson, C. (2011) Linking landscape and health: the recurring theme. *Landscape and Urban Planning* 99(3), 187–195.

Ward Thompson, C. and Aspinall, P.A. (2011) Natural environments and their impact on activity, health, and quality of life. *Applied Psychology: Health and Well-Being* 3(3), 230–260.

Warren, K. and Wapotich, L. (2011) Rachel Carson. In: Smith, T. and Knapp, C.E. (eds) *Sourcebook of Experiential Education: Key Thinkers and Their Contributions*. Routledge, New York, pp. 197–207.

Warren Wilson College (2008) Eco-Sermon Challenge. Available at: http://www.warren-wilson. edu/~advancement/Eco-Sermons.php (accessed 16 December 2012).

Watson, J.B. (1923/2012) *Behavior: An Introduction to Comparative Psychology*. Henry Holt and Company, New York.

Weigert, A. (1997) *Self, Interaction, and Natural Environment: Refocusing Our Eyesight*. State University of New York Press, Albany, New York.

Weigert, A. (2008) Pragmatic thinking about self, society, and natural environment: Mead, Carson, and beyond. *Symbolic Interaction* 31(3), 235–258.

Weinstein, N., Przybylski, A.K. and Ryan, R.M. (2009) Can nature make us more caring? Effects of immersion in nature on intrinsic aspirations and generosity. *Personality and Social Psychology Bulletin* 35(10), 1315–1329.

Weiss, D. (2003) Ecology hall of fame: Julia Butterfly Hill. Available at: http://ecotopia.org/ecology-hall-of-fame/julia-butterfly-hill/ (accessed 15 September 2013).

Wells, N.M. (2000) At home with nature: effects of 'greenness' on children's cognitive functioning. *Environment and Behavior* 32(6), 775–795.

Wells, N.M. and Evans, G.W. (2003) Nearby nature: a buffer of life stress among rural children. *Environment and Behavior* 35(3), 311–330.

Wenzel, G.W. (1999) Traditional ecological knowledge and Inuit: reflections on TEK research and ethics. *Arctic* 52(2), 113–124.

Wenzel, G.W. (2004) From TEK to IQ: Inuit Qaujimajatuqangit and Inuit cultural ecology. *Arctic Anthropology* 41(2), 238–250.

Wheatley, M.J. (2005) *Finding Our Way – Leadership for an Uncertain Time*. Berret-Koehler Publishers, San Francisco, California.

Wheatley, M.J. and Frieze, D. (2007) How large-scale change really happens – working with emergence. *The School Administrator*, Spring 2007 issue. Available at: http://www.margaretwheatley.com/articles/largescalechange.html (accessed 7 September 2013).

Wheeler, B.W., Cooper, A.R., Page, A.S. and Jago, R. (2010) Greenspace and children's physical activity: a GPS/GIS analysis of the PEACH project. *Preventive Medicine* 51(2), 148–152.

White, D.D. and Hendee, J. (2000) Primal hypotheses: the relationship between naturalness, solitude, and the wilderness experience benefits of development of self, development of community, and spiritual development. In: McCool, S.F., Cole, D.N., Borrie, W.T. and O'Loughlin, J. (comps) *Wilderness Science in a Time of Change Conference—Volume 3: Wilderness as a Place for Scientific Inquiry*, Missoula, Montana, 23–27 May 1999. *USDA Forest Service Proceedings RMRS-P-15-VOL-3*. US Department of Agriculture, Forest Service, Rocky Mountain Research Station, Ogden, Utah, pp. 23–27.

White, L. (1967) The historical roots of our ecological crisis. *Science* 155(3767), 1203–1207.

White, R. and Stoecklin, V. (1998) Children's outdoor play & learning environments: returning to nature. *Early Childhood News*, March/April 1998 issue. Available at: http://www.whitehutchinson.com/children/articles/outdoor.shtml (accessed 15 October 2013).

Wiggins, M. Jr (2012) *Kakagon and Bad River Sloughs Recognized as a Wetland of International Importance*. The Bad River Band of Lake Superior Chippewa, Odanah, Wisconsin.

Williams, D. and Stewart, S. (1998) Sense of place: an elusive concept that is finding a home in ecosystem management. *Journal of Forestry* 96(5), 18–23.

Williams, K. and Harvey, D. (2001) Transcendent experience in forest environments. *Journal of Environmental Psychology* 21(3), 249–260.

Willis, A. (2011) Re-storying wilderness and adventure therapies: healing places and selves in an era of environmental crises. *Journal of Adventure Education & Outdoor Learning* 11(2), 91–108.

Winter, D. (1996) *Ecological Psychology: Healing the Split between Planet and Self*. HarperCollins College Publishers, New York.

Wilson, E.O. (1984) *Biophilia: The Human Bond with Other Species*. Harvard University Press, Cambridge, Massachusetts.

Wilson, R. (1993) *Fostering a Sense of Wonder during the Early Childhood Years*. Greyden, Columbus, Ohio.

Wilson, R. (2012) *Nature and Young Children: Encouraging Creative Play and Learning in Natural Environments*. Routledge, Abingdon, UK.

Wilson, S.J. and Lipsey, M.W. (2000) Wilderness challenge programs for delinquent youth: a meta-analysis of outcome evaluations. *Evaluation and Program Planning* 23(1), 1–12.

Working Group on the Anthropocene, Subcommission on Quaternary Stratigraphy, International Commission on Stratigraphy (2012) What is the 'Anthropocene?' – Current definitions and status. Available at: http://quaternary.stratigraphy.org/workinggroups/anthropocene/ (accessed 12 December 2012).

World Health Organization (1948) Preamble to the Constitution of the World Health Organization as adopted by the International Health Conference. *Official Records of the World Health Organization* 2, 100.

World Health Organization (2006) Constitution of the World Health Organization – Basic Documents, 45th edition, Supplement, October 2006. Available at: http://www.who.int/governance/eb/who_constitution_en.pdf (accessed 6 April 2013).

Wuang, Y.P., Wang, C.C., Huang, M.H. and Su, C.Y. (2010) The effectiveness of simulated developmental horse-riding program in children with autism. *Adapted Physical Activity Quarterly* 27(2), 113–126.

Yale Forum on Religion and Ecology (2013) The Forum on Religion and Ecology at Yale. Available at: http://fore.research.yale.edu/ (accessed 21 September 2013).

Yankou, D. (2002) MorningStar: A place of restoration. *MorningStar Adventures* 18(2), 3.

Yeou-Lan, D.C. (1996) Conformity with nature: a theory of Chinese American elders' health promotion and illness prevention processes. *Advanced Nursing Science* 19(2), 17–26.

Young, J., Haas, E. and McGown, E. (2008) *Coyote's Guide to Connecting with Nature*. OWLink Media, Shelton, Washington.

Zabriskie, R.B. and McCormick, B.P. (2001) The influences of family leisure patterns on perceptions of family functioning. *Family Relations* 50(3), 281–289.

Zalasiewicz, J., Williams, M., Smith, A., Barry, T.L., Coe, A.L., *et al.* (2008) Are we now living in the Anthropocene? *GSA Today* 18(2), 4–8.

Zalasiewicz, J., Williams, M., Steffen, W. and Crutzen, P. (2010) The new world of the Anthropocene. *Environmental Science & Technology* 44(7), 2228–2231.

Zalasiewicz, J., Williams, M., Haywood, A. and Ellis, M. (2011) The Anthropocene: a new epoch of geological time? *Philosophical Transactions of the Royal Society A: Mathematical, Physical and Engineering Sciences* 369(1938), 835–841.

Zelenski, J.M. and Nisbet, E.K. (2014) Happiness and feeling connected: the distinct role of nature relatedness. *Environment and Behavior* 46(1), 3–23.

Zube, E.H., Pitt, D.G. and Evans, G.W. (1983) A lifespan developmental study of landscape assessment. *Journal of Environmental Psychology* 3(2), 115–128.

Index